THE ABSENT PRESENCE
of the State in Large-Scale Resource
Extraction Projects

THE ABSENT PRESENCE
of the State in Large-Scale Resource
Extraction Projects

EDITED BY NICHOLAS BAINTON
AND EMILIA E. SKRZYPEK

Australian
National
University

PRESS

ASIA-PACIFIC ENVIRONMENT MONOGRAPH 15

ANU PRESS

Published by ANU Press
The Australian National University
Acton ACT 2601, Australia
Email: anupress@anu.edu.au

Available to download for free at press.anu.edu.au

ISBN (print): 9781760464486
ISBN (online): 9781760464493

WorldCat (print): 1261481399
WorldCat (online): 1261465868

DOI: 10.22459/AP.2021

Cover design and layout by ANU Press. Cover photograph: Mining trucks lined up at a copper mine. Source: Centre for Social Responsibility in Mining, the University of Queensland.

Contents

List of Figures

Abbreviations and Currency Conversion Rates

Abbreviations

ALRA	Aboriginal Land Rights Act: the *Aboriginal Land Rights (Northern Territory) Act 1976* (Cth)
ATALA	Angore Tiddy Apa Landowners Association
ATSIC	Aboriginal and Torres Strait Islander Commission
CA	Community affairs
CLF	Community Leaders' Forum
CSG	Coal seam gas
CSR	Corporate social responsibility
EIA	Environmental impact assessment
EIS	Environmental impact statement
ESIA	Economic and social impact assessment
GCA	Gulf Communities Agreement
HGCP	Hides Gas Conditioning Plant
IBP agreement	Integrated Benefits Package agreement
LBSA	License benefit sharing agreement
LMALA	Lihir Mining Area Landowners Association
LNG	Liquefied natural gas
LSDP	Lihir Sustainable Development Plan
MCA	Minerals Council of Australia
NLC	Northern Land Council
NRLLG	Nimamar Rural Local Level Government

NSW	New South Wales
NT	Northern Territory
PJV	Porgera Joint Venture
PKKP	Puutu Kunti Kurrama People and Pinikura People
PLA/PLOA	Porgera Landowners Association
PNG	Papua New Guinea
PNGDF	Papua New Guinea Defence Force
QLD	Queensland
SIA	Social impact assessment
SML	Special Mining Lease
TO	Traditional Owners
WA	Western Australia

Currency Conversion Rates

Year	PGK (= 1AUD)	USD (= 1AUD)
2000	1.67	0.56
2007	2.42	0.88
2011	2.17	1.02
2016	2.29	0.72
2017	2.54	1.28
2018	2.29	0.70
2019	2.40	0.70

Contributors

Nicholas Bainton is an Associate Professor and Principal Research Fellow in the Centre for Social Responsibility in Mining at the University of Queensland. He has been studying the social impacts of large-scale resource extraction in Papua New Guinea for nearly two decades. He has written widely on the social and political effects of extractive capitalism in Melanesia and is the author of *The Lihir Destiny: Cultural Responses to Mining in Melanesia* (2010) and editor (with McDougall, Alexeyeff and Cox) of *Unequal Lives: Gender, Race and Class in the Western Pacific* (2021), both published by ANU Press.

John Burton is a Principal Research Fellow at the Centre for Social Responsibility in Mining at the University of Queensland. He has worked as an applied anthropologist in Papua New Guinea since the 1980s, with a focus on landowner identification studies, the social impacts of mining, and corruption risks in mining awards. The work reported on here was carried out when he was Professor and Deputy Vice President Research at Divine Word University, Papua New Guinea (2015–19), and joined Claire Levacher as a team member on the CNRT-funded project *Petites et moyennes entreprises minières en Nouvelle-Calédonie* [Small and medium mining enterprises in New Caledonia]. John's other interests include Native Title connection in North Queensland and history and identities in Torres Strait.

Martin Espig is a cultural anthropologist with an interest in environmental risk debates associated with the extractive industries and natural resource management. His research addresses the epistemic dimensions and role of science in risk controversies, social and cultural impacts of natural resource developments, and the entanglements of overlapping land uses. Martin has conducted research in the agricultural regions of Southern Queensland, and was involved in innovation projects in the mining industry in the Netherlands and Germany.

Martin is currently in New Zealand, based in a social scientist position at AgResearch, one of the country's Crown Research Institutes, where he conducts applied research on a range of topics, including responsible innovation, sustainability transitions and agricultural digitalisation.

Jo-Anne Everingham is a Senior Research Fellow at the University of Queensland's Centre for Social Responsibility in Mining who specialises in applied sociological studies of mining communities, especially in rural and regional Australia. She examines local development impacts, as well as exploring governance, social policy and management of social issues and community relations. Her recent research focuses on governance and management of social risks and impacts at various stages of the life cycle of extractive projects. Recent co-authored articles include 'Workshop processes to generate stakeholder consensus about post-mining land uses: An Australian case study' in *Journal of Environmental Planning and Management* (2021), 'The social dimensions of mineral exploration' in *SEG Discovery* (2020) and 'The governance of mining regions in Australia (2000–2012)' in *Journal of Rural Studies* (2020).

Alex Golub is an Associate Professor of Anthropology at the University of Hawai'i at Mānoa, where he studies the political anthropology of mining and the history and culture of Papua New Guinea. He is the author of *Leviathans at the Gold Mine* (Duke University Press, 2014), and editor of two special journal issues about Papua New Guinea: *The Politics of Order in Contemporary Papua New Guinea* in *Anthropological Forum* (2018) and (with Lise Dobrin) a special issue of the *Journal of Pacific History* entitled *The Legacy of Bernard Narokobi and the Melanesian Way* (2020).

Sarah Holcombe is a social anthropologist and Senior Research Fellow at the Centre for Social Responsibility in Mining at the University of Queensland. She has more than 20 years' experience in applied and academic research with Indigenous Australians. She has undertaken research, and published widely, on a diverse range of issues including human rights and intersectional challenges to implementation, and extractive industries and sustainable development. She has published in academic and non-academic outlets and is the author of *Remote Freedoms: Politics, Personhood and Human Rights in Aboriginal Central Australia* (Stanford University Press, 2018).

Julia Keenan joined the Centre for Social Responsibility in Mining at the University of Queensland in 2007, working on a range of social sustainability issues related to the extractive industries in Australia and internationally. She specialises in researching the relationship between mining and local communities, with particular focus on agreement making with Indigenous peoples, gender and community development, and methodologies for analysing and improving resource companies' social performance. She has co-authored industry guidance documents on Indigenous peoples and mining, and integrating gender into community relations work. She has also contributed to research projects examining company–community conflict, extractive industry policy, social and cumulative impact assessment, and internal management systems.

Claire Levacher is an anthropologist. She completed a PhD on indigeneity and mining governance in New Caledonia (2016) and two postdoctoral research contracts on small-scale mining in New Caledonia (New Caledonian Institute for Agronomic Research) and comparative perspective on mining encounters in New Caledonia and Canada (Laval University, Québec, QC, Canada). Apart from her work on governance, sovereignty and land issues in mining territories, she is especially interested in representations of nature, landscapes, pollution management and post-mining exploitation.

Gareth Lewis is a consultant anthropologist with more than 20 years of applied experience across Australia's Northern Territory covering Aboriginal land rights, sacred site protection and native title matters. Gareth has worked extensively at the interface between Aboriginal peoples, governments, parks agencies, the mining industry and other developers, and has held various roles including as Senior Anthropologist at both the Northern Land Council and the Aboriginal Areas Protection Authority. He has worked on Aboriginal land claims, native title claims, co-management of parks, and numerous major mining and other development projects. He has also undertaken numerous sacred site field surveys, sacred site registrations and has been involved in investigations and prosecutions for offences under the *Northern Territory Aboriginal Sacred Sites Act 1989.*

Martha Macintyre is an Honorary Principal Research Fellow at the University of Melbourne, honorary Professor at the Centre for Social Responsibility in Mining at the University of Queensland and a Fellow of the Australian Academy of Social Sciences. She has undertaken research into the social impacts of mining in New Ireland and Milne Bay Province, Papua New Guinea. Her publications include *Managing Modernity in the Western Pacific* (edited with Patterson, University of Queensland Press, 2011); *Gender Violence and Human Rights: Seeking Justice in Fiji, Papua New Guinea and Vanuatu* (edited with Biersack and Jolly, ANU Press, 2016); *Emergent Masculinities in the Pacific* (edited with Biersack, Routledge, 2017) and *Transformations of Gender in Melanesia* (edited with Spark, ANU Press, 2017).

Michael Main has a PhD in anthropology from The Australian National University. His PhD research focused on Huli people in the Papua New Guinea highlands and the impact on their lives from ExxonMobil's Papua New Guinea Liquefied Natural Gas project. Michael has a professional background in geology and environmental science, which underpins his interest and work in the anthropology of development and resource extraction.

John R. Owen is Professor and Deputy Director of the Centre for Social Responsibility in Mining at the University of Queensland. His current research interests focus on the problems of future metal supply and the implications for the communities who live and work at their source. He is the academic lead on a university–industry research consortium on mining and resettlement, and has conducted several major studies on the industry's approach to managing the complex social issues that develop around their activities. His most recent book, co-authored with Deanna Kemp, is *Extractive Relations: Countervailing Power and the Global Mining Industry*, published by Routledge (2017).

Emilia E. Skrzypek is a Senior Research Fellow in social anthropology at the University of St Andrews, and an Honorary Research Fellow at the Centre for Social Responsibility in Mining at the University of Queensland. Her work to date has largely focused on Papua New Guinea, where she investigates issues related to broadly conceived resource relations and interdependencies. She is particularly interested in stakeholder engagement and social impacts at undeveloped complex orebodies. She is the author of *Revealing the Invisible Mine: Social Complexities of an Undeveloped Mining Project* (Berghahn, 2020).

David Trigger is Emeritus Professor of Anthropology at the University of Queensland and an Adjunct Professor at the University of Western Australia. He is the principal partner in David S Trigger & Associates consulting anthropologists. His research interests encompass the different meanings attributed to land and nature across diverse sectors of society. His research on Australian society includes projects focused on a comparison of pro-development, environmentalist and Aboriginal perspectives on land and nature. His most recent works address senses of historical place that both overlap and diverge for people of diverse ancestries in northern Australia. Professor Trigger is the author of *Whitefella Comin':* *Aboriginal Responses to Colonialism in Northern Australia* (Cambridge University Press, 1992) and a wide range of scholarly articles and applied research reports.

Preface

This volume originated from a panel we convened at the combined Australian Anthropology Society, Association of Social Anthropologists and Association of Social Anthropologists Aotearoa New Zealand (AAS/ ASA/ASAANZ) conference in Adelaide in December 2017. Participants were invited to consider how ideas of the state as a central actor in resource relations are formed, negotiated and enacted, and how these relations subsequently influence events, elicit specific effects and shape outcomes at resource extraction projects. Collectively, their papers and the discussions that followed demonstrated two things: the diversity of ways in which the state is articulated and experienced in the context of resource extraction; and the opportunity and value of studying the state from the vantage point of resource extraction.

We did not begin with 'absent presence'. This concept followed rather than preceded the conference panel, emerging during the day's final discussion on the ways in which the state is present and/or absent in resource extraction contexts in both Papua New Guinea and Australia (and beyond), and the effects this produces. In the months that followed, we asked the panellists to revisit their papers and think about their material in relation to the 'absent presence of the state'. The chapters presented here are a result of this process.

We think this concept has provided fertile ground for a wider discussion about how we conceptualise and study 'the state', and the kinds of knowledge and insights this produces. We also hope that it opens up new spaces for thinking about the entanglements, uncertainties and relations that characterise resource extraction.

In focusing the analytical lens on the state, and proposing a new framework for studying the state from the vantage point of the extractive industries, this volume builds upon a theme running through the Asia-Pacific Environment Monographs (APEM) series, namely the politics of resource extraction. This volume maintains a regional focus on Papua New Guinea and Australia with the exception of the final chapter, which introduces New Caledonia as a point of comparison to test whether the notion of 'absent presence' can be applied to the study of state processes and state effects in other jurisdictions. In this sense, our volume builds upon previous works that have compared the social and cultural effects of resource use in Papua New Guinea and Australia, including James Weiner and Katie Glaskin's *Customary Land Tenure and Registration in Australia and Papua New Guinea* (2007), also published in this series. In introducing New Caledonia, our volume also complements another volume in this series that compared the 'local-level politics' of resource extraction in Papua New Guinea and New Caledonia (*Large-Scale Mines and Local-Level Politics,* Filer and Le Meur, ANU Press, 2017). Our volume extends that focus, since state actors certainly do have some presence in the local-level politics that 'surround' large-scale resource extraction projects in all jurisdictions.

This project was first conceived at the Centre for Social Responsibility in Mining at the University of Queensland (UQ), where Nicholas currently works and where Emilia was a Marie Skłodowska-Curie Research Fellow (2017–19), funded from the European Union's Horizon 2020 research and innovation program under the Marie Sklodowska-Curie grant agreement No. 753272, in a joint appointment between the University of St Andrews in Scotland and UQ, where she remains an Honorary Research Fellow. The centre has proven to be an encouraging and enabling environment for this project given the significant amount of research conducted within the centre on resource extraction in Australia and Papua New Guinea. Reflecting this research focus, when Emilia commissioned Simon Gende, an artist from Goroka in Papua New Guinea, to produce a series of mining-related paintings, he dedicated one of the images to the work of the centre, so we have included the artwork here in acknowledgement of Simon and the scholarly support provided by the centre.

Figure 0.1 Bikpela mining kamap na ol sumatin blong university blong Queensland kisim helikopta na lukluk raun long mining area: social responsibility in mining students. [Large-scale mining has come up and the students of the University of Queensland got a helicopter and are looking at the mining area: Centre for Social Responsibility in Mining students].

Artwork: Simon Gende, 2019. Emilka Skrzypek personal collection

We thank our UQ colleagues Deanna Kemp and John Owen for their encouragement and intellectual engagement, and all those hallway conversations that spur new ideas. We are grateful to Colin Filer as the APEM series editor for his editorial clarity, insight and guidance, and our peer reviewers for their helpful comments. We also express our gratitude to the University of St Andrews, whose financial support in the form of a grant from the Institutional Open Access Fund meant that we could make this volume available in this form. Finally, we extend our sincere thanks to the many different community members, and state and company actors that have engaged with our research over time and ultimately helped to inform this project and our thinking on this topic.

Nick Bainton and Emilka Skrzypek
July 2021

1

An Absent Presence: Encountering the State Through Natural Resource Extraction in Papua New Guinea and Australia

Nicholas Bainton and Emilia E. Skrzypek

Vladimir: We have to come back tomorrow.
Estragon: What for?
Vladimir: To wait for Godot.
Estragon: Ah! (Silence) He didn't come?
Vladimir: No.

Waiting for Godot (Samuel Beckett 1956)

Introduction

In Samuel Beckett's famous 'tragicomedy in two acts', the tragic tramps, Vladimir and Estragon, spend a good deal of time waiting for Godot and discussing their desire to see him. Godot never arrives, but he figures as potentially present. The absence of Godot is also about his presence and, at one level, the entire play can be read as a meditation on absence and presence. We only know of Godot's intimation, and there is little by way of actualisation; however, the promise of his presence provides the backdrop against which all other activities occur. In this respect, he is a totalising influence whose presence is everywhere and nowhere—he exerts an 'absent presence'.

The absurdity at the centre of this play reminds us that we can think of absence and presence as mutually constitutive phenomena rather than simple binary opposites, or logical antonyms of each other. Taking 'absent presence' as our point of departure, this volume focuses on the unstable and even dialectical relationship between the presence and the absence of the state in the context of natural resource extraction, and the effects this creates. This is the first volume that brings together a sustained focus on the absent presence of the state in resource extraction settings—placing this phenomenon at the very centre of analysis. The volume also provides a timely contribution to the ethnography of the state and state administration, which historically has received comparatively less attention in the Oceanic region. It details different experiences of the state in extractive settings at a time when we are witnessing increased tensions in many countries around the role of the state in managing resource extraction and the shift towards a low-carbon future. The state duty to protect the rights of citizens is matched by the corporate responsibility to respect these human rights (UN 2011). Malfunctioning or missing state-systems exacerbate the burdens of extraction on project area communities, and increase the likelihood that corporations may infringe upon the rights of these communities. This takes place amid local demands that governments enact their state responsibilities to protect the interests of the people and find more responsible and less destructive ways to harness natural resources for the benefit of society. In these contentious contexts, the absence of the state, or its particular bureaucratic functions and representatives, often creates a heightened sense of the state, with the presence of the state felt through its absence—a peculiar and paradoxical absent presence that thoroughly shapes the nature of these sites of extraction. These experienced absences tend to reinforce particular ideas of the state, signalling the interplay between the ideological and the material qualities of the state, as we shall discuss below.

The geographical focus of this volume encompasses Papua New Guinea (PNG) and Australia, with a final chapter that compares PNG and Australia with the French territory of New Caledonia—testing and demonstrating a wider application of the absent presence concept. PNG and Australia are deeply invested in the extraction and exploitation of their natural resources, and the extractive industries play an important role in national and regional economies in both. Until 1975, PNG was a colony of Australia, administered under a mandate from the United Nations. Historically and today, the relationship between these countries can be characterised as simultaneously asymmetrical and mutually dependent, entangled in a range of political and economic interdependencies, in which the industrialised extraction of natural resources plays a significant part. In both nations, the state plays a similar role in relation to the extractive industries, with responsibility for permitting and regulation, and distribution of resource-related benefits, vested in specific government agencies. While the Australian state is commonly regarded as 'mature' and 'stable', the state of PNG is more often described as 'weak' and 'ineffectual' (Dinnen 2001). The rich ethnographic material presented in this volume challenges this simple point of comparison as it applies in resource 'arenas' (Bainton and Owen 2019), demonstrating that in both jurisdictions the state is often perceived to be an absent or incomplete institution, producing common and divergent 'state effects' (Trouillot 2001). In both nations, local landowners and custodians of the land have similar expectations to benefit from extractive activities taking place on their lands, but they also experience comparable forms of depredation and dispossession, directly linked to the operation of extractive projects and the concurrent presence and absence of the state.

In this introductory chapter, we develop the idea of the absent presence of the state. We then sketch some of the main points of commonality and difference between PNG and Australia. These form the basis for comparison in this volume, including the role of the extractive industries in the development of these nation states and their identity; state sovereignty over mineral resources; and the experiences and expectations of customary landowners. In the final section, we chart the different ways in which the state is experienced in these resource settings and the various contributions of the authors. An overarching intention of this volume is to advance our understanding of the ways in which the state is experienced and instantiated through these resource 'encounters' (Pijpers and Eriksen 2018). As a whole, this volume illustrates how the concept of absent presence can be brought to life, enhancing our understanding of the state from the vantage point of resource extraction, and thus providing a specific contribution to the anthropology of the state and the anthropology of extraction.

Absence, Presence and the Absent Presence of the State

How do we study something that is absent or only partially present? How much do we need to know about a specific thing to understand the effects of its absence? More specifically, to what extent do we need to define the state before it can be studied or before we can even begin to comprehend its effects? Beckett tells us almost nothing about Godot, but the effect of his absence places him at the centre of the script—and for some readers, he is *the* central character. Conventional wisdom holds that either a thing is here or it is not, whereas we more often find that absent things leave 'traces' of their presence, so that a thing can be present in some ways while being absent in others. Philosophers and other scholars have been trying to make sense of the idea of absence, and the relationship between notions of absence and presence, for a very long time (for example, Sartre 1957; Kierkegaard 1971 [1843]; Deleuze 1990; Fuery 1995). The work of theorists such as Jacques Derrida (1976) has been instrumental in helping us to think beyond the simple static binary distinction between absence and presence, challenging conventional assumptions about 'presence', and attacking a particular view that assumes an absolute presence and an absolute absence.[1] Likewise, anthropologists and sociologists have long recognised that social relations are often performed around what is there and the *presence* of what is not (Hetherington 2004). In this regard, absence often serves as a kind of ordering device.[2] In their work on the anthropology of absence, Mikkel Bille and his co-authors emphasise the mutual interdependence between the materially present and the materially absent, suggesting that the presence or absence of phenomena—persons, things, events or places—may not necessarily depend on absolute, positive occurrences or the absolute lack thereof, but instead resides in the way

---

1   For structuralists, the concept of 'absent/presence' denotes the notion of (present) signifiers that refer to (absent) signifieds, which is a critical feature of language and representation. These ideas were especially important for media and communication theorists because they helped to emphasise the ways that 'representational absence' can become a form of presence (Bell n.d.). We also come to appreciate how certain terms, concepts, factors, questions or issues are rendered 'conspicuous by their absence', and the symbolic erasure of specific sociocultural groups in text and other mediums or in particular social contexts. A focus on absences allows us to grasp the significance of silences and how they are actively created, and pay attention to things that have been suppressed or not even allowed to exist in the first place.

2   Perhaps the most explicit examples can be drawn from the body of work on mortuary rituals and the myriad ways that people attempt to manage absences and the traces that departed kin leave on landscapes and the social realm (Hertz 1960 [1907]; Bloch and Parry 1982; Lipset and Silverman 2016).

that the experience of certain phenomena differs from expectations and preconceptions (Bille et al. 2010: 5). As such, we find a paradox in the very nature of presence and absence, where their significance is only fully realised through their mutuality.

These ideas about the relationship between absence and presence inform our approach to the ways that states are encountered through the processes of extraction. From this perspective, states can be understood as incomplete projects, which reminds us that in many settings the state cannot be taken for granted and must be actively reproduced. For example, it has been argued that in PNG the state 'is not something that has been around in the past, nor is it expected to be automatically present in the future' (Dalsgaard 2013: 36). In such contexts, the presence of the state is uncertain and incomplete. The state must be made present, and conceptualisation of state presence cannot be separated from its absence (ibid.).[3] By focusing on the absent presence of the state in resource extraction settings, we come to see the incompleteness and uncertainty that constitutes the contemporary capitalist state—as the contributors to this volume show, from the vantage point of resource extraction projects, states appear very partial. This helps us to think about the proximity of people and processes in relation to these projects and the state, and the spatial and temporal dimensions that shape these encounters. The notion of absent presence also enables us to move beyond ideas about the total absence or total presence of the state and, instead, grasp the ways that things, like states, can be simultaneously present and absent. In this respect, our approach provides a critical counterpoint to the simplistic 'big-picture' framework of the state developed by Daron Acemoglu and James Robinson (2020) in their popular political science work *The Narrow Corridor*. According to their model, states (which they prefer to call 'Leviathans') can be charted on an evolutionary scale comprising 'absent', 'despotic' and 'shackled' Leviathans, the latter representing the balance between the power of the state and society's capacity to control it. Their primary purpose is to demonstrate the range of factors that influence the trajectory of different states and how societies have failed or managed to achieve 'liberty'. But for our purposes, their model locks us into the idea

---

3   In his more recent work Dalsgaard describes how the government as a facet of the state in Melanesia 'has become locally adopted and institutionalised in multiple and sometimes contrasting ways that speak to both the varied colonial history and the cultural diversity of the region' (2019: 255).

of categorical absence, which leaves very little room for thinking about how states may be absent in some ways and present in others, at different moments in time, and for different sections of society.

In thinking about the relative absence or presence of the state in the processes of extraction, it is useful to consider the material and non-material dimensions of the state. Philip Abrams (1988) has suggested that we should distinguish between the 'state-system' and the 'state-idea'. The former refers to the palpable nexus of institutional arrangements and apparatuses centred in government, while the latter refers to the reification of this system as a substantial and totalising entity, or a thing, that is separate from society. Thus, if we think of the state as both a material force and an ideological power—as twin aspects of the same process—then it follows that we should also attend to the senses in which the state does not exist, as well as those in which it does, and the relationship between them. In other words, how the everyday materiality of the state, through its routines, rituals, representatives, activities and policies, constitutes and regulates modern subjects and shapes how people come to believe in the idea of the state and how this in turn effects everyday practice. Several contributors to this volume locate the absent presence of the state precisely in the localised interplay between state-systems and the state-idea—the way in which the state is imagined, anticipated and experienced in extractive contexts.

A decade after Abrams laid out his formula for comprehending the contemporary capitalist state, Michel-Rolph Trouillot made a similar argument. He observed that in different times and places the power of the state can seem more or less evident and encroaching. This observation is echoed in a number of chapters in this volume that reflect on spatial as well as temporal dimensions of the state: the sense of distance between the state, resource projects and impacted communities, and changing experiences of state presence at different points in time throughout the project life cycle. Trouillot reasoned that ethnographic analysis of the state must recognise that state power has 'no institutional fixity', and that 'state effects never obtain solely through national institutions or in governmental sites' (Trouillot 2001: 126). These two features, he said, are inherent in the capitalist state, and have been exacerbated by globalising processes. If the state is not limited in terms of institutions or geography, then its *presence* becomes somewhat more deceptive, and there is a 'need to theorise the state beyond the empirically obvious'. Trouillot's suggested strategy was to 'focus on the multiple sites in which state processes are recognisable through their effects'. That is, if the state is not reducible to government or a set of apparatuses, and instead can be understood as an enlarged set of practices and processes and their effects, then we need to pay attention to these practices and processes and the effects they produce (ibid.: 131).

Trouillot identified four state effects, which we regularly encounter in resource extraction settings. First, as James Scott once observed, state power relies on legibility: an illegible society is a hindrance to effective interference by the state (1998: 78). Many anthropologists have observed these legibility effects, where local communities attempt to make themselves legible to companies or the state in order to be recognised as project beneficiaries. State actors or representatives are also engaged in the project of enhancing the legibility of local communities or assigning identities to individual subjects. In the same way, companies engage in processes of 'simplification' as part of the ongoing, and always incomplete, project of legibility. Second, we also witness the so-called 'isolation effect' as community members are converted into atomised individual subjects, achieved by means of company-supplied identification numbers for its local workers or through company-sponsored census lists, the identification of 'project stakeholders', and so on. Third, the delineation of customary landowners and traditional owners who are beneficiaries of these projects resembles the 'identification effect', which leads to Trouillot's fourth and final state effect—'spatialisation', or the creation and maintenance of jurisdictions and boundaries. As governments establish lease areas and 'zones of entitlement' for resource extraction projects, this creates new forms of governance within and across these boundaries. We see an equivalence between this final effect and Peter Vandergeest and Nancy Lee Peluso's concept of 'territorialization' (1995). Such spatial strategies bring the state into being by controlling people and access to natural resources. In regards to resource extraction, territorialisation entails: 'the creation and mapping of land boundaries, the allocation of land rights to so-called private actors, and the designation of specific resource (including land) uses by both state and "private" actors according to territorial criteria' (ibid.: 415). But as the contributors to this collection show, territorialisation is often unstable or uncertain, and the people most affected by the state's resource territorialisation often undermine, disrupt or resist the state's goals and private actor interests—processes which can understood as a type of 'double movement' in property rights (Filer et al. 2020).

The contributions to this volume add several other state effects to this list, including 'order effects', 'enactment effects', 'responsibilisation effects' and 'material effects', where companies provide the services and infrastructure that states ordinarily provide to legitimise themselves. They also illustrate a 'corporate' or 'contractual' effect where neoliberal states are increasingly run like (poorly managed) companies, while multinational corporations and multilateral institutions (like the World Bank or the United Nations) can assume more state-like powers in resource extraction contexts and produce various state effects. This resembles Christian Lund's observations on the 'state-quality' of non-state institutions, where private actors 'strut in borrowed plumes' (2006: 677). As resource companies use the language of the state and its props—including deeds and contracts, stationery and stamps—this challenges the idea that states have a singular hold on the exercise of power and authority, and in some cases blurs the lines between 'state effects' and 'company effects', which often demonstrate themselves in terms of simplification, spatialisation, isolation and identification. When extractive companies use these strategies to assert exclusive claims over lease land areas, this also serves to uphold or reinforce the internal territorial strategies of the state. We might then ask whether these private sector actions constitute 'state effects' that demonstrate the presence of the state, or whether they are better conceptualised as 'company effects' that bear the hallmark of state authority but confirm private interests according to territorial criteria and signal the absence of the state.

All of this allows us to think about the state, or to recognise the presence and the influence or the effects of the state, precisely in those places that appear peripheral—like remote resource extraction enclaves in Australia and PNG.

Two States and their Resource Journeys

In most respects, PNG and Australia are profoundly different. Australia can be classified as a developed settler nation whose 'development' was based on the dispossession of its Indigenous population, and it is the dominant political and economic nation state in the Pacific region.[4] PNG is a relatively young nation. Its Indigenous population has acquired control of state institutions, and retained ownership and control of their land and resources, but the majority of the population live comparatively less developed rural lives. The country remains heavily reliant upon Australian aid and it is often described by outside commentators as a 'struggling' or 'developing' nation state that exerts limited influence beyond its shores. While the differences are significant, and we could easily expand this list, the two nation states have at least one major thing in common beyond their shared colonial history: an extraordinary commitment to resource-driven economic development, underpinned by favourable geological factors and an abundance of mineral resources in their territories. In both countries the government has strenuously advocated for extractive capitalism, or what the captains of these industries prefer to call 'resource development'.

Large-scale resource extraction looms large in PNG and Australia and both countries can be described as 'resource-dependent nations'. Resource extraction has had a profound effect on their enmeshed histories and continues to be an important part of their national identities. Mining activities shadowed the birth of these two nation states and provided the economic base to support a transition away from their respective colonial masters—although in the case of PNG, resource extraction has provided Australia with new pathways to exert power over this sovereign nation. The expansion of the resources sector in PNG provides important economic opportunities for many of its citizens and is a primary source of revenue for the state, as well as thousands of Australian businesses and skilled workers who capitalise on these opportunities. Despite these asymmetries, their economies are mutually entwined in a history of extractive ventures—through investments, trade and business opportunities, and employment—and for the foreseeable future the extractive industries will dominate the economic, political and physical landscape across these two countries.

---

4   Although in recent years the influence and authority of Australia across the Pacific region has been challenged by the growing presence of China. See, for example, Crocombe (2007) and McDougall (2019).

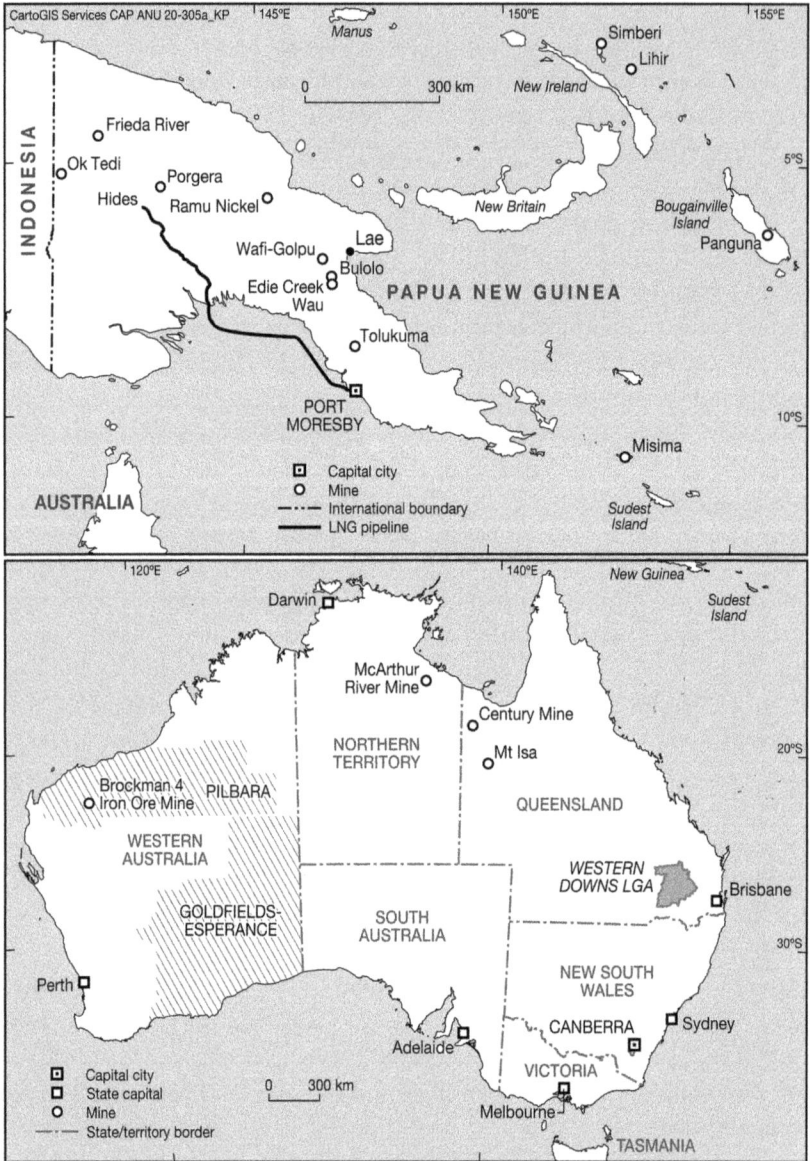

Figure 1.1 Map of Australia and Papua New Guinea with resource extraction projects and sites noted in this volume

Source: ANU Cartography

Australia

The extraction of mineral resources has long been part of the culture and development of Australia, and it has shaped the country as we now know it. It could be said that the first mining phase began some 60,000 years ago, with the first Aboriginal Australians who utilised resources of the earth for cultural and technological reasons and various forms of trade. Following European invasion in 1788, coal was soon being extracted to provide fuel for heating and cooking, and later steam locomotion. The first gold rush began in earnest in 1851 in the state of New South Wales and within three months had spread to the state of Victoria (Macintyre 2009: 87). These rushes quickly transformed the Australian colonies and fuelled the dynamic growth of towns and cities. Within just two years the number of free migrants arriving in Australia was greater than the number of convicts who had arrived in the past 70 years (ibid.). Many of these hopeful migrants, known as 'diggers', brought new skills and professions, and contributed to a burgeoning economy. The population boomed and Australia became a 'multicultural' society, of sorts. Migrants of non-European descent, especially Chinese immigrants who were particularly industrious on the goldfields, were unwelcome, signalling the xenophobic tensions that would continue to shape Australian society.[5] It is often claimed that the 'mateship' that developed between these diggers (of predominantly European descent) and their collective resistance to authority laid the foundation for a unified and unique national identity independent of British colonial rule.[6] In short, the early gold rushes transformed the colonies and helped precipitate settler Australia into nationhood.

The young colonies were quick to begin exporting minerals. During the 1850s, Victoria contributed more than one third of the world's gold output, and by the 1860s exports of copper and lead earned more than exports from wool and wheat. Following Federation in 1901, the role of mining in the economic development of Australia gradually increased. In the first half of the twentieth century, lead, zinc and copper deposits

---

5   The presence of thousands of Chinese immigrants, many of whom existed under forms of labour bondage, gave birth to the early expressions of anti-Chinese sentiment and the nation's first race-based migration restrictions, which later coalesced in the White Australia policy.

6   In 1854, 1,000 diggers assembled at Eureka in Victoria to protest against the imposition of mining licence fees. The Eureka rebellion has since become a favourite event in the national mythology of Australia.

were discovered in Queensland at Mt Isa, although their full potential was not realised until the 1950s. In the 1960s, following the Australian government's changes to the restrictions on iron ore exports, exploration commenced across the Pilbara region in Western Australia leading to the development of numerous large-scale mining operations. By the end of the twentieth century, Australia was one of the world's leading resource nations, processing and exporting huge amounts of iron ore and coal, and other valued minerals. In the early years of the twenty-first century, the industry continued to grow in size and significance. The post-2001 resources boom was unprecedented in its scope, scale and economic significance (Cleary 2011). By 2014 mining contributed almost 60 per cent of Australia's exports as resource companies scaled up their investments and their operations. What counted as a mega mine in the 1980s was just a 'normal' mine by the 2010s, and resource towns were flush with prosperity and opportunity. But, as Thomas Hyland Eriksen observed in his ethnography of a Queensland 'boomtown', the downside of such accelerated change is the rise of 'treadmill capitalism' where growth is placed ahead of other considerations (like sustainability), constant change is necessary in order to stay in the same place within the system, and, more importantly, where uncontrolled 'runaway processes' preclude political or popular governance of the system (Eriksen 2018: 104).

In recent years, the petroleum industries, comprising the extraction and processing of oil and gas, have played an equally important role in Australia. The first oil well was drilled in Western Australia in 1907, but the industry was not developed in a major way until the 1970s when massive gas and condensate discoveries (light crude oil) were made off the northwest coast of Western Australia. Over the last decade the Australian natural gas industry has grown rapidly, as part of 'global gas revolution'. Converted into liquefied natural gas (LNG), this new resource capacity primarily supplies international energy markets. In order to meet this demand, unconventional coal seam gas reserves have been rapidly developed, especially in the eastern states of Queensland and New South Wales, which has included controversial extraction techniques such as hydraulic fracturing, or 'fracking' as it is commonly known (de Rijke 2018). Fracking introduced new types of extractive infrastructure into the industrial landscape thus far dominated by large-scale projects typically centralised in single locations. Given that this gas is not located in easily accessible reservoirs, this usually results in numerous wells being drilled over large areas. Along with all of the other infrastructure that is required to transport and process coal seam gas—including gathering lines, compressor stations, well pad access tracks, and LNG processing and export plants—this means that the collective footprint of this industry is much larger and more reticular compared to other extractive industries, such as mining, which generally produces enclave spaces.

Papua New Guinea

PNG's resource journey has been somewhat similar to Australia's. Like Aboriginal people in Australia, Papua New Guineans had been extracting and utilising mineral resources for ceremonial and practical purposes for thousands of years. In the 1800s, the scale of extraction increased with European presence. When Captain John Moresby surveyed the southern coast of New Guinea in the early 1870s, he judged it was highly likely that the country contained an abundance of mineral resources. It was only a matter of years before a gold rush began, setting the future nation of PNG on a course towards resource dependency. By 1884 Britain had established a protectorate over the southeast portion of the country, which it called British New Guinea, but was later known as Papua, while the Germans claimed the northeast portion, which they called German New Guinea. Some 400 white miners had made their way from Port Moresby and north Australia to Sudest Island off the southern tip of Papua where gold was discovered in 1888, and then on to Misima Island in 1889 (Nelson 1976). In 1926 gold was found at Edie Creek in Morobe Province on the mainland, which led to the first major investment in the development of alluvial gold mining operations around Wau-Bululo in the 1930s (Waterhouse 2010).

The territories of Papua and New Guinea were combined under a single administration (or mandate) after the Second World War. In the following decades, the Australian administration came to the conclusion that the decolonisation of the territories could certainly be managed or even delayed, but definitely not prevented. Mindful that the future nation would need a stable economic base to support its independence, the administration set about mapping the natural resources of the country. In 1964 copper and gold deposits were discovered at Panguna on the island of Bougainville, and by 1967 an agreement to develop a mine had been reached and enacted by the House of Assembly. The mine was fully operational in 1972, just in time to help fund the territories' transition to independence in 1975, but not without some last-minute concessions to avert Bougainville's secession from the nation (Denoon 2000). The Panguna mine was operated by Bougainville Copper Limited Pty Ltd, which was a subsidiary of Conzinc Riotinto of Australia. The mine mobilised the largest investment in any Australian territory at the time and was therefore pivotal to the decolonisation process.

The new independent state set about growing the economy and over the following decades several large-scale gold and copper mining operations were established for this purpose. The giant Ok Tedi mine was opened on the opposite side of the country in 1984, followed by the Misima gold mine in 1989, the Porgera gold mine in 1991, the Lihir gold mine in 1997, and later the Ramu nickel mine in 2012, along with various smaller operations. The growth of this sector played out against the backdrop of the crisis that was unfolding back on Bougainville where the social and environmental impacts of the Panguna mine sparked a full-scale civil war in 1989, which forced the closure of the mine and ultimately cost some 20,000 lives (Regan 2017). Panguna represents a very special case of state presence and abandonment. The loss of revenue from Panguna increased the urgency and necessity to open new mines to make up for the shortfall—and rather than reassess the role of resource extraction, the state renewed its commitment to extractive capitalism. For example, the closure of Panguna led to a series of decisions regarding Ok Tedi's continued operations, ultimately triggering what Stuart Kirsch (2014: 133) described as 'a slow motion environmental disaster' as a result of riverine tailings disposal.[7]

As in Australia, while mineral extraction continues to play an important role in the extractive sector in PNG, the petroleum industry is growing in scale and influence, having established itself in PNG shortly after it had developed in Australia, with the first oil exports from the Southern Highlands occurring in 1992. The most significant development to date has been the massive PNG LNG project that pipes gas extracted from Hela Province in the highlands down to the huge processing facilities on the outskirts of Port Moresby, where it is then shipped to international markets (see Main, Chapter 5, this volume).

---

7   At other mines, like Misima and Lihir, deep sea tailings placement, or DSTP, has been regarded as an alternative, because tailings dams are vulnerable to earthquakes. Both riverine disposal and deep sea tailings placement practices are banned in Australia, but tailings dams are favoured because the country is much less prone to earthquakes.

Regimes of Resource Ownership

Looking at legal and policy frameworks in PNG and Australia, it would appear that Papua New Guineans are better placed to benefit from resource extraction than their Aboriginal Australian counterparts. This conventional wisdom is mostly based upon the differences between the legal regimes that operate in these countries in relation the customary ownership of land. In both countries, sovereignty over all minerals is vested in the state. In Australia landowners are the owners of the surface of the land and have no automatic right to the minerals that may be found underneath the surface.[8] While surface minerals are not mentioned in Australian legislation, in PNG customary landowners have an automatic legal right to harvest minerals *on* the surface of the land or in streams and rivers and many communities, including near large-scale resource extraction projects, engage in alluvial gold mining activities (Bainton et al. 2020).[9] In PNG the state grants leases to private companies to extract and develop these mineral resources, whereas in Australia this responsibility sits with the relevant subnational territories and state governments. In both jurisdictions access agreements must be negotiated with landowners, and compensation paid for disturbances. The major point of difference can be found in the legal recognition of customary interests in the land and the right to benefit from resource extraction. In PNG customary land rights are enshrined in the Constitution, which means that under the terms of the *Oil and Gas Act 1998* and the *Mining Act 1992*, the primary legislation governing resource extraction, customary landowners have a more direct stake in the development of any resources located within their ancestral territories. Resource companies are required to compensate the landowners for disturbance or damage to their land. They are also required to enter into benefit-sharing negotiations with landowner communities, and these communities can then expect to receive a range

8   In Australia, petroleum and mineral resources are generally the property of the Crown. There was a time when the common law position stated that the minerals belonged to the landowner, and that the minerals were regarded as an inherent product of the land itself. Common law assumed that whoever owned the land also owed the soil below and the space above the surface. The qualification was the right of the Crown to extract gold and silver, characterised as 'royal minerals'. This common law position was circumscribed with the passing of specific legislation that vested the ownership of minerals contained within the soil of private landholdings in the Crown (Hepburn 2011).

9   Section 9(2) of the *Mining Act 1992* says that: 'Any natural person who is a citizen may carry out non-mechanized mining of alluvial minerals on land owned by that natural person, provided that the mining is carried out safely and in accordance with the *Mining (Safety) Act 1977* and that the land is not the subject of a tenement (other than an exploration licence).'

of benefits from these projects (Filer 2012; Bainton and Jackson 2020). This is not to dismiss the tensions which arise between different groups with competing claims to the same areas of land, or the problems associated with 'mobilising' customary land for development (Filer 2019). However, the legal recognition of customary land rights in PNG shifts the question from recognition of those rights, to identifying specific groups in possession of those rights and the subsequent capture of benefits.

In the Australian context, the *Native Title Act 1993* dictates that Aboriginal people must prove a connection to their ancestral lands in order to exercise their interests, which is a rather difficult task for many groups given the onerous legal hurdles associated with native title (Weiner and Glaskin 2007; Glaskin 2017). This situation is compounded by the history of Aboriginal dispossession, including the founding fiction of terra nullius, which essentially held that the continent was 'no man's land', and the fact that so many groups were forcibly removed from their land to reserves and missions. Other groups were massacred en masse in the frontier wars, which makes it doubly difficult for their descendants to demonstrate lines of ancestral continuity. As Sarah Holcombe and Gareth Lewis discuss in their chapters for this volume, this early history of violence and dispossession set the tone for the relationship between Aboriginal communities, resource companies and the Australian state, and continues to impact the ways that some Aboriginal groups in Australia relate to the state both as an idea and as a system. Aboriginal groups who are granted recognition of their traditional ownership of specific areas are entitled to the right to negotiate with any entities seeking to develop the minerals resources on their land, but it doesn't grant them veto rights. The exception to this rule can be found in the earlier *Aboriginal Land Rights Act 1976*, which grants veto rights to owners of 'Aboriginal Land' in the Northern Territory. Similar to PNG, in some cases this can result in negotiated agreements for the provision of specific benefits and payments (O'Faircheallaigh 2016).

The question of ownership extends beyond land and resources. Following global neoliberal trends, the Australian government has steadily reduced its stock of state-owned enterprises, which are sometimes termed 'government business enterprises'. While the government has retained ownership of various energy and infrastructure companies, there are no state-owned extractive companies operating in Australia. In PNG, the processes of privatisation have been less extensive. The state has maintained ownership of companies providing essential services and established several enterprises to consolidate and manage its interests in the business of hard rock mining and petroleum development.[10] Among other things, this has enabled the state to become a major partner in the PNG LNG project, and to assume ownership of the Ok Tedi and Tolukuma mines. However, the acquisition of these mines took place in rather special circumstances and should not be confused with the government's statutory right to acquire a minority stake in joint ventures with foreign investors. While the latter may sometimes appear to outside observers as a 'commitment to nationalisation', it is better interpreted as a commitment to an option for 'participation'. The growing public and political discourse about the need to actually nationalise the resource sector is partly influenced by ideas about resource dependency that were prominent in the early independence period (Amarshi et al. 1979), and continuing concerns about the ways that powerful foreign states and companies maintain their economic and political advantage by exploiting the raw materials of less developed countries. There is a popular expectation (in some quarters at least) that the state should acquire ownership of all resource projects so that PNG can reverse these trends and reap a greater share of the financial benefits derived from the exploitation of its natural resources. But this pathway to development also entails a fundamental conflict of interest as the state becomes both a shareholder and a regulator of these projects. In the case of the Ok Tedi Mine, the state made regulatory decisions that sought to minimise operational expenditures and maximise economic returns as both a shareholder and a tax collector. Many of these decisions were made at the expense of people living downstream from the mine. This kind of situation emphasises the very selective presence of the state and embeds the state in a web of conflicting interests. This includes conflicts emerging from the state's dual role as a regulator and an investor, and extends to internal conflicts between different state functions and institutions.

10   In 1975 the Mineral Resources Development Company (MRDC) was established as a 100 per cent state-owned company. Originally established to acquire and manage state and landowner interests in mining and petroleum projects, including management of equity funds for landowner companies, in 2007 the state's equity interests were then transferred to Petromin Holdings Limited, and the focus of MRDC was limited to landowner interests. In 2015, the government passed the Kumul Minerals Act, consolidating the state's oil, gas and mining interests, and Petromin was restructured into two companies—Kumul Mineral Holdings Limited and Kumul Petroleum Holdings Limited.

Resource Development

While both countries can be considered resource-rich nations, PNG is often described as suffering from the so-called 'resource curse' or the 'paradox of plenty', which generally holds that countries with an abundance of natural resources tend to experience less economic growth and worse development outcomes (Gilberthorpe and Rajak 2017).[11] This term is less often used to describe the Australian experience of extractive-led development, although some commentators would certainly not agree that Australia has avoided the resource curse (i.e. Cleary 2011). In PNG, the curse is often thought to be linked to faulty and ineffective governance, or 'maladministration' (Burton 1998), where the state is unable to curb the excesses of the extractive industries or convert natural resources into sustainable forms of development. Anthropologists and other social scientists working in the country who have written about the 'resource curse' have generally focused on the ways in which it is manifest through increased social conflicts and environmental impacts and the forms of structural violence that are created and intensified by large-scale resource extraction projects (see Filer 1990; Haley and May 2007; Banks 2008; Gilberthorpe and Rajak 2017; Allen 2018). In Australia, on the other hand, analytical focus is usually directed at the macro-scale, which might go some way towards explaining why the country is not generally thought to be afflicted by this curse. But when we shift our focus to the local scale, especially to the Indigenous estate, the picture begins to look rather different and in many cases pre-existing forms of discrimination and structurally unjust forms of political economy have been exacerbated by resource extraction (Langton and Mazel 2008; Altman and Martin 2009). It now becomes clear that the negative force of the resource curse should not be understood as a purely economic issue, but it can 'also be portrayed in political, cultural, or environmental terms, which means that the resource curse may have several dimensions in the same country or region' (Filer and Macintyre 2006: 217).

The presence of the resource curse can be reconceptualised in terms of the absence of development that resource extraction promises to deliver—yet another demonstration of the absent presence of the state. Continuing with the metaphor of Godot is apposite, as some communities have

---

11   The concept of the resource curse was first elaborated by British economist Richard Auty (1993). See also Karl (1997) and Ross (2003). See also Ross' (2015) examination of a relationship between 'resource curse' and state governance.

maintained their enthusiasm for extractive-led development despite the known impacts of extraction and the fact that the anticipated benefits, or the government and the developers, have simply never arrived. Many of these communities find themselves anxiously 'waiting for company' (Dwyer and Minnegal 1998). Several authors in this volume describe situations where communities are waiting for certain forms of development, which, like Godot, ultimately remain unknown and uncertain. In both national settings this absence can be partly understood as result of the failure of the state to bring 'development' (economic opportunities and services) to many rural areas. In the Australian context this reflects the extent to which most Aboriginal people have been marginalised from the mainstream economy and the opportunities enjoyed by the majority of the population and the desire for self-determination (and the ways in which resource extraction is imagined as a possible pathway to such ends). In PNG this enthusiasm is expressed through a particular set of beliefs, which Colin Filer has called 'the ideology of landownership'. This ideology generally holds that the road to development can be found in the compensation and benefits that are provided by resource companies to customary landowners in return for the extraction of the natural resources located within their ancestral lands (Filer 1997). Of course, commitment to this ideology does not mean that people are not wary of the impacts of extraction, but it does highlight a particular expression of local sovereignty and understanding of development. In both settings the prospect of resource extraction has generated strong contestation between groups who want these activities to proceed in the hope that this will improve their circumstances and enhance their autonomy, and those who remain unconvinced by the 'value proposition' of extractive capitalism.

In Australia this resistance is often connected to the broader Aboriginal land rights movement, but it also finds expression among a coalition of environmental activists and pastoralists in, for example, the 'Lock the Gate Alliance' that seeks to prevent further fracking operations (de Rijke 2013; and Espig, Chapter 6, this volume). Enthusiasm and resistance at the local level are also mirrored at the national level. Back in 2017 when he was still the Treasurer of Australia, Scott Morrison, who is now the country's prime minister, brought a lump of coal into the House of Representatives as a political stunt to indicate the government's support for the coal industry, coal-fired power stations and a carbon-intensive economy (Murphy 2017). Morrison's support for the extractive industries reached new heights in 2019 when he signalled his intent to clamp down on individuals and organisations that protest against the impacts of extraction.[12] The government's position has generated much unrest, and in the wake of the 2019 and 2020 mega fires across large parts of Australia, there have been increased calls to stop all new coal mining developments to help prevent further overheating of the globe. The sense of the failure of the state to respond to the emergency caused by the fires was made worse by the prime minister's initial absence, and then his widely debated 'inappropriate' and 'inadequate' presence in the affected areas (Brett 2020). While the current conservative government in Australia wants to remove the barriers to expand global extractive investment in Australia, and has proposed a 'gas-fired recovery' to lift Australia out its COVID-19 induced economic recession, the PNG government has sent a more nationalistic message about containing mineral wealth within the country. In his 2019 'manifesto' to the nation, the current prime minister of PNG, James Marape, pledged that he would 'take back PNG' and turn the country into the 'richest black Christian nation on earth'. While these political slogans encapsulated a very broad agenda for reform (and represent a direct response to past government failings), his manifesto contained specific commitments to ensure that Papua New Guineans will obtain a much larger share of the benefits arising from exploitation of the nation's natural resources, positioning extraction at the forefront of the future prosperity of the nation. These sentiments are best encapsulated in the tagline used by the government regulator of the industry, the Mineral Resources Authority: 'minerals for life'. However, in the event, the reality of extractive capitalism frequently belies the promises of 'resource development'. At this point, the experiences of customary landowners in PNG begin to converge with the experiences of traditional owners in Australia.

Implicating the State in the Extraction of Resources

The state is often perceived by different groups as habitually failing to be present in meaningful ways in resource extraction contexts. Of course, it can also be argued that this failure has occurred in many 'remote' areas

---

12   In a press release titled 'Australian mining: Building a better future', the Minerals Council of Australia (the peak representative body of the Australian mining industry) welcomed this political support (MCA 2019).

where resource projects are not present. And, in the PNG context, the state's absence is just as marked in areas where logging companies or palm oil companies have large-scale operations as it is in areas where mining or petroleum companies operate (e.g. Tammisto 2016). Nevertheless, in many cases, including a number of examples presented in this volume, the state is perceived as being more present for some groups and less present for others, raising questions about the state's complex and multifaceted obligations towards local communities, foreign investors and broadly conceived goals of national development. In both PNG and Australia it is common to hear complaints from different quarters that the state has abandoned its responsibilities in relation to resource extraction; overlooked or overridden its primary duties to its citizens; forgotten that some portions of the population even exist; left corporate entities and local communities to their own devices; that it is working against the growth of the resources sector and blocking industrial progress or, on the other hand, that it is too closely aligned with the interests of multinational capital, or has outsourced its responsibilities to the private sector, and so on.

Rather than trying to unpack all these conflicting expectations and critiques, it is useful to first consider what is significant and unique about resource extraction and how the state is implicated, experienced and brought into being through these activities. Several structural features of resource extraction are important in this regard. First of all, states reinforce the primacy of their sovereign authority through their exclusive claim over the mineral resources in the land. The state's supreme authority is symbolically upheld through formal legislative frameworks that govern the ownership and extraction of these resources. States open up land to facilitate development, and in some contexts, where states are struggling under macroeconomic constraints, the revenues generated by resource extraction allow states to continue functioning as states. Perversely, in countries like PNG, economic dependence upon extraction, inaction and limited capacity also weaken the ability of the state to exercise its sovereign power, contributing to the experience of a 'resource curse'. Similar charges have been made against the Australian state, which remains heavily dependent on mineral exports and has been captured by the interests of the industry at all jurisdictional levels. Second, large-scale resource extraction generates profound transformations. It extends capitalist development into places where the capitalist economy is not institutionalised or places where it is characterised by petty commodity production or agrarian

21

capitalism, such as the development of large-scale mines in frontier zones in PNG. In these instances, it introduces a form of capitalism that is characterised by accumulation on a large scale combined with significant forms of dispossession and disruption. It is not uncommon for conflicts between local and national interests to play out in those arenas, and for national political leaders to invoke the 'greater national good' to justify locally experienced impacts of extraction and the creation of 'sacrifice zones'. Third, large-scale resource extraction also changes the dynamics of existing capitalist arrangements by introducing new forms of capitalism into areas that are already incorporated into the global capitalist economy, such as the development of coal seam gas operations in the pastoral regions of Australia. It ultimately affects the geography of capitalist development and accumulation in ways that are related to the nature and geography of natural resources, and the geography of prior political and economic arrangements. And as we discuss below, these structural features point to a temporal dimension as the state's presence or absence alternates through different phases of the resource project cycle. As the authors in this volume show, presence often turns to absence once approvals and licences have been granted.

The rapid and unruly socioeconomic change generated by these projects constitute forms of immanent development in the sense conveyed by Cowen and Shenton (1996). Under these conditions, there is a heightened need for state-systems to manage or offset the effects of these transformations—and it is here that we frequently encounter the absence of the state. Indeed, local communities often demand an increase in the presence of the state and regard resource extraction as a way of securing state services, including law and order, health and education services, and public infrastructure. But in lieu of a more active role, states have tended to delegate greater regulatory responsibility and discretion to resource companies—which may constitute a form of 'indirect rule', where the state is anchored in project locations through the delegation of its functions to extractive corporations.

Large-scale resource extraction is also implicated in a range of intentional interventions that are geared towards particular ends. These interventions are mainly directed towards the management or governance of people, policies, landscapes and resources in ways that secure resource investments and extend extractive capitalism, as well as other sociopolitical projects. Some of these interventions are designed to secure local support for these ventures, which can take the form of community development programs and investment in other forms of 'intentional development' that are designed to mitigate the immanent processes associated with resource extraction (Banks et al. 2017). Here we begin to see a convergence of interest between states and private companies, both in terms of function (as companies assume state responsibilities) and style (creeping bureaucratisation and managerialism). In this respect, extractive companies are involved in a form of 'trusteeship' that is more typically associated with the state (Cowan and Shenton 1996). By assuming some responsibilities for the local population, resource companies also play a role in defining aspects of a desirable future for that population. More importantly, evidence from both countries suggests that resource companies invest considerable energy and resources into steering populations towards that future, framing it in terms of 'progress' and enhanced personal and communal opportunities.

States also seek specific goals through resource extraction. In most jurisdictions, including PNG and Australia, the state holds sovereignty over mineral resources and notionally claims that these resources will be used for the collective benefit of the population. It is important to recall that states are not passive custodians of the common wealth. They may intervene through policies and legislation that enable the expansion of extractive capitalism within their national territory, or they may seek to constrain extractive activities in certain areas, such as conservation zones. The state issues permits to companies to develop these resources, and it is expected to perform a regulatory function to ensure that these activities occur within prescribed parameters. The state may be a direct financier or shareholder in some resource operations, meaning that in addition to its regulatory role it has a direct stake in the commercial success of individual ventures (which can constitute a contradictory role). And in many instances the state—through the relevant branches of government—will enter into various agreements with the parties to these projects and make specific commitments as part of the overall compensation and benefits sharing process.

There can also be an inter-state dimension to resource extraction, both in terms of international trade in resources, and attracting and managing foreign direct investment, as well as using extraction and trade in resources to achieve economic and political aims in international arenas as well as domestically. For example, the complete absence of the PNG state at the Ramu nickel mine in PNG has opened up a space for the Chinese state (as the owner of the company developing this resource) to exercise forms of 'soft power' within this sovereign state as part of its (somewhat unsuccessful) attempt to retrofit this project into its Belt and Road Initiative (Smith 2018). Although states pursue a range of political and economic goals through resource extraction, these may not always align with corporate interests, and it is for this reason that states sometimes grant tax holidays, stabilisation agreements, and subsidies, or provide supporting infrastructure to reduce these differences and make certain investments more attractive, allowing capital to accumulate in some areas and not others. The general point is that in the context of resource extraction we find that the state is thoroughly implicated, but it is also selectively present—at different times, and in different places, it will be more or less present and more or less aligned with the interests of different groups. Depending upon the vantage point, the state will cast a longer or shorter shadow and its presence or its absence will be felt in different ways.

Encountering the Absent Present State

For ease of navigation we have organised the chapters in this volume in terms of regional focus, connected via the two chapters concerned with gas extraction in PNG and Australia. The final chapter by John Burton and Claire Levacher introduces New Caledonia as a point of comparison to PNG and Australia. They explore whether the notion of 'absent presence' can be applied to a national context where the state is of a different kind to that in PNG and Australia, and most notably, as they argue, where it cannot absent itself. The volume concludes with a short afterword reflecting on the uncertainty of the state.

Between them, the contributing authors discuss cases of proposed mining ventures, existing large-scale mining operations and the extraction of natural gas (both coal seam gas and LNG). While these two national settings are distinctive in their own right, several interconnecting themes can be found, including particular types of engagement with the state, and specific experiences of the absence, the presence and the absent presence of the state in extractive contexts. Although these three phenomena are found in each, they have different weight in every chapter.

The most prominent theme running through the volume, and the most common way in which the absence of the state is manifest, is the failure of the state to fulfil its responsibilities (both stated and perceived). These specific failures, silences and expressions of unreliability create a range of effects that evidence the presence of an absent state. Although the details of these effects differ between the cases, state effects are nevertheless present in all the chapters and constitute a second core interconnecting theme. These two themes speak to spatial and temporal dimensions of the absent presence of the state, and the materiality and morality of this absence—all of which are explored in the remainder of this introductory chapter, and in the chapters that follow.

Failure as Presence

In one way or another, each chapter speaks to the responsibilities of the state in extractive contexts, and the state's performance against those obligations. In democratic theory, the modern state legitimates itself through a simple bargain: citizens pay tax and agree to obey its laws, and in return, the state protects its people and keeps them safe. The inability of states to keep that pact in respect of resource extraction appears to be increasing, especially in Australia where commitment to the coal industry will exacerbate climate change and constitutes an existential threat to the security of the nation (Brett 2020). Among other things, state responsibilities encompass regulatory oversight of extractive processes; protection from human rights abuses and other forms of harm; the monitoring and management of the social and environmental risks and impacts of extraction; implementation of agreements and other commitments including the distribution of benefits to project area landowners, and the delivery of services and programs or the construction of public infrastructure; participation in governance processes surrounding resource extraction including involvement in forums, working groups and reporting initiatives; the general maintenance of 'order' in these extractive sites; and so on. These and others are intertwined and actively negotiated in the context of resource extraction.

While failures of the state can be taken as a demonstration of state absence or impotence, they can also be reconceptualised and experienced as a form of presence, what we term an absent presence. Such failure often informs experiences and engagement with the state, and these failures are often revealed through a series of state effects. Each of the contributing authors observes moments where the state is encountered through a series of lacks and problems, unreliability, ineffectiveness or deception. These encounters are neither straightforward nor uniform, and paradoxically, these failures and absences often lead to an enhanced sense of state presence.

Locating the state in contemporary resource extraction contexts is not always a simple task. A number of contributors describe instances of what David Szablowski (2007) calls the 'selective absence of the state', or what others in this volume have inversely termed a 'selective' or a 'strategic presence'. For example, both Gareth Lewis and Sarah Holcombe explain how the legal environment in Australia's Northern Territory allows companies to operate with minimum regulatory oversight. Lewis describes the systematic marginalisation of Aboriginal people around the McArthur River Mine and argues that, from a local Aboriginal perspective, the state has essentially absented itself from the project, failing to enforce existing legislation designed to protect the rights and interests of Aboriginal people. Holcombe uses a case study of a junior mining company seeking to operate on Aboriginal lands to show how the state has strategically acted to pursue extractive capitalism, and how the state has been significantly more present in its role as an enabler for the industry than a protector for local Aboriginal communities and environments. In a similar vein, Martin Espig describes a 'partially absent' state in the context of coal seam gas developments in western Queensland, where the rapid development of multiple gas projects outpaced the regulating departments of the state and pushed them beyond their capacity once developments fully commenced. These 'runaway processes' (Eriksen 2018) reveal the contradictions at the heart of contemporary extractive capitalism, and the double bind that encompasses the state, suspended between short-term growth imperatives and long-term goals of stability and sustainability.

When PNG gained independence from Australia, 'many of the new state's claims to legitimacy were based on the promise that all Papua New Guineans could expect development to come their way' (Jorgensen 2007: 58). The inability to deliver upon this promise has seriously eroded the legitimacy of the state in the eyes of many citizens, which partly explains why so many rural communities hope that extractive companies might come along and fill this gap, and the willingness of the government to allow this to happen. In PNG and Australia, the state habitually reduces delivery of services around mining projects. As resource companies 'have become increasingly involved in "community development" roles, often promoting the newly acquired corporate language of "sustainability" and "partnerships"' (Banks 2008: 24), this extended corporate role often results in the withdrawal of the state from some of its functions and responsibilities, in what is effectively a privatisation of state functions. For instance, in the Lihir islands, Nicholas Bainton and Martha Macintyre have observed how the state has consistently failed to fulfil its commitments under the terms of the agreements that it has signed for the development and operation of the gold mine. The gap created by absent state actors has produced a far more intense relationship between the community and the company characterised by 'mutual incomprehension' and antagonism, and a very high degree of local dependency (Bainton 2021). In the absence of the state, the community have 'manoeuvred' the company into the space ordinarily occupied by the state, forcing the company to provide social services and infrastructure that are the responsibility of the government.

Alex Golub's historical anthropology of the Porgera mine reminds us that we cannot take collective actors like 'the mine', 'the state' or 'the landowners' for granted, and that an accurate account of history must also focus on networks of people who appear in different times and different situations as personating one or the other of these entities. From this vantage point, we can begin to really grasp the interpenetration between collective actors and how the absence or the presence of the state is experienced over time.

In her work on kinship and law, Marilyn Strathern theorised the differences between 'categorical' and 'interpersonal' relations, demonstrating that while these different relations animate each other, these types of relationships are not always present at the same time (Strathern 2005). In PNG, the state exercises a substantial categorical presence through its legislative and permitting powers, but this may not equate to a relationship. Emilia Skrzypek describes how the absence of an interpersonal relationship between community members and state actors at the Frieda River Project in the remote Sepik region led local landowners to feel like they could never really get to know the state. Instead, they invested their hopes and energy in the relationships they developed with company personnel, which they came to consider as more important, more consequential and more effective in pursuit of their development aspirations. Not enacted through relations, and not made visible through delivery of services, the state constitutes an ambiguous, uncertain and absent presence at Frieda River where local communities look to the company to fulfil what would traditionally be considered roles and responsibilities of the state, and to facilitate the missing relationship with state representatives. Across the mountain range, at the PNG LNG project, local landowners staged an attack on project facilities in an attempt to use the company as a proxy to get the attention of the government and voice their anger over unpaid royalties and unfulfilled promises of development. Michael Main argues that their anger was really directed at the state and not the company, who they regarded as 'a pawn in a larger game of corruption that was being played out by various levels of government and the judiciary'. Other chapters also speak to the lack of interaction between communities and state representatives and the effects of these non-existent relationships.

While the chapters by Skrzypek and Main describe instances where the state fails to assert effective presence, Jo-Anne Everingham and her co-authors describe an instance where the state did assert a presence but failed to maintain it. Their chapter documents the changing presence of the state at the large Century zinc mine in the Gulf of Carpentaria in northern Queensland since the negotiation of the Gulf Communities Agreement began in 1992. Like many resource extraction projects in Australia, the Century Mine evidences the tensions between the state's responsibilities to ensure economic development on the one hand, and the well-being of regional populations, especially Aboriginal communities, on the other hand. The negotiation of the Gulf Communities Agreement was the high point of the mine; it set out a development vision that aimed to reduce welfare dependency among Aboriginal people in the region. A range of state actors engaged intensively with Aboriginal groups at the outset of the project to help establish the various representative committees and bodies that would form the institutional framework of the agreement and play a central role in implementing its objectives. But as priorities changed within the Government of Queensland, combined with considerable turnover in government personnel during the operational years of the mine, the state's actions were often at a remove from the region, important commitments were not fulfilled and local dependency upon the mine increased. The authors of this chapter argue that rather than seeing such resource contexts as 'stateless spaces', where the state plays a limited role, they can be understood as spaces where the state takes 'action at a distance' to pursue its ends indirectly, through the actions of others, in this case the mining company.

State Effects

In mapping the effects and the presence of the state, anthropologists have generally looked for signs of administrative bureaucracies, regulation, service provision and other technologies of governance that support larger order-making agendas. Thinking of the state in terms of these order-making functions leads us to consider those places where the state has been unable to impose its order. Golub's historical account of law and order in Enga Province, provides a case in point. The presence of gold, and in later years the development of the Porgera mine, altered the history of policing in the Porgera valley, challenging the state's monopoly over violence, and law and order. The understaffed and under-resourced police force contrasts with the company's well-resourced private security force. However, as Golub explains, the picture is more complicated as many police officers were also members of the mine's security team, and the latter often supplied resources to the former. As a result, law and order has gradually been recast in terms of 'security', as order is upheld by privately contracted security firms.

The limited presence of the state in those rural areas where mines are often located has turned these projects into new sites of 'governmentality', which can be defined as the ensemble of disciplining processes, institutions and tactics that attempt to shape human conduct—or how conduct is governed (Dean 1999). As companies are forced into closer contact with surrounding communities they often assume some of the responsibilities of the state including the provision of social services and infrastructure, development and implementation of programs, and greater levels of monitoring and self-regulation. In assuming some degree of responsibility for classic state functions, resource companies shape local desires and configure habits, aspirations and beliefs. They may act with a notional 'will to improve' (Li 2007) the local population, but they do so in ways that are rationally calculated towards the larger goals of extractive capitalism. In doing so, the actions of resource companies constitute these locales as authoritative and powerful, as different agents are assembled with specific powers, and particular domains—like lease land areas, group identification or the receipt of benefits—are constituted as governable and administrable (Dean 1999: 29). As sites of governmentality, extractive projects produce 'extractive subjectivities'. This occurs by reworking forms of exclusion and inclusion in ways that serve the interests of extractive companies, by reshaping lifeworlds in areas targeted for investment, and by transforming state–society–market relations, with companies assuming state-like functions and subjecting them to capitalist logics (Fredricksen and Himley 2019: 59). This ultimately results in new 'extractive subjects' coming into being.

These new forms of governmentality often expose the shortcomings of the state, and the inability of companies to meet local expectations for development. In Australia, Espig describes instances where, let down by the state and increasingly distrustful of its ability to regulate and monitor the industry and associated risks, citizens stepped in and took it upon themselves to hold companies accountable for their actions. This is representative of a process addressed by several other authors in this volume, where the absence of the state incentivises people to deal directly with the company, facilitating new forms of relations, interdependencies and interactions. As the state retreats from certain spaces, companies and community members can become 'responsiblised' as they assume greater levels of responsibility for managing various health and social risks related to these developments. Drawing from the work of Ulrich Beck (2009: 193), Espig argues that these runaway developments have resulted in a situation where 'no individual or institution seems to be held specifically accountable for anything'. This resonates with a point made elsewhere by Filer and his co-authors that in PNG, and from the perspective of the mining industry, the resource curse takes the form of a 'responsibility vacuum', which makes partnerships with host governments and local communities 'difficult, dangerous or downright impossible' (2008: 175).

State effects also assume a material form. For example, the original agreements for the Lihir gold mine contained a commitment by the state and the company to build and maintain a road around the main island where the mine is located. After some 20 years of mining operations, the road remained unsealed. While the company eventually took steps to seal half of the road (fulfilling part of its original commitment), the state has not fulfilled its part. The materiality of the road, and indeed the mine itself and the various other pieces of public infrastructure that are promised as benefits of extraction, invokes the dynamic relationship between the presence and absence of the state. Penny Harvey makes a similar observation in her ethnography of roads in the Peruvian Andes. She engages the state ethnographically through the material 'traces' and 'effects' which open out, rather than close down, further perspectives on the structures and practices through which specific modes of state power are effected and reproduced (Harvey 2005: 131). Like the Andean roads, the local Lihir ring road produces numerous state effects. The road is a contested space. It has become the site of political struggle between the company and the state as they both shift the responsibility for completion. It is a point of conflict between the various levels of government who are signatories to the original agreement, and it is a source of acrimony as the community demand that the company fill the gap left by the state and the potholes left by its trucks. As 'state effect', the road provides a complex material site through which people experience and negotiate their relationships with power (ibid.: 134). Standing on the incomplete road, the state then appears in a multivalent form—as the object of people's desire, and the source of bitterness and the feeling of abandonment.

A Multivalent Presence

Anthropologists often study states from the 'margins' (Das and Poole 2004), or indirectly through the prism of nation and nationalism and related discourses (e.g. Herzfeld 1987; Handler 1988). Ethnographers tend to conduct much less fieldwork in the 'halls of power', where state political power is articulated and directly wielded.[13] In a geographical sense, resource extraction sites are often peripheral places, distant from metropolitan centres where state power is concentrated, adding to the complex task of locating the state in extractive contexts. However, the processes of resource extraction occur both at the points of extraction as well as in major capital cities and industrial hubs, where companies are headquartered in close proximity to financial markets and the functions of the state that enable extractive capitalism. The various processes and effects of extraction also take place between these spaces, and beyond.

Many of the communities who reside around resource extraction projects, or who desire or fear them, inhabit 'marginal spaces' characterised by an absence of roads, schools, hospitals and other services that are signs of the presence of the state. Thus, the geographical distance often translates into physical and moral absence of the state and can be used to contextualise specific kinds of state behaviour, such as the withdrawal of government functions. As the contributing authors show, the state as a macro-actor is rarely held responsible for its promises and commitments in extractive contexts, which is partly due to the combination of physical distance and

---

13 Exceptions from the region include Santos da Costa (2018).

ideas that circulate about the 'natural' marginality of these places and the people who inhabit them, especially Indigenous people. This also reinforces the value of thinking about the nature of the state as a conglomerate of actors and institutions, rather than a collective actor. Skrzypek recalls a government representative explicitly telling community members at the Frieda River Project: 'Before, you all lived in the bush and you had nothing. The government is too far away, it will not help you. If it was not for the company you would still be in the bush with no development, nothing. If you don't support the company, you will go back'.

The geographic distance between centres of national and regional governance and the often-remote sites of extraction means that communities in those locations find themselves subjected to decisions and policies made away from local contexts. Lewis describes the experience of those communities affected by the McArthur River Mine who live a long way from centres of state power. 'Absent decision makers' have long 'wielded power from afar', which has contributed to their experience of the state as a distant and disinterested institution on the one hand, and as a powerful and malevolent presence on the other hand. He adds the neoliberal state to a long list of 'broken promise men' who have altered the course of the mining project and Aboriginal people's lives over time. An even more extreme case of wielding power from afar is described by Burton and Levacher in the context of New Caledonia where the industry was directly impacted by events and policy changes decided in France, literally halfway around the world. The task of locating the state in resource extraction contexts gets even more complicated if we consider other kinds of peripheralities at play, particularly those associated with asymmetries of power, representation and decision making experienced by Indigenous people. Taking stock of legislative systems, procedures and effects, both Lewis and Holcombe locate Australia's Aboriginal communities on a kind of political periphery, drawing links between geographic remoteness and a history of violence and political and socioeconomic peripheralisation which combine to push Aboriginal communities to marginalised and compromised positions vis-à-vis the state and multinational corporations.

The temporal dimensions of absent presence also emerge across the chapters. Writing about the temporality of the government in PNG, Dalsgaard (2013) describes how the state is made present in time, for example through rhythms, schedules and engagements. In both national contexts discussed here, the state as a regulator influences temporalities of resource extraction projects and is present in time through, for example, specific initiatives and interventions (as discussed by Golub), negotiated agreements (as illustrated by Everingham et al. and Bainton and Macintyre); and changes in the regulatory and governance environment, including those related to the critical question of national sovereignty, within which the industry operates (Burton and Levacher). The presence of the state in time can also be experienced through its failure to be present at a time when this presence is expected. For example by delaying the distribution of royalties to affected communities (as discussed by Main), or where sluggish state processes are outpaced by a fast-moving industry, increasing potentially risky gaps in industry permitting and oversight (as elaborated by Espig).

While these examples show how the state can be made present in time, it is also present (or absent) over time. Most of the chapters in this volume take a broader historical view to show how the state was present (and absent) prior to extraction, and how and if this changed, or remained the same, in the context of extraction. Thus the presence and the absence of the state should be considered across the entire mine life cycle. For example, Everingham and her co-authors also trace the government's receding involvement over time at the Century Mine. And at the contentious PNG LNG project, Main locates experiences of the state in a local ontology that interprets the state's action and inaction in terms of ancestral prophecies of abundance and a historical trajectory of decline. The spatial and temporal configuration and experience of such encounters with the state, and its arbitrary exercise of power, can also be thought of as maps of frustration or horizons of disillusioned expectations. These are spaces where the legal claims and commitments of the state—and the assurances of the state—become unsettled by the practices of state representatives and government departments and their emissaries. Where state-systems and the actions of its representatives and agencies fail to dissolve the tension, people may take it upon themselves to address the dissonance.

Lewis provides one of the most 'colourful' examples of local protest over the failure of the government to make itself present in meaningful ways. In order to voice their opposition to the McArthur River Mine and gain support for their cause, community members launched an emotive exhibition of photographs and artwork in Darwin. The exhibition drew attention to the intrinsic cultural and spiritual value of their land, seemingly disregarded by the state in its ardent support for the mine. Their artwork stood as a moral critique of the state's absence, or its uneven presence, and the effects of state sponsored extractive capitalism. In a similar vein, Main shows how Huli landowners attempted to hold the government responsible for unfulfilled promises of benefits and development at the PNG LNG project. He argues that local landowners have interpreted a range of events and processes, including natural phenomena such as a recent devastating earthquake, as a direct effect of what they regard as the 'immoral absence of the state'. He writes that Huli people saw the immorality of the state as a 'material threat to the future aspirations of Huli' and their prophetic entitlement to benefit from their resources.

Conclusion

In this introductory chapter we have claimed that resource extraction provides a useful starting point for studying the state and its effects. To paraphrase Clifford Geertz (1973: 22) who once argued that anthropologists don't simply study villages, 'they study *in* villages', anthropologists do well to not only study resource projects, but can also study the state from the particular vantage point of resource extraction. The ethnographic examples from PNG and Australia offered here illustrate the illegibility and obscurity of the state. From the perspective of these resource extraction sites, the state is often barely perceptible. It is bamboozling, deceptive, dangerous and two-faced, self-interested and sometimes completely unknowable. Material and ideological aspects of the state are mutually reinforcing in these resource domains, where the state looms as an eternally incomplete and uncertain project having to constantly reproduce itself.

The chapters in this volume show that although states might be imagined, or contested, they have real effects in people's lives, especially in the settings described here. Where gaps emerge as a result of the failure of the state to enact its responsibilities or, to paraphrase Skrzypek, where dissonance appears in the networks which inevitably emerge around resource extraction projects, something or someone will either choose or be persuaded to fill this space. An anthropology of absent presence which we propose here offers a unique perspective on the state in relation to resource extraction. Not just because it captures the gaps and unruliness that is thought to reside in resource frontier zones, but because it suggests that incompleteness, partial presence, and frustrated forms of order-making are a necessary entailment of the state and the way it is experienced and encountered in the context of resource extraction.

References

Abrams, P., 1988. 'Notes on the Difficulty of Studying the State (1977).' *Journal of Historical Sociology* 1: 58–89. doi.org/10.1111/j.1467-6443.1988.tb00004.x

Acemoglu, D. and J.A. Robinson, 2020. *The Narrow Corridor: States, Societies, and the Fate of Liberty.* New York: Penguin Books.

Allen, M.G., 2018. *Resource Extraction and Contentious States: Mining and the Politics of Scale in the Pacific Islands*: Singapore: Springer Singapore.

Altman, J.C. and D.F. Martin (eds), 2009. *Power, Culture, Economy: Indigenous Australians and Mining.* Canberra: ANU E Press (Centre for Aboriginal Economic Policy Research Monograph 30). doi.org/10.22459/CAEPR30.08.2009

Amarshi, A., K.R. Good and R. Mortimer, 1979. *Development and Dependency: The Political Economy of Papua New Guinea.* Melbourne; New York: Oxford University Press.

Auty, R., 1993. *Sustaining Development in Mineral Economies: The Resource-Curse Thesis.* London: Routledge.

Bainton, N., 2021. 'Menacing the Mine: Double Asymmetry and Mutual Incomprehension in the Lihir Islands.' In N. Bainton, D. McDougall, K. Alexeyeff and J. Cox (eds), *Unequal Lives: Gender, Race and Class in the Western Pacific.* Canberra: ANU Press. doi.org/10.22459/UE.2020.14

Bainton, N. and R.T. Jackson, 2020. 'Adding and Sustaining Benefits: Large-Scale Mining and Landowner Business Development in Papua New Guinea.' *Extractive Industries and Society* 7: 366–375. doi.org/10.1016/j.exis.2019.10.005

Bainton, N. and J.R. Owen, 2019. 'Zones of Entanglement: Researching Mining Arenas in Melanesia and Beyond.' *Extractive Industries and Society* 6: 767–774. doi.org/10.1016/j.exis.2018.08.012

Bainton, N., J.R. Owen, S. Kenema and J. Burton, 2020. 'Land, Labour and Capital: Small and Large-Scale Miners in Papua New Guinea.' *Resources Policy* 68: 101805. doi.org/10.1016/j.resourpol.2020.101805

Banks, G., 2008. 'Understanding "Resource" Conflicts in Papua New Guinea.' *Asia Pacific Viewpoint* 49: 23–34. doi.org/10.1111/j.1467-8373.2008.00358.x

Banks, G., D. Kuir-Ayius, D. Kombako and B. Sagir, 2017. 'Dissecting Corporate Community Development in the Large-Scale Melanesian Mining Sector.' In C. Filer and P.-Y. Le Meur (eds), *Large-Scale Mines and Local-Level Politics: Between New Caledonia and Papua New Guinea*. Canberra: ANU Press (Asia-Pacific Environment Monograph 12). doi.org/10.22459/lmlp.10.2017.07

Beck, U., 2009. *World at Risk*. Cambridge: Polity Press.

Beckett, S., 1956. *Waiting for Godot: A Tragicomedy in Two Acts*. London: Faber & Faber.

Bell, A., n.d. 'Absence/Presence.' Chicago School of Media Theory blog. Viewed 15 September 2020 at: lucian.uchicago.edu/blogs/mediatheory/keywords/absence-presence/

Bille, M., F. Hastrup and T.F. Soerensen (eds), 2010. *An Anthropology of Absence: Materializations of Transcendence and Loss*. New York: Springer.

Bloch, M. and J. Parry (eds), 1982. *Death and the Regeneration of Life*. Cambridge: Cambridge University Press.

Brett, J., 2020. 'The Coal Curse: Resources, Climate and Australia's Future.' *Quarterly Essay* 78: 1–76.

Burton, J., 1998. 'Mining and Maladministration in Papua New Guinea.' In P. Lamour (ed.), *Governance and Reform in the South Pacific*. Canberra: The Australian National University, National Centre for Development Studies (Pacific Policy Paper 23).

Cleary, P., 2011. *Too Much Luck: The Mining Boom and Australia's Future*. Collingwood (Vic.): Black Inc.

Cowen, M. and R.W. Shenton, 1996. *Doctrines of Development*. New York: Routledge.

Crocombe, R., 2007. *Asia in the Pacific: Replacing the West*. Suva: IPS Publications.

Dalsgaard, S., 2013. 'The Time of the State and the Temporality of the *Gavman* in Manus Province, Papua New Guinea.' *Social Analysis* 57: 34–49. doi.org/10.3167/sa.2013.570103

———, 2019. '"Seeing" Papua New Guinea.' *Social Analysis* 63: 44–63. doi.org/10.3167/sa.2013.570103

Das, V. and D. Poole (eds), 2004. *Anthropology in the Margins Comparative Ethnographies*. Santa Fe (NM): School of American Research Press.

Dean, M., 1999. *Governmentality: Power and Rule in Modern Society*. London: Sage.

Deleuze, G., 1990. *The Logic of Sense* (ed. C.V. Boundas, trans. M. Lester). New York: Columbia University Press.

Denoon, D., 2000. *Getting Under the Skin: The Bougainville Copper Agreement and the Creation of the Panguna Mine*. Melbourne: Melbourne University Press.

De Rijke, K., 2013. 'The Agri-Gas Fields of Australia: Black Soil, Food, and Unconventional Gas.' *Culture, Agriculture, Food and Environment* 35(1): 41–53. doi.org/10.1111/cuag.12004

———, 2018. 'Drilling Down Comparatively: Resource Histories, Subterranean Unconventional Gas and Diverging Social Responses in Two Australian Regions.' In R.J. Pijpers and T.H. Eriksen (eds), *Mining Encounters: Extractive Industries in an Overheated World*. London: Pluto Press. doi.org/10.2307/j.ctv893jxv.10

Derrida, J., 1976. *Of Grammatology*. Baltimore (MD): Johns Hopkins University Press.

Dinnen, S., 2001. *Law and Order in a Weak State: Crime and Politics in Papua New Guinea*. Honolulu: University of Hawai'i Press. doi.org/10.1515/9780824863296

Dwyer, P.D. and M. Minnegal, 1998. 'Waiting for Company: Ethos and Environment Among Kubo of Papua New Guinea.' *Journal of the Royal Anthropological Institute* 4: 23–42. doi.org/10.2307/3034426

Eriksen, T.H., 2018. *Boomtown: Runaway Globalisation on the Queensland Coast*. London: Pluto Press. doi.org/10.2307/3034426

Filer, C., 1990. 'The Bougainville Rebellion, the Mining Industry and the Process of Social Disintegration in Papua New Guinea.' *Canberra Anthropology* 13(1): 1–39. doi.org/10.1080/03149099009508487

———, 1997. 'Compensation, Rent and Power in Papua New Guinea.' In S. Toft (ed.), *Compensation for Resource Development in Papua New Guinea*. Boroko: Law Reform Commission (Monograph 6). Canberra: The Australian National University, National Centre for Development Studies (Pacific Policy Paper 24).

———, 2012. 'The Development Forum in Papua New Guinea: Evaluating Outcomes for Local Communities.' In M. Langton and J. Longbottom (eds), *Community Futures, Legal Architecture: Foundations for Indigenous Peoples in the Global Mining Boom*. Abingdon (UK): Routledge.

————, 2019. 'Two Steps Forward, Two Steps Back: The Mobilisation of Customary Land in Papua New Guinea.' Canberra: The Australian National University, Crawford School of Public Policy, Development Policy Centre (Discussion Paper 86). doi.org/10.2139/ssrn.3502585

Filer, C., J. Burton and G. Banks, 2008. 'The Fragmentation of Responsibilities in the Melanesian Mining Sector.' In C. O'Faircheallaigh and S. Ali (eds), *Earth Matters: Indigenous Peoples, The Extractive Industries and Corporate Social Responsibility*. Sheffield: Greanleaf Publishing.

Filer, C. and M. Macintyre, 2006. 'Grass Roots and Deep Holes: Community Responses to Mining in Melanesia.' *Contemporary Pacific* 18: 215–232. doi.org/10.1353/cp.2006.0012

Filer, C., S. Mahanty and L. Potter, 2020. 'The FPIC Principle Meets Land Struggles in Cambodia, Indonesia and Papua New Guinea.' *Land* 9(3): 67. doi.org/10.3390/land9030067

Frederiksen, T. and M. Himley, 2019. 'Tactics of Dispossession: Access, Power, and Subjectivity at the Extractive Frontier.' *Transactions of the Institute of British Geographers* 45: 50–64. doi.org/10.1111/tran.12329

Fuery, P., 1995. *The Theory of Absence: Subjectivity, Signification and Desire.* Westport (CT): Greenwood Press.

Geertz, C., 1973. *The Interpretation of Cultures: Selected Essays.* New York: Basic Books.

Gilberthorpe, E. and D. Rajak, 2017. 'The Anthropology of Extraction: Critical Perspectives on the Resource Curse.' *Journal of Development Studies* 53: 186–204. doi.org/10.1080/00220388.2016.1160064

Glaskin, K., 2017. *Crosscurrents: Law and Society in a Native Title Claim to Land and Sea.* Crawley: University of Western Australia Publishing.

Haley, N. and R.J. May (eds), 2007. *Conflict and Resource Development in the Southern Highlands of Papua New Guinea.* Canberra: ANU E Press. doi.org/10.22459/crd.11.2007

Handler, R., 1988. *Nationalism and the Politics of Culture in Quebec.* Madison: University of Wisconsin Press.

Harvey, P., 2005. 'The Materiality of State-Effects: An Ethnography of a Road in the Peruvian Andes.' In C. Krohn-Hansen, K.G. Nustad and B. Kapferer (eds), *State Formation: Anthropological Perspectives.* London: Pluto Press. doi.org/10.2307/j.ctt18fs5n7.10

Hepburn, S., 2011. 'Not Quite the Castle: Why Miners Have a Right to What's Under Your Land.' The Conversation blog, 15 November. Viewed 15 September 2020 at: theconversation.com/not-quite-the-castle-why-miners-have-a-right-to-whats-under-your-land-4176

Hertz, R., 1960 [1907]. *Death and the Right Hand* (trans. R. Needham). Aberdeen: The University Press.

Herzfeld, M., 1987. *Anthropology Through the Looking-Glass: Critical Ethnography in the Margins of Europe*. Cambridge: Cambridge University Press. doi.org/10.1017/cbo9780511607769

Hetherington, K., 2004. 'Secondhandedness: Consumption, Disposal, and Absent Presence.' *Environment and Planning D: Society and Space* 22: 157–173. doi.org/10.1068/d315t

Jorgensen, D., 2007. 'Clan-Finding, Clan-Making and the Politics of Identity in a Papua New Guinea Mining Project.' In J.F. Weiner and K. Glaskin (eds), *Customary Land Tenure and Registration in Australia and Papua New Guinea: Anthropological Perspectives*. Canberra: ANU E Press (Asia-Pacific Environment Monograph 3). doi.org/10.22459/cltrapng.06.2007.04

Karl, T.L., 1997. *The Paradox of Plenty: Oil Booms and Petro-States*. Berkeley: University of California Press. doi.org/10.1525/9780520918696

Kierkegaard, S., 1971 [1843]. *Either/Or*. Princeton (NJ): Princeton University Press.

Kirsch, S., 2014. *Mining Capitalism: The Relationship Between Corporations and Their Critics*. Oakland: University of California Press.

Langton, M. and O. Mazel, 2008. 'Poverty in the Midst of Plenty: Aboriginal People, the "Resource Curse" and Australia's Mining Boom.' *Journal of Energy and Natural Resources Law* 26: 31–65. doi.org/10.1080/02646811.2008.11435177

Li, T.M., 2007. *The Will to Improve: Governmentality, Development, and the Practice of Politics*. Durham; London: Duke University Press.

Lipset, D. and E. Silverman (eds), 2016. *Mortuary Dialogues, Death Ritual and the Reproduction of Moral Community in Pacific Modernities*. New York: Berghahn Books. doi.org/10.2307/j.ctvpj7hc4

Lund, C., 2006. 'Twilight Institutions: An Introduction.' *Development and Change*, 37(4): 673–684.

Macintyre, S., 2009. *A Concise History of Australia.* Cambridge: Cambridge University Press.

MCA (Minerals Council of Australia), 2019. 'Australian mining: Building a better future.' Canberra: Minerals Council of Australia. Viewed June 2021 at: www.minerals.org.au/news/australian-mining-building-better-future

McDougall, D., 2019. 'Singing the Same Tune? Scott Morrison and Australia's Pacific Family.' Australian Outlook blog, 13 June. Viewed 15 September 2020 at: www.internationalaffairs.org.au/australianoutlook/singing-same-tune-scott-morrison-australias-pacific-family/

Moresby, J., 1876. *Discoveries and Surveys in New Guinea and the D'Entrecasteaux Islands: A Cruise in Polynesia and Visits to the Pearl-Shelling Stations in Torres Straits of H.M.S. Basilisk.* London: Murray.

Murphy, K., 2017. 'Scott Morrison Brings Coal to Question Time: What Fresh Idiocy is This?' *The Guardian*, 9 February.

Nelson, H., 1976. *Black, White and Gold: Goldmining in Papua New Guinea 1987–1930.* Canberra: Australian National University Press. hdl.handle.net/1885/114813

O'Faircheallaigh, C., 2016. *Negotiations in the Indigenous World: Aboriginal Peoples and the Extractive Industry in Australia and Canada.* London: Routledge. doi.org/10.4324/9781315717951

Pijpers, R.J. and T.H. Eriksen (eds), 2018. *Mining Encounters: Extractive Industries in an Overheated World.* London: Pluto Press. doi.org/10.2307/j.ctv893jxv

Regan, A., 2017. 'Bougainville: Origins of the Conflict, and Debating the Future of Large-Scale Mining.' In C. Filer and P.-Y. Le Meur (eds), *Large-Scale Mines and Local-Level Politics: Between New Caledonia and Papua New Guinea.* Canberra: ANU Press (Asia-Pacific Environment Monograph 12). doi.org/10.22459/lmlp.10.2017.12

Ross, M., 2003. 'The Natural Resource Curse: How Wealth Can Make You Poor.' In I. Bannon and P. Collier (eds), *Natural Resources and Violent Conflict: Options and Actions.* Washington (DC): World Bank.

———, 2015. 'What Have We Learned About the Resource Curse?' *Annual Review of Political Science* 18: 239–259. doi.org/10.1146/annurev-polisci-052213-040359

Santos da Costa, P., 2018. Re-Designing the Nation: Politics and Christianity in Papua New Guinea's National Parliament. St Andrews: University of St Andrews (PhD thesis).

Sartre, J.-P., 1957. *Being and Nothingness: An Essay on Phenomenological Ontology* (trans. H.E. Barnes). London: Methuen.

Scott, J.C., 1998. *Seeing Like a State: How Certain Schemes to Improve the Human Condition Have Failed.* New Haven (CT): Yale University Press. doi.org/10.2307/j.ctvxkn7ds

Smith, G., 2018. 'Just the Nickel, Please: China's New Power in the Pacific.' *Griffith Review* 59: 191–202.

Strathern, M., 2005. *Kinship, Law and the Unexpected: Relatives Are Always a Surprise.* Cambridge: Cambridge University Press.

Szablowski, D., 2007. *Transnational Law and Local Struggles: Mining, Communities and the World Bank.* Portland (OR): Hart Publishing.

Tammisto, T., 2016. 'Enacting the absent state: State-formation on the oil-palm frontier of Pomio (Papua New Guinea).' *Paideuma: Mitteilungen zur Kulturkunde* 62: 51–68.

Trouillot, M., 2001. 'The Anthropology of the State in the Age of Globalization: Close Encounters of the Deceptive Kind.' *Current Anthropology* 42: 125–138. doi.org/10.1086/318437

UN (United Nations), 2011. 'Guiding Principles on Business and Human Rights: Implementing the United Nations "Protect, Respect and Remedy" Framework.' New York and Geneva: UN. doi.org/10.1017/9781316677117.043

Vandergeest, P. and N.L. Peluso, 1995. 'Territorialization and State Power in Thailand.' *Theory and Society* 24(3): 385–426.

Waterhouse, M., 2010. *Not a Poor Man's Field: The New Guinea Goldfields to 1942: An Australian Colonial History.* Ultimo (NSW): Halstead Press.

Weiner, J.F. and K. Glaskin (eds), 2007. *Customary Land Tenure and Registration in Australia and Papua New Guinea: Anthropological Perspectives.* Canberra: ANU E Press (Asia-Pacific Environment Monograph 3). doi.org/10.22459/cltrapng.06.2007.01

2

Categorical Dissonance: Experiencing *Gavman* at the Frieda River Project in Papua New Guinea

Emilia E. Skrzypek

Introduction

Much of the thinking about roles and responsibilities in resource extraction projects is framed around the notion of stakeholder categories. These categories group people of a given type under a particular label, and award them with prescriptive roles and responsibilities. Like many countries around the world, in Papua New Guinea (PNG) provisions in national legislation make the state the owner of the country's mineral resources and the regulator of the extractive sector. In order to capitalise on these resources, the government enters into agreements with mining companies, which have the financial and technological capacity to operate resource extraction projects, and the host area communities who claim customary ownership over the land. The country's first large-scale mining project, the Panguna mine on the island of Bougainville, began its operations in 1972, with the Ok Tedi mine in the country's Western Province following 12 years later. In the 1980s and 1990s respectively, the two mines were sites of dramatic events that sent shockwaves through the sector. A breakdown of social relations at Panguna resulted in an armed conflict, claiming thousands of lives (May and Spriggs 1990), while mismanagement at the Ok Tedi mine led to a natural disaster and one of the most famous legal challenges in the history of the industry (Banks and Ballard 1997). The legacy of those events forced governments and companies in and beyond PNG to acknowledge landowners as the third key stakeholder category in resource extraction projects (Imbun 2007), where stakeholders are defined as 'groups and individuals who can affect, or are affected by, the achievement of an organisation's missions' (Freeman 2010 [1984]: 48).[1] This triad model, which recognises the state, the company and local communities as the three key stakeholder groups, continues to serve as a 'provisional analytical device' (Ballard and Banks 2003: 289) in industry as well as academia, amid mounting acknowledgement that 'it has not generally served to capture much of the complexity of the relationships that form around mining as a site' (ibid.).

As the sense of a mining community broadened, the significance of additional stakeholders such as NGOs, media, legal agencies and global financial institutions rapidly expanded, forming what Ballard and Banks (2003: 304) glossed 'the fourth estate'—included in resource relations 'by virtue of connections to or alliances with (rather than membership of) one of the three principal stakeholder categories'. More recently, Filer and Le Meur proposed extending the triad model to include the fourth element, which they called 'society'. They also reframed the original three stakeholder categories as spaces rather than conceptual entities and argued that:

> there is no large-scale mining project that involves just one company, one government agency or even one mine-affected community. Instead, there is a corporate space, a government space and a community space, each of which is liable to be occupied by a number of organisations or agencies in different relationships with each other, some of which count as political relationships, and within each of these organisations or agencies there are more political relationships. (2017: 22)

This view is reflective of a growing recognition in the anthropology of mining that, held too tightly, categories can hinder rather than aid our understanding of complexities of stakeholder engagement in resource extraction projects. Instead, the discipline has turned much of its focus to nuanced processes and enactments of categories as they are taking place, in project locations (e.g. Golub 2014; Welker 2014). Inevitably, the closer

---

1 This development was then reflected in the establishment of the Development Forum process for the Porgera mine—PNG's next major development project (Filer 2008).

we get to the ambiguities and complexities of on-the-ground relations, the more distant and blurry the categorical prescriptions become. Yet, it would be dangerous to neglect categorical prescriptions from any systematic analysis, as they frame and influence on-the-ground interactions and are indispensable to our understanding of 'stakeholder relations'. Here, I focus on the contested spaces in between stakeholder categories and their enactments in the project location. Categorical models tend to make assumptions about what the latter ought to look like. They often contain specific instructions to facilitate types of interactions and relations (e.g. through legal frameworks, employment contracts and bureaucratic procedures). But, in a version of a looping effect described by Douglas (1986), categorical models are themselves influenced by ways in which they are experienced and imagined in resource project localities.

Categorical Dissonance

Building on Le Meur and Filer's take on categories as complex spaces rather than discreet entities, this chapter explores what I term here as 'categorical dissonance' which contrasts expectations of perceived roles and responsibilities with actual enactments and effects of those roles. The concept pivots on cognitive dissonance theory, first proposed by Festinger in the late 1950s, which 'suggests that when individuals hold two or more cognitions that are contradictory, they will feel an unpleasant state— dissonance—until they are able to resolve this state by altering their cognitions' (Hinojosa et al. 2016: 171). I use it here to investigate the role of the state in stakeholder engagement at the Frieda River Project in PNG: to explore ways in which the absence of the state was experienced in the project location, and describe how the company developing the project and members of local communities responded and altered their cognition and actions to redress the imbalance these absences introduced into the process of planning for a mine at Frieda.

Two interconnected types of categorical dissonance will be described here. The first is dissonance within categories, found in dynamic spaces between categorical prescriptions of stakeholder roles and responsibilities, and their enactment in project contexts. In her study of kinship, law and the unexpected, Marilyn Strathern theorised two kinds of relations. The first kind, she wrote, are categorical relations, connections acquired 'through a logic or power of articulation that acquires its own conceptual momentum' (2005: 7). The second are interpersonal relations taken as connections between persons, which develop over time 'inflected in a precise and particular history' (ibid.). Together, the two create 'a tool for social living' as 'it is through interacting with persons that diverse interactions and further connections become intellectually conceivable, while it is through creating concepts and categories that connections come to have a social life of their own' (ibid.: 8). This chapter uses Strathern's framework to contrast the state's categorical presence with its interpersonal absences at the Frieda River Project. It describes how members of one of the project's so-called impact communities experienced the categorical prescriptions of state legislation which they could neither understand nor leverage, and how they became disillusioned with enactments of the state and the lack of interpersonal relationships with government agents. Instead, they vested their hopes in the relationships formed with mining company personnel, in anticipation of using those relations to leverage positive outcomes and bring civic infrastructure and services to the Frieda River area.

The second type is dissonance between categories, which emerges where one or more stakeholder groups fail to meet the expectations of their role leading to absences in their respective functions. Where this affects functions deemed necessary for the shared project, and where such absence arises in a way that threatens it, another actor (or actors) with the capacity to do so, will likely be forced to fill the missing role to address the disharmony and ensure continuity of the process. Although such 'absences' can be found in all stakeholder groups, this chapter focuses specifically on the state's performance against its categorical role at the Frieda River Project. It describes how the company developing the project has attempted to mediate the failures of the government, or its profound absence, and found itself under pressure to take on service delivery functions in the project area.

To demonstrate the multiple dimensions of the experiences of the state on the ground, in a project location, the chapter triangulates findings of a village-based ethnography with material collected among on-site company personnel and data from interviews conducted with government representatives during their rare visits to the area. It follows stakeholders engaged in the process of planning for a mine at Frieda River in their use of categorical terminology, referring broadly to the government, the company, and 'local' or 'impact' communities as key stakeholder groups at the Frieda River Project. It shows how the Paiyamo community of

Paupe, who live in the vicinity of the project, strived for a relationship with the state and strategically engaged the company to fill some of the voids created by the gap between expectations and reality of government involvement at the Frieda River Project. An important part of this process was the anticipation that if the mine begins operations, the government's interest and its active presence in these communities will follow. This chapter follows local use of the Tok Pisin term *gavman*, translated here as government, to describe the state, its institutions and people who enact it at any given point in time—expanding on the view of the state as a discrete polity made up of organising rules and principles and possessing the power to enforce those over particular peoples and territory, to include people who enact it and enforce its policies in the project location. It is based on ethnographic material collected during 12 months of fieldwork conducted in the Frieda area in 2011–12 and again in 2018.

Emergence of the State

The image of the Australian flag being lowered and the Papua New Guinean one being raised signalling the birth of an independent PNG is widely used in descriptions of the moment in which PNG became an independent nation state. The red, black and gold bird of paradise flag on a mast at sunrise on 16 September 1975 was heralded in the national media as a symbol of a new dawn, uniting the independent nation state around imagery filled with hope and anticipation of a bright future. The public rhetoric surrounding independence created an idea of a nation state akin to Benedict Anderson's (1983: 6) model of an 'imagined community'—a group of people, spread over a particular geographic area, who will likely never meet 'yet in the minds of each lives the image of their communion' (see also Main, Chapter 5, this volume). Accounts of the time leading up to the event speak of high levels of excitement, particularly and understandably in the capital district where official celebrations were taking place, attended by powerful political figures from across the country and abroad.

In the mid-1970s, the newly formed nationalist discourse pivoted on the powerful idea that independent PNG will also be a 'developed' PNG. Despite early national optimism, in the years following independence the newly established government struggled to fulfil its development aspirations. Even though the experience of colonialism was brief for a large proportion of the population, it left a lasting legacy of expectations and shaped the ways in which citizens of the newly independent nation conceptualised the state and the role of the *gavman*. Accounts of the emergence of the 'state' in PNG describe the state, seen as regulator and provider of infrastructure and services, as very much a colonial creation and describe how, particularly in the years between the Second World War and independence, the colonial administration played a key role in 'development' (May 2004). As it withdrew, the colonial administration left voids in the provision of services, and the maintenance and development of infrastructure. While the country's public service was functional, its capacity to deliver services and respond to the needs of the country's growing population was relatively limited (Gelu 2010) meaning that gradually, in the years following independence, in large parts of the country civic infrastructure wilted and access to government services diminished.

Like the experience of colonial administration before it, the experience of the state of PNG was unevenly shared across the nation following independence. To an extent this is still the case today, with people in many rural parts of the country expressing feelings of being forgotten by the government and excluded from national political representation and capital flows. In 1975 'many of the new state's claims to legitimacy were based on the promise that all Papua New Guineans could expect development to come their way' (Jorgensen 2007: 58). By the 1980s tensions arose between the goals of 'development for all', set out in the country's largely egalitarian Eight Point Plan (1975), and the realisation that development, out of its very nature, produces inequality (Rivers 1999; Stead 2016). The mining industry seemed to offer an ideal opportunity to bring revenues to the national economy and development into local areas (Gilberthorpe and Banks 2012). The government embraced the promise of mining-induced national development. By the end of the 1970s, PNG became a 'mining resource frontier' (Hyndman 2001). Over time, mining become a central feature of the country's growing economy, capitalising on what Ballard and Banks (2003) referred to as the 'mining exploration bonanza' of the 1980s and early 1990s. By then, at least 10 large mineral prospects had been identified in PNG, and national leaders restated the promise that 'mining revenues will be the catalyst for an ambitious program of rural development and that capital works programs and government services will enable subsistence villagers to enjoy the benefits of the cash economy, hitherto largely confined to the urban minority' (Ashton, 1990). By the end of the twentieth century, 156 mining exploration licences had either already been awarded or were awaiting grants, and four major operations (Ok Tedi, Lihir, Misima and Porgera) and one medium-sized mine (Tolukuma) operated in PNG (Hirsch 2001).

The 1990s saw intensification of exploration activities in many parts of the country, including the copper and gold prospects in the Frieda River area. The companies' increasing presence on the ground and their interactions with members of host communities established exploration and mining companies as powerful social, economic and development agents across the nation. In light of the ongoing decline in the provision of government services, and increasing disillusionment with the government's capacity to maintain and expand civic infrastructure, local communities looked to mining and exploration companies to fulfil those roles—with the companies' ability to secure local consent for their operations, known in the industry as a 'social license to operate' (SLTO),[2] increasingly influenced by their performance against those expectations. This anticipation was shared by many local and provincial governments that linked their own ability to cater to the needs of their communities to their success in attracting and retaining mining interests in their area. For some people, the enabling legislative climate created by the state in the hope of encouraging growth of the industry was the state's way of helping its citizens.

While people hoped that the companies would bring development to their areas in the form of infrastructure, services, and employment and business opportunities, in the decades following independence the government was very much hoping for the same. Unsurprisingly, over the years, companies tended to resist taking on what they saw as the role of the government. Nevertheless, faced with the necessity to secure support for their projects from the host communities, and confronted with the weak institutions of the state, many companies have found themselves with no choice but

2   Although it adapts technical and tangible language associated with licencing and permitting, in reality SLTO is 'an intangible construct associated with acceptance, approval, consent, demands, expectations and reputation' (Parsons et al. 2014: 84). The term has been criticised on the grounds of limiting, not expanding, critical debates about entanglements and regimes of responsibility and accountability between companies and society (Owen and Kemp 2014). Despite risks associated with the way in which the concept 'constructs its notion of approval, based on the absence of disapproval' (Owen 2016: 103), it has become a popular shorthand to describe ways in which companies deliver benefits to communities in exchange for the latter not interrupting their local operations.

to either fill these gaps (see also Bainton and Macintyre, Chapter 4, this volume), or facilitate the government in fulfilling its role, and most often both—as the remainder of this chapter will show.

The Frieda River Project

As the global geographies of demand and supply in mineral resources shift (Bebbington et al. 2008), deposits that have been long considered either too dangerous or economically unviable are having their viability reassessed. Frieda River's copper and gold deposit falls into this category. It is what can be described as a 'complex ore body', where this complexity includes a range of geographical, geological and non-technical factors, such as political uncertainties and social and cultural complexities (Valenta et al. 2019). Frieda River is not a mine, but an advanced resource exploration project.

Even by Papua New Guinean standards, the Frieda River Project is geographically remote. Located near the border between East Sepik and Sandaun provinces, some 70 km south of the Sepik River, the site is not accessible by road from either of the provincial capitals, Vanimo or Wewak, but it can be reached by air, using a helicopter or a lightweight plane, or by water—following the Sepik from Pagwi in the East Sepik or from Green River in Sandaun. Despite its geographic isolation (see Figure 2.1), the area's rich copper and gold deposits mean that it is of great interest to the mining industry, and has been for a number of decades. Mineral prospects were first identified at Frieda during a regional mapping exercise in 1966. In 1969 a Prospecting Authority (today known as Exploration License 58) was granted to an Australian mining company, Mount Isa Mines Ltd, and its subsidiary, Carpentaria Exploration Company, allowing for exploration work at Frieda to begin. Exploration has been ongoing since, but noticeably intensified in 1996 when, in the middle of the global resource high, Highlands Gold Properties Pty Ltd completed the first full feasibility study for the Frieda River Project. The second feasibility study was completed in 2013 by Xstrata Frieda River, a subsidiary of Xstrata Copper, which at the time was the fourth largest copper producer in the world. Xstrata withdrew from the project a short time later.

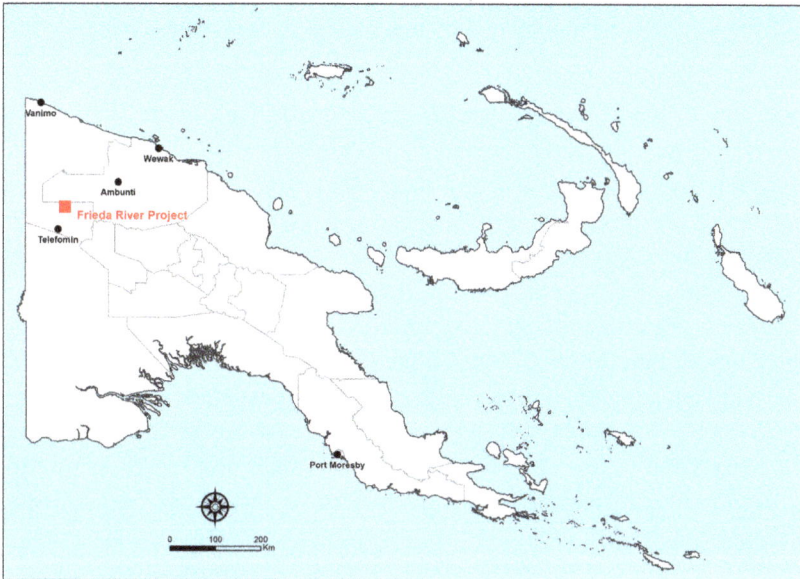

Figure 2.1 Location of the Frieda River Project
Source: Map created by Kamila Svobodova

The Frieda River Project is now co-owned by Brisbane-based PanAust (80 per cent)[3] and a Papua New Guinean incorporated company, Highlands Pacific Ltd (20 per cent). The current owners see Frieda River as 'a core component of a transformative nation-building opportunity—which the company has dubbed the Sepik Development Project' that promises to 'deliver direct and indirect local employment, create business opportunities, attract foreign investment and boost trade and commerce' (PanAust 2018). Company materials publicly promoting the project claim that:

> host communities, especially in rural areas, will benefit from access to improved transport, telecommunications, health, education and government services that will support a higher quality of life and greater social participation. More broadly, training and employment of Papua New Guineans will provide the skills and capacity to support the nation's future development and prosperity. (ibid.)

---

3   In 2015, soon after it acquired shares in the Frieda River Project, a Chinese company—Guangdong Rising Assets Management, acquired PanAust. It is now fully Chinese-owned, but continues to operate from Brisbane, and remains listed on the Australian Stock Exchange. The majority of the on-site personnel are either Papua New Guinean or Australian.

The promotional materials do not specify who will deliver those benefits, only that they will be substantial.

Frieda River Joint Venture partners were, at the time of writing, in the process of making adjustments to their feasibility study with a view of submitting a Special Mining Lease (SML) application to the state government, as required by the Mining Act—the main instrument of mining industry governance in PNG. The licence is granted on the basis of the application and conditional to successful completion of a development forum—a process of tripartite consultations between government, company and community representatives developed by the government 'for addressing issues of community participation' and distribution of landowner benefits in resource extraction projects (IIED 2002: 211, see also Filer 2008). Granting of the SML opens the door for companies to proceed to the construction and operations phases of the mineral project life cycle.

Since the turn of the century, on-site exploration activities at Frieda were tailored towards servicing the demands of the consecutive feasibility studies, including but not limited to hydrological testing, drilling programs and technical scoping for the project's numerous designs. They also included environmental surveys, social mapping and landownership demarcation exercises, and community engagement activities. At the time of research, the company maintained a presence of operations and community affairs personnel on site. In 2011–12, the community affairs team consisted of three superintendents and seven to eight officers, who were all accommodated in one of the two semi-permanent residential camps—the Base Camp and the Airstrip Camp. In 2018 the team was smaller, comprising of two superintendents and six officers, working on a roster. On both occasions, the team was led by a manager based in the company's office in Brisbane. While the on-site teams interacted with representatives of provincial and district governments, particularly during their visits to the site, the relationship with the national government was largely handled by the company's Brisbane-based general manager for 'external affairs'.

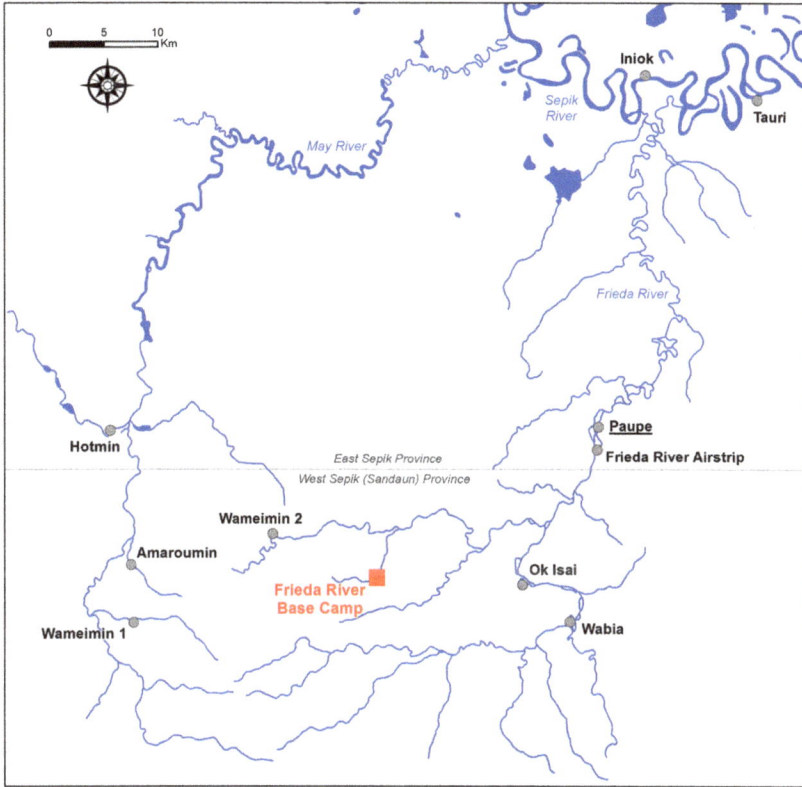

Figure 2.2 Frieda River Project area
Source: Map created by Kamila Svobodova

Following a landowner identification exercise for an area under Exploration License 58, requested by the government and completed by the company in 1995–96, seven communities were awarded a status of what are now known as 'impact communities' of the Frieda River Project: Telefolmin villages of Ok Isai and Wabia; Mianmin villages of Wameimin 1 and 2, Amaroumin, and Hotmin; and a Paiyamo village of Paupe, where I was based for most of the time I spent at Frieda (see Figure 2.2). Located in the East Sepik Province, a short boat ride away from the Sandaun provincial border, and a few minutes' walk away from the Airstrip Camp, Paupe is a truly beautiful place nestled in the rainforest. At first glance, there is little to signify Paupe's status as an impact community to a large, advanced resource exploration project, based upon the visible signs of development that we have come to expect in 'mining impact communities'. In the absence of actual mining, Paupe is really an 'exploration impact community'. Most of its members continue to live in temporary houses made of bush materials, and subsistence agriculture remains the main economic activity in the village. Saying that, during the time I have spent in the area, there have been some noticeable changes in community infrastructure and services. In that time, and with the company's assistance, Paupe gained new school buildings and a community clinic, complete with staff housing. A gravel service road now runs through the community linking the Airstrip Camp with Paupe, and further with a water port on the banks of the Sepik River. The service road was built by the company to enable transport of cargo between supply barges moored on the Sepik and the company camp. In 2018 my hosts told me, with more than a note of excitement in their voices, that when the road was being built a PMV (public motor vehicle, a popular bus substitute in PNG) travelled through Paupe, bringing workers to site in the morning, and returning them to the Airstrip Camp in the afternoon. Although the PMV runs ceased once the construction was completed, the image of a bus passing through the community stayed with the Paiyamo, evoking imagery of the kinds of services they will have access to once the project proceeds from exploration and permitting to the construction and operations phases of its life cycle.

The Absent Presence of the State

The Paiyamo's first encounters with state administration occurred relatively late in the country's colonial history, with the first government patrol officer (or 'kiap') reaching the area in 1962. The kiap brought together Paiyamo speakers dispersed around the region, and established the foundations of the current village of Paupe. In local narratives, this marks a moment of transition from past-focused time of the ancestors, to a future-driven time of 'white men' and 'development'. The Paiyamo rarely speak about their first experiences of the government. But when probed they talk about establishment of the village, advent of local administrative roles like *luluais* (village heads or chiefs) and *tultuls* (their assistants), introduction of village books, and first encounters with organised law and order structures which saw a group of local men arrested and taken to jail in Ambunti, the district headquarters, in the late 1960s. From its early days, the relationship between the Paiyamo and the *gavman* was weak, and limited to sporadic and brief encounters with colonial patrols. Even now, the Paiyamo feel very little obligation towards the state and ascribe much greater significance to another arrival—which followed a few years after the first kiap moored his boat on the banks of the river—the mining company. Over time, members of local communities came to see the mining industry as a 'modern road to cargo' (see also Bainton 2010) and began to conceptualise the future mine as a portal through which they will gain access to 'development'.

For all intents and purposes, the way the Paiyamo see it, they do not have a relationship with the government independently of the Frieda River Project, and they look to the company for what would traditionally be seen as the role(s) and responsibilities of the state. 'We must support the company and work with them. It is the company that will bring services to us. It will not be the government who will come and give medications to our children. If the company was not there, the government would not remember about us', I often heard them say. The self-description as 'people forgotten by the government' featured strongly in such statements, as the Paiyamo turned to the company for goods and services, which they came to associate with the promise of development and modernity they read from the imagery of the anticipated copper and gold mine in their area.

Despite their brief and limited encounters with the colonial administration, like many people all over the country, they too spoke of a decline in government presence in the years since independence. 'These are no longer the times when Australians were looking after us and when there was enough of everything for everyone. We should be happy we have the company here because the government has completely forgotten about us', I was told by a community leader in Paupe. In the few encounters I witnessed between Paiyamo people and government representatives, the latter actively encouraged the communities' dependency on the company for both civic infrastructure and basic services, laying bare the government's own expectations of the company to deliver development to what it saw as some of the most geographically remote and hard-to-access parts of the country. On one occasion in 2012, during a brief visit to Paupe, I heard a member of the provincial administration say to the community:

> Before, you all lived in the bush and you had nothing. The government is too far away, it will not help you. If it was not for the company you would still be in the bush with no development, nothing. If you don't support the company, you will go back there.

This is not to say that the government was completely absent from the project and the area. National legislation meant that, in categorical terms, the state's presence was very much felt at Frieda. It was demonstrated through prescriptions of government forms, laws, procedures and—last but not least—the sovereign power held by the government to grant the SML for the project. The company would often bring the state into conversations by invoking government process and legislation to explain its own decisions and actions, and to settle disputes on issues such as compensation rates. From the perspective of local communities, the government processes were slow, and were often blamed for holding back the mine. This meant that there was also a temporal dimension to the categorical presence of the government at the Frieda River Project (cf. Dalsgaard 2013).

In many ways, the categorical presence of the state at Frieda, where subsistence and artisanal mining activities kept people largely outside the formal economy and taxation system, and where most law and order issues were settled using prescriptions of customary rather than common law, was felt more strongly in the context of the project than in any other area of people's lives—the project serving as a tool for local anchoring of the state at Frieda. But this categorical presence of the state was not equal to a relationship—which the Paiyamo were very quick to remark upon when I asked. It lacked the interpersonal component and access to government representatives with whom consequential relations could be leveraged and built. In that regard, the interpersonal absences, lack of direct interactions with government officials, and poor delivery of government services in the area contributed to the feeling that the state was absent from Frieda.[4] These feelings were exacerbated by the community members' very limited understanding of government processes and systems—a form of institutional uncertainty about the rules of the new game (see also Minnegal and Dwyer 2017). On more than one occasion I heard people ask 'How can we participate fully if we do not understand?' This led to high levels of frustration among the Paiyamo who, faced with

---

4   Since independence, Paupe has had a local government councillor. Formally the holder of this elected position served as an intermediary between the community and the state, together with the village court magistrate. Generally, the Paiyamo respected their local councillor and his role in resolving community matters. However, they had little confidence in the subsequent post-holders' ability to leverage relations with 'the state'. Limited interaction and communication between the councillor and higher levels of government, and the company's emphasis on district, provincial and state representatives meant that while respected within the community, the councillor was not considered a fully-fledged 'government representative' in Paupe.

state regulations and bureaucratic prescriptions that were inflexible and non-negotiable, felt they could not really get to know the government. Further, as their local epistemology treats knowledge as a relational endeavour, they took the government's lack of interest in entering into interpersonal relationships with members of the community as a sign of the state's unwillingness or lack of interest in getting to know them.

Local ideas of personhood play an important role in the processes described here. The Paiyamo judge knowledge and relations against the transformative effects these have on the world. For them, personal relations award meaning and significance to past and current events, and create opportunities for the future. Over time, my hosts in Paupe came to see the (future) Frieda River Mine as an effect of knowledge and relations, which they actively navigated hoping to reveal the long-awaited mine in their area (Skrzypek 2020). Which is why the almost exclusively categorical presence of the state, which they did not know and felt did not want to know them, and with which no two-way relationship could be developed, was so frustrating. In the absence of interpersonal relations with the government, members of Paupe community did not identify with it in a way in which they came to identify with the company developing the project, and did not form relations with government representatives in the way they entered into complex relations with company employees. There was a strong categorical relationship between the Paiyamo and the company, grounded in the national Constitution and prescriptions of legally recognised customary land ownership. In this relationship, the company (the developer) must secure and maintain access to land owned by the Paiyamo (the landowners) in order to develop infrastructure essential for the project's activities and future operations (Bainton and Banks 2018). However, to my hosts in Paupe, it was the interpersonal relations they entered into with the company's on-site personnel that made the categorical relations meaningful and gave them effect, with the ultimate goal of leveraging those relations to precipitate a mine at Frieda. On my most recent visit to the area I was told by a company representative that now, even when the government representatives visit the villages, they are given less priority than the company personnel, the latter seen as more capable to impact change and address their desires and concerns. In his view, 'people no longer believe the *gavman* will help them'.

Community Leaders' Forum

The problematic relationship, or the lack of association, between government representatives and community members had an effect on the company and the way in which it approached the task of 'stakeholder engagement' at the project. Company representatives frequently raised concerns about the government's limited capacity to proactively engage with the project, and the country's political instability, which, in their view, weakened the state's position as the regulator of the industry. 'At the national level, elections are coming up next year and anything can happen', I was told in 2011. 'Personnel changes may mean setbacks, interests may change. The bureaucracy is slow. Laws can change all around the country. But all companies want security. They invest the money, they want to see it back'. Interestingly, this concern with the country's political instability was shared by many government representatives, including those based at the Mineral Resources Authority (MRA)—a government agency which acts on behalf of the national government to enforce rules and regulations at exploration and mining projects, who expressed frustration with the effect political processes had on their own ability to fulfil their administrative roles and responsibilities in the context of the Frieda River Project.

Rather unsurprisingly, company employees felt uncomfortable with refracted responsibilities and the pressure they have found themselves under to fulfil what they saw as the role of the state. 'The government doesn't play its role. It thinks that since we are here we can provide all', a community affairs officer (CA) told me, following a community meeting in one of the impact villages. 'But we cannot be seen as the provider of government services in the area—this is just not right', he added, signalling that the absence of state function was not only a practical but also a moral issue. In an attempt to manage the absence of the government representatives at Frieda, and to fill the void in interpersonal relations between the government and local communities, the company management revived an old and for some time abandoned model of a 'community leaders forum' (CLF)—a common practice in the 1990s in the Project Coordination Branch of what was then the Department of Minerals and Energy. The revived forum was designed specifically to, in the words of one CA officer, 'connect landowners and the government representatives with the company present'.[5] The quarterly meetings brought government representatives, company personnel and community leaders together in the company's Base Camp to disseminate and discuss project-related information and, as it was described to me by one of the participants, 'to agree on common issues and allow everyone to move forward'. The meetings were intended as mechanisms for transparency, disseminating information at the same time to the different stakeholder groups, and ensuring accountability in decision making and acting on items agreed upon during the meetings. For many community leaders the forum did indeed offer a rare chance to interact with government representatives, although the forum did not address the bigger issue of the absence of state functions.

Generally, the list of government invitees included district representatives from Ambunti-Dreikikir and Telefomin, and an MRA representative. Depending on specific agenda items, representatives of relevant provincial government divisions would sometimes also be asked to attend. The costs and logistics of their attendance at CLF were covered by the company, who flew them into the area on either the company's fixed-wing charter or a chopper, and provided them with meals and accommodation in the Base Camp. This was a common arrangement for government representatives visiting the site. Often, on top of transport and accommodation, the representatives expected and received a daily stipend, to provide an additional incentive for travel to site. During the CLF, government representatives were encouraged to provide ward reports, speak about district priorities and provide updates on broadly conceived government issues. The invited representatives did not always come (on occasions leading the to the CLF being postponed or cancelled) and when they did, the level of their active participation in the meetings varied. Despite this, the company as well as community members spoke highly of the meetings and considered them to be the single most important stakeholder engagement mechanism at the project. A number of people I talked to in 2018 went a step further and described CLF as currently 'the main political body' for the Frieda River Project.

---

5   The company also meets with the two provincial governments and the MRA at the Joint Provincial Consultative Committee (JPCC) meetings. Held quarterly, usually shortly after the CLF, the JPCC meetings have been described to me as the 'main vehicles for formal interaction between the company and government stakeholders'. JPCC meetings are held away from Frieda and do not include community participation.

While stakeholders on all sides told me that the forum created a valuable platform for discussion, agreeing on action items and allocating responsibilities for their delivery, community members felt that of the three broad stakeholder groups they had the least influence over forum proceedings. To quote one community leader and a regular CLF attendee:

> we develop agendas. The company asks us to. But the agenda does not have power—our points get missed out, rejected. Why? Because this is our thinking. But it is good to bring everyone in a meeting. We come from different places we need to get together.

Echoing this sentiment, another community representative told me:

> the CLF discussion is generally good. But sometimes we just sit and listen to what the company has to say. This has been changing as we give feedback, but there is room for improvement. For example, the company and the government—they need to change how they listen to us.

When probed about the last part of his statement, that particular person explained to me that issues brought to the table by the company take priority over issues presented by community, the latter often falling off the table altogether. This is reflective of a conflict of interpretation about the nature of this type of forum among the different participating groups and individuals. While community representatives tend to see them as opportunities for discussion, expressing discontent and resolving conflicts, companies more often see this as the opportunity to 'share information' and 'conduct awareness', and such processes tend to entail a much more one-sided flow of information.

There was also the issue of efficacy of agreements made during the CLFs. 'Sometimes, decisions are made at the CLF regarding certain projects, but nothing happens—no power, talk only', I heard in Paupe. Company representatives I spoke to about this told me that the meetings were established to move the project along a particular trajectory, set out by the requirements of government regulation and the company's own business processes and procedures. Where the CA officers reject agenda items, it was because they felt that they were either not appropriate for the forum, or wrongly timed—with the most time-sensitive and pressing issues needing to take precedence. While they acknowledged the great importance of having good relationships with the impact communities and their leaders, the CAs were open about the fact that the directives they received clearly prioritised moving the project towards an SML application. Ultimately, the forum was aimed at servicing the needs of the project, guided by the business rationality of the company. During my conversations with the CA officers, they spoke at length about how important it was that the government plays its role in the CLF and in the process of planning for a mine at Frieda, and how careful they had to be not to become a de facto government in the area. They also talked about the difficult task of trying to secure government's active involvement and engagement with the local communities, especially in the context of poor accountability and evasion of responsibility for their well-being (cf. Golub and Rhee 2013). Community members as well as company staff described numerous instances of government representatives making a promise and committing to a task or a project, which was never delivered.

For the CA officers it was important that the government representatives made promises directly to community leaders. This way, if the promises were not kept, the government, not the company (or the CA officers, for that matter) were held accountable for the failures—a coping strategy in a difficult work context. In a mirror process, it was not uncommon for government representatives to exercise a 'strategic absence' and avoid the meetings in order to avoid accountability, especially if they felt they lacked the capacity or were, for any other reason, unable to perform tasks and duties required of them. Reflecting on the impact of the CLF on the relationship between communities and the government, one CA officer told me: 'for people this is the only time they see the government. They out their complaints, demands and expectations. But do not actually expect the government to fulfil them'. 'No funding is their favourite excuse', I heard some time later in Paupe. But by far the most common reaction to my questions about promises made by the government representatives during the CLF and the state's involvement in delivery of services was a shrug of shoulders and the word *maski* (Tok Pisin for never mind, forget about it) uttered under the breath—signalling a certain resignation to the absence of the state.

Refracted Responsibilities

Aside from trying to facilitate a relationship between the government and local communities, the company actively worked on establishing its own relationship with government agencies and representatives. It resisted taking on service delivery responsibilities, and what it saw as the role of a 'quasi-government', but the CA officers felt that they did not have much choice if they wanted to maintain good relations with the impact communities, and to secure and maintain their support for the project. Over the years, the company made a concerted effort to collaborate with the government on facilitating access to basic services in the impact communities. For example, PanAust assisted the Telefolmin District Health division with delivery of quarterly health patrols in communities. This was one of few success stories. On many other occasions, company collaboration with government agencies encountered significant problems, like when the CA officers worked with communities to finance and erect a community clinic in Paupe. A government community health worker was brought in on a company charter plane to run and manage the facility. Although he was on the government payroll, he often did not receive payment for his work. He continued to fulfil his duties but was growing increasingly hesitant to do so. This was not the only instance where the government did not keep its side of the deal. As one CA officer explained to me 'school funding faced [a] similar challenge. And then service providers—the health worker, the teachers—turn to the company for help'.[6]

While the high-level conversations between the company management and the national government happened away from Frieda, as the face of the company, time and time again the CA officers found themselves put on the spot during on-site meetings and gatherings. They spent a lot of their time managing community expectations, drawing lines in the sand to delineate their role from that of the government and encouraging communities to demand that the government provides them with basic services and fulfil its obligations to communities. 'The government thinks that we can solve all its problems'; 'The government is always two steps behind, not keeping up the pace'; 'The government relies too heavily on the company to provide development assistance. It is not taking initiative'; 'The government never played its role properly. As a result, the communities too are not being proactive with alternatives. They are just waiting for the company and the mine'. These are just some of the CA voices I heard which expressed frustration with the lack of active state involvement in services provision at Frieda.

---

6   It is not uncommon in PNG for schoolteachers and health workers not to get paid, especially in remote areas.

Many company and government members I spoke to about the government's dependency on the company to fulfil some of its roles and responsibilities referred to the lack of government capacity to provide infrastructure and services to communities in the Frieda area, rather than the government's bad intentions or lack of willingness to do so. Some of my company interlocutors linked this to the weakness of the national government and saw this incapacity being passed down through all levels of state governance:

> The governments know that it is their job, but they lack the resources, the manpower. They always thank us. They are very appreciative. They know that we are doing what they should be doing. But they lack the capacity and resources and as a result the company finds that it must provide services it really shouldn't have to.

Others were less sympathetic, of the opinion that the government purposefully took a step back and let the company fill the gap in providing services and development assistance to the local communities: 'The [company] feels compelled—it has no choice.' Sitting down for an interview with another, cup of tea in hand, I recorded him express a frustration echoed in many of the interviews I conducted with the company's on-site personnel: 'Government participation at early stages is lacking. CAs are caught in between. They cannot speak for the government. They should not have to.' Trapped in the state's strategic absence, CA officers have become reluctant policy brokers.

A number of district and provincial government representatives admitted that they depended on the company's assistance to bring development not only to the impact communities, but also to the two affected provinces. 'We are very remote here, we don't have development. With the mine, the company will bring the money for development', one of them told me. 'We understand the problems associated with mining projects and the impact it will have on the [Sepik] River. But if we don't get major development in the province, where do we get the money to help people?', I heard another official ask on a different occasion. Representatives of the national government shared this anticipation and also hoped to precipitate development of a mine at Frieda. As one MRA officer explained to me:

The national government has one hope—for the project to develop as a mine. We are asking, preaching, hoping—when will it happen? When will the company submit the SML application? We are waiting happy to grant the lease. We are waiting for the economic impact and development.[7]

Mind the Gap(s)

Despite being host to numerous mining projects, and notwithstanding the industry's dominant presence in the national economy, PNG offers a challenging operating environment for mining companies. Upon arrival, companies find themselves caught up between what Emma Gilberthorpe and Glenn Banks (2012: 188) identified as the 'weak state' and the 'powerful local' and become 'entangled in a range of pre-existing local, regional and national political plays' over which they have little control. The weak state is incapable of regulating not just the corporation involved (see Wesley-Smith 1990; Toft 1997; and Walton 2015), but also the landowning communities with which companies find themselves entangled. As a result, I was told at the PNG Chamber of Mines and Petroleum—the peak representative body for the extractive industries—companies in PNG find themselves 'operating against a backdrop of bad governance and ever growing community expectations'. In academic as well as public discourse, the country is often described in terms of a failed state, affected by the so-called resource curse (Ballard and Banks 2003), which has been linked to factors such as government's incapacity to distribute benefits to the people (Jackson 1991, and Main, Chapter 5, this volume); underperformance of state institutions (Genasci and Pray 2008); disconnect between national-level decision makers and society (Collier 2008); and the government's preoccupation with distant opportunities over locally defined social and economic goals (Connell and Howitt 1991). These macro-level assessments (see Gilberthorpe and Papyrakis 2015) judge the government's performance against the

---

7   This is not to suggest that there were no people in both provinces who opposed the Frieda River Project. While their views were reflected by the provincial representatives, their arguments were generally overshadowed by the desperate need for basic services and infrastructure in the two provinces, emphasised by the government's self-admitted reliance on the industry to deliver 'development' to the region. The level of opposition increased in 2018 when the developer announced major changes in the design of the project, and further in 2019 following the company's publication of an Environmental Impact Statement for the Frieda River Project (see Skrzypek 2020).

expectation of state function, and the promise it made at independence to strategically engage the mining industry with the aim of securing positive development outcomes for the country and its people.

At a micro-level, my hosts in Paupe tended to describe their relationship with the government in terms of absences and deficiencies. The fact that Frieda River is not an operating mine but an advanced mineral exploration project means that the area has not yet experienced the entrenched negative localised impacts present at the many mines operating across the country.[8] However, the way in which the government is experienced by community and company stakeholders in the project location already points to performance and capacity gaps commonly associated with the resource curse, judged against categorical prescriptions of the role of the state as a regulator for the industry, and provider of services, infrastructure and effective governance for its people. The Paiyamo talked about the government that has forgotten them, that never came to see them. They also spoke of the government's failure to deliver and maintain basic services and the ease with which it made promises, which, the Paiyamo learned over time, it almost never kept. Desire for services and infrastructure and high expectations of 'development' which formed at Frieda in the decades since independence meant that there were numerous gaps between what they understood the role of the government to be, and the government's performance against this role. The way in which the categorical presence of the government in the process of planning for a mine at Frieda contrasted with its interpersonal absence further exacerbated the situation, creating in Paupe a sense of the government's presence most strongly experienced through its absence. The Paiyamo saw the mining company as a way of filling this gap, at least insofar as it demonstrated through securing access to what they came to see as indicators of development—services, infrastructure, learning and earning opportunities. They considered their ability to form interpersonal relations with company representatives as a crucially important part of this process, with the ultimate goal of leveraging those relations to precipitate a mine at Frieda. Despite their difficult experiences of *gavman* to date they hoped that, once the mine starts operating, the *gavman* will step up and become more actively involved in 'looking after its people and communities'. In spite of decades of engagement and exploration, they have not yet succeeded in this task.

---

8   See, for example, Golub (2014) on Porgera; Bainton (2010) on Lihir; and Kirsch (2014) on Ok Tedi.

Conclusions

Stakeholder categories are based on a series of assumptions and expectations about their nature and effect. Governance mechanisms commonly used to guide stakeholder engagement in resource extraction projects further reinforce such categorical prescriptions. On face value at least, categorical models of stakeholder relations are grounded in a belief that, in entering into a relationship, the stakeholders accept roles assigned to them by the model, and further that they have the capacity required to enact them and achieve the desired effect. For example, the popular triad stakeholder model reflects a belief that, as a regulator and a main service provider, the state can and will regulate the industry, and provide services for its people. Furthermore, engrained in categorical models is the implication that stakeholders representing different categories have the means of holding the others accountable for their performance against their designated roles and responsibilities. To that end, such models tend to reflect an erroneous assumption that power inequalities can be treated as externalities, painting a deceptively apolitical picture of highly politicised relations and dependencies.

Conceptually, the models stipulate that if all stakeholders enact their categorical roles and meet conditions of their categorical relations, the process will succeed. In the ethnographic example presented here, this success is anticipated to take form of the Frieda River Mine. However, like any attempt at categorisation and standardisation, such models 'assume a set of conditions that do not exist in most of the world' (Newell 2005: 556). Once the categorical stakeholder model is enacted in a real-life scenario of a resource exploration (or extraction) project, categories are exposed to power inequalities, different world-views, political relations, personal interpretations and a range of other socioeconomic forces, which the conceptual models push aside or hide. What emerges is a form of dissonance which questions prescriptive categories on the basis of enactment and effect of those; and forces some or all of the stakeholders to revise their understandings of their roles and responsibilities at the project.

Cognitive dissonance theory helps us explore how people deal with uncertainty, making it useful for thinking through impacts and implications of resource exploration projects characterised by prolepsis, anticipation and desires to effect and affect the future (see Minnegal and Dwyer 2017; and Skrzypek 2020). In cognitive dissonance theory, when there is an inconsistency between attitudes or behaviours, something must change to eradicate the disharmony (Festinger 1957). In a similar way, categorical dissonance stipulates that where there emerges inconsistency between expectations of roles and their enactments, something must change to alleviate the tension and eliminate the risk of project failure. At the Frieda River Project, members of local communities saw the potential future mine as a pathway to development, a 'modern road to cargo'. They associated the mine with industrial development, but even more so with civic infrastructure and widespread access to education, services and earning opportunities. Their expectations were high, having grown over time, fuelled by generous promises made by early exploration companies who sought to secure access to their land for exploration and, when the time comes, mining purposes. Government representatives visiting the area not only encouraged but actively partook in this imagery of mining as a pathway to development and a road to wealth and cargo. As described early on in this chapter, the country has a long history of deficiencies in service delivery and high levels of dependency on the extractive industries for both national economic growth and localised forms of development. In anticipation of a large-scale mine, in the Frieda River area, as in many other resource-rich parts of the country, the government has largely abdicated its service and infrastructure delivery responsibilities, refracting them instead onto resource extraction companies.

The ethnographic material presented here described dissonance created by expectations of the state's function and the government's performance against that role; the gap formed by its categorical presence and interpersonal absence from the project area; and the company's response to local communities' and the government's expectations of service delivery assistance. The company resisted becoming a 'quasi-government' and felt that was not part of its role as the project's key investor and operator. However, in order to maintain local communities' and the government's support, its representatives felt that they had no choice but to alter what they saw as the boundaries of the company's role and, consequently, behaviour at the Frieda River Project. The company took on some of the state's service delivery responsibilities, and facilitated the relationship between the government and communities, thus mediating the failure and absence of the government from the local nexus of relations, and from the service delivery landscape. It did not do so because it wanted to—the company's ideas about community development and engagement programs saw their function as 'in extension' rather than 'in place' of government provision of basic services. The company took on elements of the role of the state because this allowed it to advance its interests and fulfil its primary role—as a commercial business and project developer.

Such corrective behaviour associated with categorical dissonance brings to light critical limitations of oversimplified stakeholder models, which assume conditions that do not exist in the lived experiences of stakeholders in resource contexts. In this case, the company's behaviour responded to state absences and pointed to gaps between prescriptions and expectations of the state role at the project, and the government's performance against those. However, rather than exposing the shortcomings of the model, and holding the government accountable for its failure to fulfil its obligations, the company provided it with a self-correction mechanism. Its actions meant that the government's demonstrated absence and underperformance against its prescribed stakeholder role was mediated, as the developer filled the very gaps that could have threatened the future of the project. As a result, while the company and the Paiyamo faced ongoing tension and conflict over the allocation of responsibilities, viewed from a distance, the demands of the categorical model were met, and the project could continue moving forward towards the desired effect—the Frieda River Mine.

Acknowledgements

The work presented here was made possible by funding from the European and Social Research Council (UK)—award number ES/1904107/1, and the European Union's Horizon 2020 research and innovation program under the Marie Sklodowska-Curie grant agreement No. 753272.

References

Anderson, B., 1983. *Imagined Communities: Reflections on the Origin and Spread of Nationalism*. London: Verso Books.

Ashton, C., 1990. 'Papua New Guinea: A Broken-Backed State.' *Quadrant* 34(7/8): 25–30.

Bainton, N.A., 2010. *The Lihir Destiny: Cultural Responses to Mining in Melanesia*. Canberra: ANU E Press (Asia-Pacific Environment Monograph 5). doi.org/10.22459/ld.10.2010

Bainton, N. and G. Banks, 2018. 'Land and Access: A Framework for Analysing Mining, Migration and Development in Melanesia.' *Sustainable Development* 26: 450–460. doi.org/10.1002/sd.1890

Ballard, C. and G. Banks, 2003. 'Resource Wars: The Anthropology of Mining.' *Annual Review of Anthropology* 32: 287–313. doi.org/10.1146/annurev. anthro.32.061002.093116

Banks, G.A. and C. Ballard, (eds), 1997. *The Ok Tedi Settlement: Issues, Outcomes and Implications.* Canberra: The Australian National University, National Centre for Development Studies (Pacific Policy Paper 27).

Bebbington, A., L. Hinojosa, D.H. Bebbington, M.L. Burneo and X. Warnaars, 2008. 'Contention and Ambiguity: Mining and the Possibilities of Development.' *Development and Change* 39(6): 887–914. doi.org/10.1111/ j.1467-7660.2008.00517.x

Collier, P., 2008. 'Laws and Codes for the Resource Curse.' *Yale Human Rights and Development Law Journal* 11: 9–28.

Connell, J. and R. Howitt, 1991. 'Mining, Dispossession, and Development.' In J. Connell and R. Howitt (eds), *Mining and Indigenous Peoples in Australasia.* Sydney: Sydney University Press.

Dalsgaard, S., 2013. 'The Time of the State and the Temporality of the *Gavman* in Manus Province, Papua New Guinea.' *Social Analysis* 57: 34–49. doi.org/ 10.3167/sa.2013.570103

Douglas, M., 1986. *How Institutions Think.* New York: Syracuse University Press.

Festinger, L., 1957. *A Theory of Cognitive Dissonance.* Stanford (CA): Stanford University Press.

Filer, C., 2008. 'Development Forum in Papua New Guinea: Upsides and Downsides.' *Journal of Energy & Natural Resources Law* 26: 120–150. doi.org/ 10.1080/02646811.2008.11435180

Filer, C. and P.Y. Le Meur, 2017. 'Large-Scale Mines and Local-Level Politics.' In C. Filer and P.Y. Le Meur (eds), *Large-Scale Mines and Local-Level Politics: Between New Caledonia and Papua New Guinea.* Canberra: ANU Press (Asia-Pacific Environment Monograph 12). doi.org/10.22459/LMLP.10.2017.01

Freeman, R.E., 2010 [1984]. *Strategic Management: A Stakeholder Approach.* Cambridge: Cambridge University Press.

Gelu, A., 2010. 'Politics and Governance.' In T. Webster and L. Duncan (eds), *Papua New Guinea's Development Performance 1975–2008*. Port Moresby: National Research Institute (Monograph 41).

Genasci, M. and S. Pray, 2008. 'Extracting Accountability: The Implications of the Resource Curse for CSR Theory and Practice.' *Yale Human Rights and Development Law Journal* 11: 37–58.

Gilberthorpe, E. and G. Banks, 2012. 'Development on Whose Terms? CSR Discourse and Social Realities in Papua New Guinea's Extractive Industries Sector.' *Resources Policy* 37: 185–193. doi.org/10.1016/j.resourpol.2011.09.005

Gilberthorpe, E. and E. Papyrakis, 2015. 'The Extractive Industries and Development: The Resource Curse at the Micro, Meso and Macro Levels.' *Extractive Industries and Society* 2: 381–390. doi.org/10.1016/j.exis.2015.02.008

Golub, A., 2014. *Leviathans at the Gold Mine: Creating Indigenous and Corporate Actors in Papua New Guinea*. Durham (NC): Duke University Press.

Golub, A. and M. Rhee, 2013. 'Traction: The Role of Executives in Localising Global Mining and Petroleum Industries in Papua New Guinea.' *Paideuma* 59: 215–236.

Hinojosa, A.S., W.L. Gardner, H.J. Walker, C. Cogliser and D. Gullifor, 2016. 'A Review of Cognitive Dissonance Theory in Management Research: Opportunities for Further Development.' *Journal of Management* 43: 170–199. doi.org/10.1177/0149206316668236

Hirsch, E., 2001. 'New Boundaries of Influence in Highland Papua: "Culture", Mining and Ritual Conversions.' *Oceania* 71: 298–312. doi.org/10.1002/j.1834-4461.2001.tb02755.x

Hyndman, D., 2001. 'Academic Responsibilities and Representation of the Ok Tedi Crisis in Postcolonial Papua New Guinea.' *Contemporary Pacific* 13: 33–54. doi.org/10.1353/cp.2001.0014

IIED (International Institute for Environment and Development), 2002. *Breaking New Ground: Mining, Minerals and Sustainable Development*. London: IIED, Mining Minerals and Sustainable Development Project.

Imbun, B.Y., 2007. 'Cannot Manage without the "Significant Other": Mining, Corporate Social Responsibility and Local Communities in Papua New Guinea.' *Journal of Business Ethics* 73: 177–192. doi.org/10.1007/s10551-006-9189-z

Jackson, R., 1991. 'Not Without Influence–Villages, Mining Companies and Government in PNG.' In J. Connell and R. Howitt (eds), *Mining and Indigenous Peoples in Australasia*. Sydney: Sydney University Press.

Jorgensen, D., 2007. 'Clan-Finding, Clan-Making and the Politics of Identity in a Papua New Guinea Mining Project.' In J.F. Weiner and K. Glaskin (eds), *Customary Land Tenure and Registration in Australia and Papua New Guinea: Anthropological Perspectives*. Canberra: ANU E Press (Asia-Pacific Environment Monograph 3). doi.org/10.22459/cltrapng.06.2007.04

Kirsch, S., 2014. *Mining Capitalism: The Relationship Between Corporations and Their Critics*. Oakland: University of California Press.

May, R.J., 2004. *State and Society in Papua New Guinea: The First Twenty-Five Years*. Canberra: ANU E Press. doi.org/10.22459/SSPNG.05.2004

May, R.J and M. Spriggs (eds), 1990. *The Bougainville Crisis*. Bathurst (NSW): Crawford House Press.

Minnegal, M. and P.D. Dwyer, 2017. *Navigating the Future: An Ethnography of Change in Papua New Guinea*. Canberra: ANU Press (Asia-Pacific Environment Monograph 11). doi.org/10.22459/NTF.06.2017

Newell, P., 2005. 'Citizenship, Accountability and Community: The Limits of the CSR Agenda.' *International Affairs* 81(3): 541–557. doi.org/10.1111/j.1468-2346.2005.00468.x

Owen, J.R., 2016. 'Social License and the Fear of Mineras Interruptus.' *Geoforum* 77: 102–105. doi.org/10.1016/j.geoforum.2016.10.014

Owen, J.R. and D. Kemp, 2014. 'Mining and Community Relations: Mapping the Internal Dimensions of Practice.' *Extractive Industries and Society* 1: 12–19. doi.org/10.1016/j.exis.2013.12.004

PanAust, 2018. 'A Nation-Building Development Pathway for the Frieda River Project.' Press release, 11 December. Viewed 16 September 2020 at: panaust. com.au/sites/default/files/201812_A%20nation-building%20development %20pathway%20for%20the%20Frieda%20River%20Project_0.pdf

Parsons, R., J. Lacey and K. Moffat, 2014. 'Maintaining Legitimacy of a Contested Practice: How the Minerals Industry Understands Social Licence to Operate.' *Resources Policy* 41: 83–90. doi.org/10.1016/j.resourpol.2014.04.002

Rivers, J., 1999. 'Formulating Basic Policy for Community Relations Programs.' Canberra: The Australian National University, Research School of Pacific and Asian Studies, State, Society and Governance in Melanesia Project (Discussion Paper 99/1).

Skrzypek, E., 2020. *Revealing the Invisible Mine: Social Complexities of an Undeveloped Mining Project.* New York: Berghahn.

Stead, V.C., 2016. *Becoming Landowners: Entanglements of Custom and Modernity in Papua New Guinea and Timor-Leste.* Honolulu: University of Hawai'i Press. doi.org/10.21313/hawaii/9780824856663.001.0001

Strathern, M., 2005. *Kinship, Law and the Unexpected: Relatives Are Always a Surprise.* Cambridge: Cambridge University Press.

Toft, S. (ed.), 1997. *Compensation for Resource Development in Papua New Guinea.* Boroko: Law Reform Commission (Monograph 6). Canberra: The Australian National University, National Centre for Development Studies (Pacific Policy Paper 24).

Valenta, R.K., D. Kemp, J.R. Owen, G.D. Corder and E. Lèbre, 2019. 'Re-Thinking Complex Orebodies: Consequences for the Future World Supply of Copper.' *Journal of Cleaner Production* 220: 816–826. doi.org/10.1016/j.jclepro.2019.02.146

Walton, G., 2015. 'Defining Corruption where the State is Weak: The Case of Papua New Guinea.' *Journal of Development Studies* 51: 15–31. doi.org/10.1080/00220388.2014.925541

Welker, M., 2014. *Enacting the Corporation: An American Mining Firm in Post-Authoritarian Indonesia.* Berkeley: University of California Press. doi.org/10.1525/9780520957954

Wesley-Smith, T., 1990. 'The Politics of Access: Mining Companies, the State, and Landowners in Papua New Guinea.' *Political Science* 42(2): 1–19. doi.org/10.1177/003231879004200201

3

'Restraint without Control': Law and Order in Porgera and Enga Province, 1950–2015

Alex Golub

The clans are fighting a mile away.
They paint their faces, wear bits of grass and bark
to tell us this is happening at an earlier time.
Do not believe it, this is wishful thinking;
they wear suits and conquer the centuries
by aircraft. In capital cities
they shake each others hands, careless
of what smears them.
They are what old men fearing, feared to predict.

'Lambchops in PNG' (Markham 1985)

Introduction

On 6 November 1884, Commodore James Erskine hoisted the Union Jack over Port Moresby and declared British New Guinea to be a protectorate of the British empire. At the ceremony, he promised the Motu-speaking men assembled there that:

> Evil disposed men will not be permitted to occupy your Country, to seize your lands, or to take away from your homes ... Your lands will be secured to you. Your wives and children will be protected. Should any injury be done to you, you will immediately inform

> Her Majesty's Officers who will reside amongst you, and they will
> hear your complaints and do justice … You will all keep peace
> among yourselves, and if you have disputes with each other you
> will bring them before the Queen's Officers, who will settle them
> for you without bloodshed. (Biskup et al. 1968: 56–7)

Wolfers describes Erskine's policy as one of 'restraint without control'
(1975: 13). On the one hand, the British administration sought to make
the colony 'attractive to white settlers and entrepreneurs exercising their
undoubted rights' (MacWilliam 2013: 24–5) to make a profit in Papua,
assuming that the 'immanent, spontaneous development led by private
firms and expatriate planters' would transform British New Guinea into
'a profitable (colonial) possession' (ibid.: 17), to the presumed benefit
of everyone, including Melanesians. On the other hand, as a sign of
its enlightened paternalism, 'the protectorate administration … was
empowered by law only to control the entry of foreigners to the area, and
to legislate only for its non-indigenous population' (Wolfers 1975: 14)
and would restrain 'the depredations of the British fortune- soul-
and adventure-seekers' (ibid.: 14–15) without unduly controlling the
Indigenous population.

Erskine promised, in essence, that the presence of the state would benefit
the people of what is now Papua New Guinea (PNG). This chapter
uses historical anthropology (Sahlins 1985) to describe the historical
transformations of Erskine's promise and to analyse the role it plays
in PNG today. In particular, it examines the history of 'restraint' and
'control' in Enga Province, PNG, from 1950 to 2015. Erskine's promise,
although made far from Enga and decades before government would
reach the highlands, is key to understanding the post–Second World War
history of PNG because it was the historical precedent that grounded
the state's legitimacy for both the colonial and postcolonial governments
of PNG. This chapter describes how violence, policing and governance
changed over time in central Enga, where Wabag, the provincial capital,
is located. It then examines continuities and differences between central
Enga and the Porgera valley in the western half of the province where a
large gold mine operated by the Porgera Joint Venture (PJV) is located.
The chapter argues that in contemporary Enga the relationship between
restraint and control is the reverse of Erskine's vision. Rather than the
government protecting Melanesians from global capitalism, in Porgera
today the government protects global capitalism from Melanesians.
Where once the government aspired to restrain foreign business without

controlling the population, today the government exercises little control over mine employees even as it seeks to restrain Papua New Guineans who act against the PJV's interests.

This chapter takes up this volume's focus on the state's absent presence by examining the role restraint and control have played in realising Erskine's promise. Erskine's vision is essentially a nineteenth-century liberal one of restrained government: that a limited government would protect lives and properties while allowing the market to create a spontaneous order which enabled prosperity and growth. Present to protect Melanesians from dangers of alcohol and indentured servitude, it would restrain itself from imposing in their customary life, but preserve their ownership of land and guarantee their personal safety. Thus the state was present in some areas of activities but absent in others.

The governments which followed sought to fulfil Erskine's promise, but used different means to achieve it. Both the postwar Australian administration and the independent state of PNG pursued a developmentalist agenda (Ferns 2020) which actively intervened in life in Enga. Theirs was an order-making project (Golub 2018) that sought to routinise and regulate life. Most obviously, 'control' meant controlling outbreaks of violence and bringing peace. But it also manifested itself in subtler ways. The construction and maintenance of schools, roads and hospitals required transformations in senses of time and labour. Protecting the rights of women meant intervening in the 'private' domain of the family. Agricultural officers would teach people new ways of farming, transforming nutrition and diet. Control, not restraint, was the means by which the government sought to fulfil Erskine's promise. As we shall see, in Enga the government has failed to achieve many of the goals it has set for itself. In this sense, it is an 'absent presence' in that its promises to be present, to control and regulate life, are too often honoured in the breach rather than the observance.

But this chapter challenges the concept of the state as an 'absent presence' as well. While broad historical narratives must necessarily involve collective actors such as 'the state' and 'society' or 'the mine' and 'the community', this chapter will argue that an accurate understanding of history cannot take these actors for granted. Using the approach I have followed elsewhere (Golub 2014), I will demonstrate how history is not really made up of these collective actors, but rather of networks of people who appear in different times and different situations as personating one or the other of these entities. A full history of Enga requires understanding how at times 'the government' and 'clans' overlap, just as 'the police' and 'mine security' interpenetrate.

Decomposing these corporate actors contributes not only to the history of PNG, but to broader social scientific work on 'the resource curse' and 'resource conflict'. It is easy to write particularistic histories of individual resource projects, just as it is easy to produce highly abstract, schematic models of resource conflicts. But it is much more difficult to produce 'middle range' (Tilly 2010) theories of resource conflict which engage in comparison and make generalisations about cases without losing track of the historical uniqueness of each one. This chapter aims to show how a historical anthropology which decomposes corporate actors might help make this middle range theory possible. In doing so, it attempts to answer Le Billon's call for '"thick" historical and geographical contextualisation, relating the past to the present, as well as resource locales to places of belonging and spaces of social relations' (Le Billon 2014: 57–8).

The main body of this chapter is split into two primary sections. In the first, I describe state governance and tribal fighting in central Enga Province from 1950 to the mid-1990s. In doing so, I take issue with one common narrative of the history of Enga and of PNG more generally. I will call this a 'degenerationist' narrative (Golub 2014: 169–70) similar to the 'discourse of abjection' that Bashkow notes (Bashkow 2006: 231). This is, in essence, a narrative of successful control, a claim that Australia developed PNG and brought peace and government to the country, but that after independence the country went backwards. A good example of this narrative can be found in the comments of the YouTube video 'Tribal Conflict in Papua New Guinea | On The Frontline':

> PNG under Australian administration, life was pretty close to normal. Tribal conflicts were extinguished quickly. Small police stations scattered throughout the highlands kept an eye out for trouble. Riot squad was stationed in the highlands and a backup unit in the capital, Port Moresby. Back up squad could be deployed at a moment's notice to hot spots. All weapons were seized/destroyed. Those involved were sent to prison to face judgement. Police regularly visited places searching /seizing/ destroying confiscated weapons. Locals started to respect law and order, mediation was encouraged. After independence, the png government closed all regional stations, not sure what happened to the riot squad. PNG government past/present offer very little to regional areas unfortunately. (comment on ICRC 2017)

In the first section of this chapter, I will claim that key aspects of this narrative are incorrect. The Australian administration never devoted the resources necessary to police Enga adequately. It is true that there was a decrease in violence in Enga after the administration arrived, but this was due to Engans, not the administration, who 'self-pacified' for roughly a decade between 1955 and 1965 due to a temporary enthusiasm for state governance. Disillusion with the administration and other factors led to a growth in fighting *prior* to independence. In Enga, the breakdown in law and order was not something that Papua New Guineans created, it was something they inherited from the Australians. During the independence period, tribal conflict increased as different groups attempted to control the spoils system of governance that took hold in the province. Policing in Enga, before and after independence, was often both illiberal and ineffective. Restraint without control has been a hallmark of governance in Enga during both the Australian and independence period. In its unfulfilled promises, the state has been both present and absent in Enga.

Using this discussion of central Enga as context, the second section of this chapter examines the history of Porgera from the mid-1980s to 2015. This section examines the growth of violence and disorder that accompanied the creation and operation of the Porgera Gold Mine. It tests two more hypotheses that circulate in the public sphere about Porgera. The first one, which might be associated with mine management, claims that Porgera is a 'rough neighbourhood' and that violence and disorder are cultural norms for which the mine cannot be blamed. Second, there is the 'resource conflict' hypothesis, which argues that in Porgera, as in so many other resource projects, it is the presence of the mine which induces conflict. The evidence in this section shows that both hypotheses explain some of the historical evidence for Porgera, but are insufficient alone. The 'rough neighbourhood' model does not capture the way that social change—including that created by the mine—has amplified the negative aspects of Engan culture. The 'resource conflict' model, on the other hand, cannot see how police and PJV violence is a transformation of the illiberal policing tactics which have been the historical norm in the valley. Disorder in Porgera is not new: it is an amplified version of existing trends in Enga.

Finally, this chapter intentionally dwells on the some of the most troubling aspects of life in Enga so as to understand them and, hopefully, form some small part of a solution to them. I want to note at the beginning of this chapter, therefore, that there are many dedicated police officers and mine employees who do their best to act ethically, just as there are many Papua New Guineans who abhor war and seek peace. There have also been numerous attempts at conflict mediation over the decades in Enga and Porgera as well. For reasons of length and analytic focus this chapter does not focus on these people, but I want to emphasise at the outset that I recognise their existence and contributions. While I focus on the negative parts of life in the province, the reader should not develop an overly negative view of Enga and Porgera.

Enga and the 'Pax Australiana'

Enga Province is part of the Highlands Region of PNG (Figure 3.1). This chapter focuses on two areas of the province. First I examine what I call 'central Enga', or the area around Wabag, the capital of the province. Second, I will focus on the Porgera Valley, which is in the far west of the province. Unusually for PNG, Engans consider themselves to be part of a single ethnolinguistic group (the largest in the country) whose boundaries are coterminous with their province—a unique political and ethnic unit. Porgerans, who speak a language called 'Ipili' by linguists, are often seen as a separate group. In fact, linguists and anthropologists view this area to be characterised by interconnection than clear separation. 'Engan' appears to be a series of closely related dialect chains which stretch across the province rather than a single language. Porgerans and their 'Engan' neighbours in Laiagam have more in common linguistically than people from Laiagam would have with Engan speakers on the other side of the province. Historically, the area was characterised by a complex system of migrating and intermarried social groups rather than bounded and localised ethnic groups and these connections continue to this day. In fact, the ethnonyms 'Porgeran' and 'Engan' developed in the 1950s and are not endogenous to the area. This is not to gainsay the identity claims of any particular group. On the contrary I will use terms like 'Engan' and 'Porgerans' throughout this chapter. But it is important to understand that behind this useful shorthand is a more complicated ethnolinguistic situation which will become relevant in different contexts in the chapter, especially as I tack between the history of central Enga and Porgera.

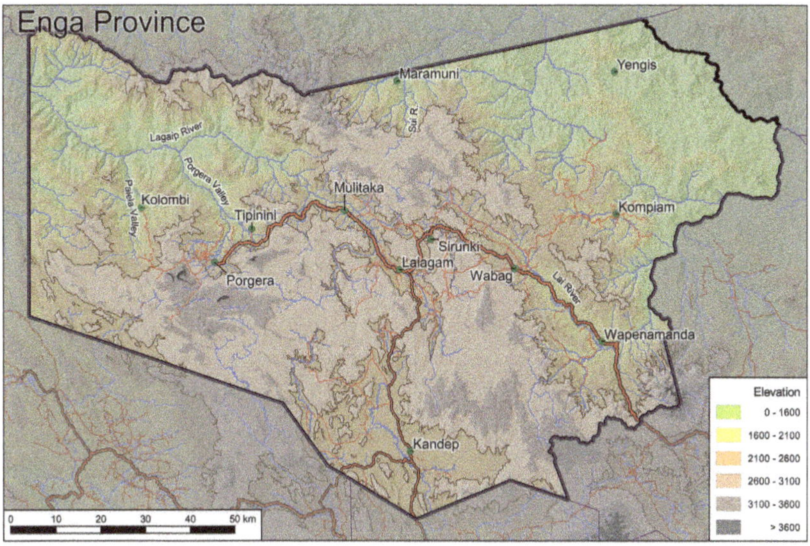

Figure 3.1 Enga Province

Source: Map by Nelson Minar with data from OpenStreetMap, Humanitarian Data Exchange, OpenDEM, and SRTM. openstreetmap.org/copyright

While a few early exploratory patrols by miners and government officers began entering Enga in the early 1930s, it was not really until after the Second World War that outside powers seriously attempted governing what is now Enga Province. In the late 1940s Enga fell within the Trust Territory of New Guinea, which was administered by the Australian government on behalf of the United Nations. The Australian territory of Papua would later merge with the Trust Territory of New Guinea to form the Independent State of Papua New Guinea in 1975, which was also governed by Australia, but under a different administration.

The nature of Erskine's promise had shifted in the postwar period. Administration policy in this period was overseen by Paul Hasluck, who sought 'uniform development' to bring the highlands region the same level of administration of government services as the coast, which had a much longer history of entanglement with the global system (Hasluck 1976; MacWilliam 2013). The administration sought to control and develop areas, not merely to restrain bad behaviour and trust in spontaneous development. The result was what some authors have seen as 'acceptable colonialism' or 'administrative colonialism' (Hawksley 2007), a form of colonialism that would lead to development. In terms of policing, this meant pacification and the creation of law and order, a situation frequently described in the literature of this period as the 'Pax Australiana' on par with the Pax Britannica or Pax Romana of the past.

And yet, from the beginning the New Guinea administration was remarkably understaffed. In 1948–49 the territory was roughly 250,000 square kilometres and home to 1 million people (UN 1951: 8) but was policed by just 22 white officers and 1,236 New Guinean policemen (Kituai 1998: 45). This was an all-time high. In 1927–28 there were just 10 white officers and 588 police in New Guinea (ibid.). So, despite Erskine's promise, the Australian government never produced a police force sufficient to protect Papua New Guinean property and safety. At the same time, it could not project sufficient force to dispossess Papua New Guineans of their lands and property. The relatively benign nature of Australian colonialism in PNG was as much a result of a lack of capacity as it was of good intentions. The state was 'absent' in PNG long before independence.

Enga was originally part of the Western Highlands province and was declared 'uncontrolled' (later this category was changed to 'restricted') and off-limits to foreigners on the assumption that they could both harm and be harmed by highlanders if they ventured into the area. The government centre of this section of the Western Highlands was Wabag, an area that had first served as a staging ground for the large 'Hagen-Sepik Patrol' of 1938–39 (Gammage 1998). In 1941 it was repurposed as a patrol post with an expatriate 'kiap', or patrol officer, stationed there (Lakau 1988: 84). Land was not alienated there for a government station until 1949 (ibid.). Government during this period took the form of patrols from Wabag to neighbouring areas, during which government patrols would attempt to census the population, resolve disputes peacefully and administer basic healthcare. Over time the government would open additional patrol posts and derestrict the areas around the patrol posts until, segment by segment, they had the entire highlands under their control.

Conflict was a part of the situation which Australian officials encountered. All sources agree that in precolonial Enga, violence and fighting were common but not highly culturally valued by most people. Wiessner and colleagues summarise this view when they write that 'in Enga oral history, there was never a time before war' but that 'historical traditions do not glorify war, but describe it as an unfortunate last resort when problems could not be solved by other means' (Wiessner et al. 2016: 164).

A common cause of war was 'to re-establish the balance of power and respect between clans after insult or injury so that exchange could flow' (ibid.: 165). Warfare was a topic of ambivalence, in that it was destructive and inhibited the flow of wealth, but it also provided an opportunity for young men to demonstrate their bravery and for the clan to reinforce its solidarity. Confusingly, the Enga unit of customary social organisation known as *tata* is often translated into English as 'clan' even though the conflict between them is often referred to as 'tribal' fighting.

Overall, casualties were low and there was not a great deal of training for warriors, songs glorifying combat and death, and so forth. As in much of the highlands, prosperity and wealth were highly valued, not one's reputation as a warrior. It is no surprise, then, that the arrival of the government in the 1950s began a relatively speedy process of 'pacification', in which fighting declined markedly and people became fascinated with the arrival of white men and what the administration could mean for their history. In a remarkable piece of early opinion polling, Spruth found that in 1968 '97 per cent of people agreed with the statement, "I think the government should stay here always," (katawaka katapyali lao masilya)' (Spruth 1968: 6). Peace was a central benefit of the arrival of government. 'The Enga people feel good about and recognise the value of the peace bought by the administration. Particularly in private interviews this was seen as one of the most important things that administration had done and the door through which many other good things have come' Spruth wrote, and '99% of the people believe that fighting would break out all over if the government was to depart' (ibid.: 7). A similar poll showed that 42 per cent of people felt that the main work of the government was to 'keep peace' (the runner-up, at 21 per cent, was 'to get money') (Lutz and Rivers 1968: 6).

Most writers of this period agree that it was Engans, not kiaps, who stopped the fighting. Wormsley and Toke noted in the 1980s that 'increasing violence is the trend and the period following pacification was the anomaly' (Wormsley and Toke 1985: 63). According to them, it was a 'fiction' that 'pacification was achieved at the hands of the kiaps and that fighting ceased. Reality suggests that Enga took a step back, viewed what the government offered, and agreed to give the government an opportunity to deal with disputes' (ibid.). When it was discovered that 'what the government offered was inadequate … the Enga … simply reverted to their own tried and true techniques of dispute settlement' (ibid.).

Gordon and Kipalan agree: 'Law and Order [sic] was achieved, despite acts of intimidation against them, through the active cooperation and collaboration of the Engas' (1982: 332).

Why did Engans embrace the administration? Access to wealth and prosperity were key. As we have seen, Engans do not value warfare and aggression. Peace was in accordance with Engan values as an end in itself. Another central Engan value is generating wealth and prosperity through trade and exchange with wealthy external sources, such as trading partners in the large-scale exchange ceremonials known as *tee*. Peace furthered this cause. As Gordon and Kipalan point out, the most desired forms of wealth were manufactured goods, whose distribution was controlled by the administration. They quote someone (identified only as 'an old Enga') as saying:

> The first 'kiaps' gave beads, salt, steel-axes—everyone wanted it so they all followed the 'kiap' and stopped fighting. They stopped fighting because the 'kiap' told them to and they obeyed because they did not want to lose the source of these things. (Gordon and Kipalan 1982: 333)

In sum, peace was in accordance with Engan values and interests, and kiaps had a near monopoly on manufactured goods.

This golden age of harmony did not last long. Spruth and Lutz's early opinion polling work was presented in 1968. And yet, 'by the early 1970s, the Australian administration believed it was facing a rapidly deteriorating law-and-order situation' (Gordon and Meggitt 1985: 27). From February 1972 to February 1973, 17 'large riots' occurred in Enga which killed 20, wounded 'hundreds' and did 'untold damage' to 'gardens, cash crops, and buildings' (MacKellar 1975: 215).[1] A sign of the seriousness of the situation was the commissioning on 20 December 1972 of a special committee to study tribal fighting in the highlands (GPNG 1973: 1). As the decade continued, Engans had adapted to the incursions of the mobile squads of police by learning to exploit the patterns of response 'including counteracting tear gas canisters by throwing them back at the police or covering them in mud' (Gordon and Meggitt 1985: 31).

---

1   Throughout this chapter I use terms such as 'riots' or 'mobs' following the terminology of the authors I am quoting. These are not terms of art for me.

In 1979 the Enga provincial government declared a state of emergency after a policeman was killed in revenge after an Engan died in prison (ibid.: 34). By this time, one report noted that:

> tribal fighting on a vast scale has completely paralysed the Enga ... the field service has all but disintegrated in this province, and the police there have for the best part of three years been almost powerless. They repeatedly express themselves unable to find or arrest the accused, or to bring witnesses to court. In these circumstances the Law does not run. (Gordon and Meggitt 1985: 18–19)

Given these facts, it is hard to see a decline in law and order from the 'Australian' period to independence. Self-government was achieved in PNG on 1 December 1973. Nineteen days later, the Paney Commission on tribal fighting was formed. It appears that tribal fighting was escalating even as Australia was preparing to withdraw from the country. One can hardly say, then, that Australia had Enga under control when it handed the province over to the newly independent state of PNG. Nor could one say that police in this period had a high standard of conduct from which the post-independence police force would fall. There have been numerous government attempts at various culturally appropriate grassroots projects to restore law and order throughout the province (e.g. Young 2004). However, this should not give the impression that policing as a whole in Enga has always strictly adhered to international best practices in human rights—while Edward LiPuma is doubtless right that 'there is no ... Australian Cortez' and that 'few peoples of Melanesia have felt the exercise of power in its most crushing, overt, agentive mode' (2000: 24).

The Australian administration relied on illiberal modes of justice, just as the PNG government would. Despite Erskine's promise, Native Regulation Acts were in force in both Papua and New Guinea which deprived Indigenous people of many basic human rights. The 'traditional method of stopping fighting' used by police was by 'taking pigs and women as hostages', at least until courts declared such activity unconstitutional in 1979 (Gordon and Meggitt 1985: 31). The Paney report on tribal fighting recommended group punishment (GPNG 1973: 30) and a majority of its members recommend hanging as punishment for murder to disincentivise violence (ibid.: 15). Other suggested innovations included 'reversing the onus of proof' (ibid.: 29) (that is, assuming people to be guilty of tribal fighting until proved innocent), and group punishment (making all members of a group guilty for the action of one of its members) (ibid.: 29).

These recommendations were implemented in the *Inter-Group Fighting Act 1977*, which allowed the government to declare areas 'fight zones', and to make any attempt to prepare for fighting or to incite fighting in such zones illegal (Gordon and Meggitt 1985: 31–2). The law was later overturned as unconstitutional.

Clans and the Government after Independence

In the late 1970s and throughout the 1980s conflict continued in Enga Province even as the province elected its first leaders and created an Indigenous provincial civil service. Daniel Kumbon writes that during the 1980s 'there was a widespread breakdown in law and order. Tribal warfare, armed robbery, rape and other social problems reached high levels as government services deteriorated. Intense politics and deep-rooted corruption took centre stage' (Kumbon 2017: 41). A large grant from the World Bank funded and integrated development programs in Enga which included a report on 'law and order'. In the course of three years it 'observed and recorded the details of eighty-five tribal fights' and developed 'incomplete profiles on another twenty fights' (Wormsley and Toke 1985: 49). Enga also developed a reputation for violence, fighting and corruption in this period. The Law and Order Project reports 49 headlines appeared in the press between January 1981 and June 1985 which featured headlines such as 'clansman axed to death', 'chopped to death', 'police harass Enga villagers' and 'police burn down huts in Enga' (Wormsley and Toke 1985: 21). It should be noted that Enga was not alone in seeing a rise of crime in this period. Law and order were issues around the country, a situation reflected in the fact that a national law and order study was also commissioned and completed in this period (Clifford et al. 1984).

There were many causes of conflict in this period, only some of which were novel. Historical enmities remained. Wormsley and Toke, the authors of the Law and Order report, note that 'one fight is estimated possibly to be as long-lived as one hundred sixty years, based on oral histories and conflicting claims to occupation' (1985: 59). Kundapen Talyaga, an Engan scholar and public servant, argues that the presence of alcohol and the degeneration of traditional forms of control resulted in more fighting (Talyaga 1982). Wormsley and Toke point out that these incidents always take place in the context of clan affiliations and past clan histories (1985: 61). As a result a single drunken incident can reignite pre-existing grudges and mobilise an entire clan. Indeed, given how ineffective the government has historically been at settling disputes, 'Engas conceive of fights not as problems (or crimes) but as a solution to other problems and not as threats to order but attempts to restore order' (Wormsley and Toke 1985: 61). 'Engans,' the report notes, 'recognise impotence when they see it' (ibid.).

Changing relations to wealth also shifted in this period, and this in turn shaped conflict. As Engans became integrated into the global economy the government lost its monopoly on imported goods, giving it less ability to control fighting by controlling the flow of wealth. Moreover, the greatest source of wealth was control of government funds, resources and positions, which were provided by the funding the province received directly from the national government. By the early 1990s, Harry Derkley could write that:

> Provincial government ... functions as an arena where the furious
> contest between clans for the spoils of government—jobs, cash,
> houses, vehicles—continues unabated. It is often only the threat
> of total loss of control of the source of these goods by the ultimate
> sanction—suspension of provincial government—which serves to
> preserve a facade of legality and adherence to 'proper procedure'.
> (Derkley 1997: 142)

Derkley's comments are consistent with contemporary observers, who described democratic politics as a contest between the haves and have-nots, or rather 'those who eat and those who don't eat' (Lakane and Gibbs 2003; Kanaparo n.d.). In a serious sense, in post-independence PNG the government did not govern Engans, but rather the other way around. Gordon and Kipalan write that 'in Enga it appears that the state is being held at spear-point by pre-capitalist forces' (Gordon and Kipalan 1982: 334).

In 1973 Enga was established as an 'area authority', a transitional administrative structure which was governed by a white colonial official and a selected group of Engans nominated by Indigenous Local Government Counselors and approved by the administration (Scott and Pitzz 1982: 280). The next year Enga was established as a district separate from Western Highlands. This district then became a province of the Independent State of Papua New Guinea in 1975. Leadership was still provided by the area authority figures, who in 1978 were appointed as the Interim Provincial Government until 1980, when the first elections were held (Scott and Pitzz 1982: 280).

It was in this period that concerns about 'corruption' increased in Enga. In 1984 PNG's auditor-general found 'gross financial mismanagement' in the province's accounting systems, and the premier of the province, Danley Tindiwi, was jailed for misappropriation (Dorney 2000: 238) and the provincial government was suspended from February 1984 to March 1986 (GPNG 1987: 25). In May 1986 a new provincial assembly was sworn in, with Ned Laina as the new premier. In retribution, supporters of the losing candidate in the election burned down the administrative headquarters along with eight motor vehicles, resulting in a half million kina worth of damages (Dorney 2000: 254). In the 1987 national election, Malipu Balakau unseated Paul Torato, only to be shot in June 1988 in what many considered a payback killing. In the rioting and looting which followed, the Bromley and Manton supermarket was burned down in Wabag, an example of the way that commercial property could be threatened by tribal rivalries (Derkley 1997: 142). In 1990 Tindiwi re-won the premiership, only to have provincial government suspended for corruption and financial mismanagement again on 12 March 1993 (GPNG 1994). On 26 March, the provincial capital—which had been rebuilt, at the cost of PGK5 million (Dorney 2000: 260)—was burned down again by Tindiwi's supporters. In 1997 Peter Ipatas was elected governor of Enga Province, and gained sufficient control to win four more five-year terms, bringing a measure of stability to the province—in 2002 when hand grenades were used to detonate the helicopter fuel in which ballots were soaked, the provincial headquarters was merely damaged, not destroyed.

The thrust of Engan history is clearly that it is not the case that a weak state struggles to govern a strong society in Enga. Rather, the state was a prize over which a society of clans competed. Indeed, the first full elections were held in Enga in 1980, and in 1984 the government began a series of crises from which it did not fully recover. In this situation, it is hard to see 'the state' as an enduring, agentive institution at all.

Wormsley and Toke are clear on this point when they state 'the government is not in control of law and order' (1985: 69, emphasis in original). During this period police were able to restrain fighting but not control it. That is to say, they could mobilise sufficient force to disperse fighters but could not stop tribal fighting in the long run. A state of emergency was declared in 1979 which allowed increased use of force, and this was followed up by a special police operation in 1982. Both achieved short-term success in ending immediate conflict, but did not bring lasting order in the province. Policing continued to be both illiberal and ineffective. 'Police do burn houses. Police also "confiscate" property. Police harass and physically assault people not only in tribal fights but in the streets and market place of Wabag' (ibid.: 68, emphasis in original). Enga politicians, like their colonial predecessors, did not emphasise human rights. A 'senior provincial politician' told the authors of the Law and Order study that 'were it within his power he would remove all the special police [who might be witnesses] from Enga. Then when a tribal fight erupted he would '"wipe them out"' (ibid.: 69).

Violence and Policing in Porgera

Porgera's history is both similar to and different from that of central Enga. Like the Wabag area, it appears that fighting was part of pre-contact life in Porgera. In the diary of his 1939 patrol, one of the first Australian explorers in Porgera noted that 'today's journey was through a devastated tribal no-man's land, the scene of recent tribal fighting. All houses had been burnt and gardens razed to the ground. Several square miles had been devastated' (Black in Jacka 2019: 4–5). Jacka reports that prior to contact people in Porgera felt like they were '"living in a fence" owing to restricted social mobility arising from inter-group conflict' (Jacka 2016: 134). This is consistent with reports I received that the arrival of government and missionaries 'opened the roads' and allowed people to move freely without fear (Golub 2001: 225). In his review of government patrol reports from the late 1940s to the 1960s, Glenn Banks concludes that 'an outstanding feature of the Ipili ... is the amount of fighting that was taking place' (Banks 1997: 93). My own review of these reports (Golub 2014: 80–7) reached a similar conclusion.

While both central Enga and Porgera experienced violence pre-contact, Porgera's experience of governance was very different. Despite the brevity of Australian administration in Wabag, formal administrative authority in Porgera was even shorter: The government established the first airstrip and government station in the valley in 1962 and independence occurred in 1975, making Porgera's entire 'colonial period' 13 years long—including two years of self-government between 1973 and 1975! It is hard to sustain degenerationist narratives about Porgera given such a short timeline.

In this regard, Porgera was similar to Kandep, Kompiam and other rural areas of the province: it took longer for the effects of the administration to be felt in remote areas of the province.

Porgera was different from other peripheral areas, however, because of the presence of gold, which attracted outsiders and provided a source of wealth. Alluvial gold was discovered in Porgera by a government exploratory patrol in 1939. After their war service ended, in 1948, a handful of expatriates moved to Porgera, creating a small but robust artisanal mining operation in which Porgerans began as workers but soon began operating their own claims. During this period the valley was patrolled occasionally by kiaps stationed at Wabag and, after 1952, at Laiagam. Thus unlike central Enga, Porgera experienced 'development' (i.e. income from mining) before it received 'government'. In this frontier zone, Porgerans learned that the government was not an all-powerful force, but rather just one factor, albeit a potent one, in valley life.

In 1962, Porgera was derestricted and an airstrip was put in. Originally, police in Porgera were stationed in a building adjoining the main government office at Yandiakale, at a strength of less than 10 men (Kelly Talia and Geoffrey Puge, personal communication, 28 April 2019). Although it is difficult to estimate the population of Porgera and Paiela, the sister valley which the Porgera government station was also responsible for, it is clear that the police must have been extremely underpowered, even given the fact that they had firearms and Porgerans did not.

During the colonial period, as in the rest of Enga, the government was viewed favourably as a source of wealth and prosperity. When the Australian flag was lowered at an independence ceremony in 1975, a man climbed up the flagpole and attempted to raise it again (Biersack, personal communication, April 1999). This was a period when alluvial gold workings were common and Porgerans earned good money as artisanal miners relative to the rest of the highlands. A permanent government presence meant that the government was willing to issue exploration leases for potential large-scale mines. This resulted in additional wealth flowing into the valley. In Wabag, hopes for quick and extreme wealth and development were disappointed as worries about government corruption and warfare became widespread. In Porgera, on the other hand, these hopes were realised, at least to a certain extent—given the sometimes millennial expectations of Porgerans, the mining economy was at best a partial success.

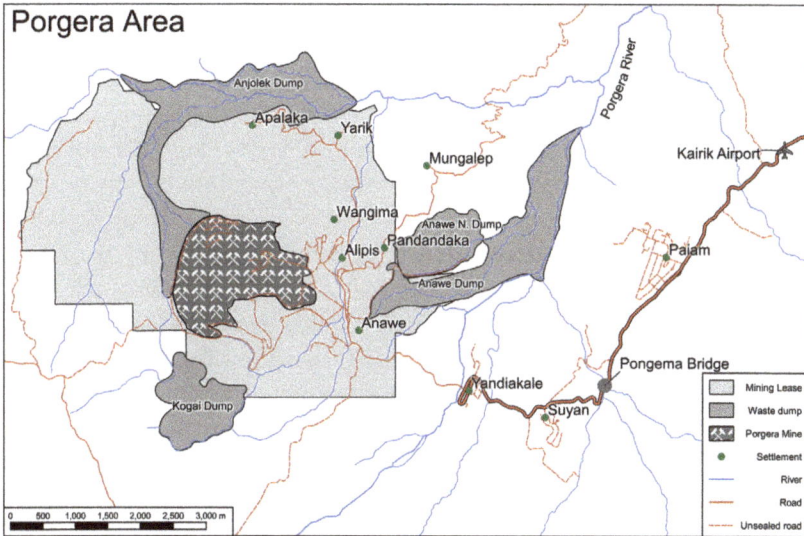

Figure 3.2 Porgera mining area

Source: Map by Nelson Minar with data from OpenStreetMap, Humanitarian Data Exchange, OpenDEM, and SRTM. openstreetmap.org/copyright

Wealth flowed into Porgera because of business interests, not the government. In fact, if anything, the reverse was true: government followed extractive industry and worked, to various degrees of success, to enable it. It is clear that Porgera was a place the government had its eye on because gold mining was happening there. However, it could not be said that it provided much initial security.

It is not really until the mid-1980s that the first historical records of the valley are available. They clearly demonstrate that the government was unable to control the valley's population. In 1987, a group of Ipili broke into the prospecting camp where Placer (the company then looking for gold) and its contractors has their base of operation. The Ipili stole the safe from Coya Constructions, an Australia-based contractor. The safe contained PGK69,000 in cash as well as the passports of expatriate Coya employees. The response of the state was to burn down houses at Alipis, an area near the prospecting camp, in the course of three days. One of the heist's leaders went to jail, while the other reappeared and was employed by the PJV just a few months later (Golub 2001: 322).

During the construction period when the mine was being built, the fighting continued. From 1989 to 1992, '100 percent of Porgera Census Division's population was affected by tribal fighting' (Bonnell 1999: 64).

89

In the 1980s, the Enga Law and Order Project noted only one conflict which involved a gun. In the 1990s fighters in Enga began using automatic weapons in conflicts, despite there previously having been a ban on them. Wiessner writes that this move was based on dissatisfaction with the government's ability to govern and a desire to settle scores for one's self (Wiessner and Pupu 2012: 1251). Bonnell reports that it was in 1992 that guns first began to be used in the valley in fights (Bonnell 1999: 64–5). In 1990 the Westpac bank agency was robbed and there was a 'riot' in 1990 on New Year's Day in which the IPI Supa Stoa, the most modern-looking store in the valley, was looted (ibid.: 66–7). There were also 'riots' connected to a union strike in May 1991 (ibid.: 67). Weissner and Pupu (2012: 1251) note a skyrocketing amount of violence following the introduction of guns and record over 250 conflicts (which Wiessner calls 'wars' and which other have called 'fights') between 1991 and 2005. During this period, mean fatalities per war were 17.8, almost five times higher than the mean of 3.7 per fight which Wiessner and Tumu's ethnohistorical work uncovered. This was the context against which the Porgera mine would begin operation.

Policing in Porgera followed a similar pattern to that of the rest of the province: the illiberal and often ineffective use of force. As Glenn Banks diplomatically noted, 'the activities of the police often undermine parallel work of corporate community affairs staff' (Banks 2000: 255). Police burning of houses also occurred in this period. For instance, in 1990, 'the police reportedly burned and looted approximately 39 houses in the Mungalep area after the official opening of the Porgera mine on 20 October' (Dinnen 2001: 213). In 1993 a policeman was killed while trying to intervene in a tribal fight in Paiam. In retaliation, 'mobile squads and other police' burned down 'more than 200 homes and buildings' (Banks 2000: 260).

As the mine's presence increased in the valley, government policing was increased. In late 1986 a police station was built in Porgera with three officers and five constables (Banks 2000: 256) at Pandandaka, close to the exploration camp and the future mine site. The Pandandaka police station was moved to make way for the mine, initially to containers at the Anawe boom gate (Bonnell 1999: 60–1) and then in 1990 to an 'impressive concrete building' at the Yandiakale. The multilateral legal agreements which created the mine specified that policing should be increased in Porgera and that the police headquarters in Mulitaka, the regional hub, should be upgraded as well. Unfortunately, the new police station was destroyed by a huge landslip in 1991 (Derkley 1997). As a result, the centre of governance was moved from geologically unstable Yandiakale to Paiam. In 1992 police stations were built in Paiam and Suyan, and Paiam increasingly became the centre for law and order infrastructure in the valley (Bonnell 1999: 60–1).

In addition, the government used special militarised police forces to (unsuccessfully) secure law and order in the valley. As a result, in the early 1990s Porgera was policed by three disciplined forces: the police, a mobile squad, and a rapid deployment unit, or 'RDU'. Mobile squads are widely used around PNG. These squads were modelled on earlier forces used in Fiji to control riots in urban areas, and were first used in PNG's islands provinces to quell anti-colonial movements such as Mataungan in East New Britain (Mapusia 1986: 65). In the highlands, they were used to intervene in tribal fights in the 1970s, and continue to be used to this day. The RDU was a special force designed by the national government of PNG, with the support and encouragement of the Papua New Guinea Chamber of Mining and Petroleum. They were stationed near each major resource development in order to secure them (Banks 2000; Dinnen 2001). Like the regular police force, whose housing and uniforms are provided by the mine, the RDU was stationed at the PJV camp in Tipinini (Bonnell 1999: 61).

The RDU was perhaps more a source of disorder than of order. On 14 October 1992, members of the RDU confronted a group of young men at Anawe, and the confrontation escalated until a man was shot and killed. The RDU then fled the scene, trying to head east back to their base camp at Tipinini. They were pursued by a large group (perhaps as many as 1,000 people) and sought shelter at Suyan camp before leaving and continuing back to Tipinini. There, the mob destroyed mine property worth PGK1 million (Banks 2000: 256–9). Bonnell notes that 'the only surprising thing about this serious attack on Suyan was that it had not happened sooner' (1999: 67), arguing that:

> even with all the new security arrangements which now give Suyan the appearance and ambience of a prison, it would be hard to say that the risk of such an attack has been removed. There are plenty of Porgerans with grievances against the mine and vast numbers of idle outsiders ready to fuel these grievances or join in demonstrations. The rent-a-crowd mob have nothing to lose. (Bonnell 1999: 68)

After it sparked this violent conflict, the RDU program was cancelled.

While these forces received various amounts of PJV support, they were much smaller and more poorly equipped than the PJV's own security forces. In 1992, Porgera had eight police, and the PJV employed 210 security guards, 140 of which were on-shift at any time. This included a dog squad, as well as 'static guards, escort teams, investigators, and supervisors. These figures included a group of 40 special constables, known as the Mobile Reaction Force, trained to respond to riots and low-level armed-offender incidents' (Dinnen 2001: 213). On the surface, then, it would appear that the PJV had a large and well-resourced security program while the police are under-resourced. But the picture is more complicated than this. In 1996, for example, 56 members of PJV's Loss Control Department (the department which includes security) were police reservists (Banks and Bonnell 1997: 10). Mine security literally were the police, or at least reservists (see Bainton and Macintyre, Chapter 4, this volume).

In central Enga the major source of wealth was the government, which could be controlled by clans through elections. In Porgera, in contrast, the mine was a source of wealth but the PJV's security were capable of restraining Porgerans who sought to compromise its operations. For instance, the PJV rebuilt and refortified Suyan camp after the disturbance there, despite the fact that it cost PGK1 million to do so (Banks 2000: 256–9). Unlike the government, however, the PJV had no charge to control Porgerans or maintain public order. Restraint, rather than control, was the goal of mine management.

Porgera mine security used the same illiberal force to restrain Porgerans which the police had used throughout Enga's history. A 2005 report claimed that PJV security and police had shot 10 people to death (Akali Tange Association 2005). By 2015, the mine had indemnified 130 women for sexual violence committed against them by mine employees (Jungk et al. 2018: 24). As of September 2018, 821 additional people had filed human rights abuses claims against the mine using its Porgera Remedy Framework (Jungk et al. 2018: 28). Rosenau reports that 'local PJV security staff and contracted Reserve Police personnel were acting without due regard for six of the seven major categories specified in the VPSHR [voluntary principles on security and human rights]—the principle document that defines guidelines for extractive companies when seeking to protect their facilities' (Rosenau et al. 2009: 27). In the early 2000s, of the 65 hours of training received by Reserve Police hired by the mine, only 40 minutes focused on 'wider relations with the local community' (Rosenau et al. 2009: 29).

Call-Out Operations and States of Emergency: 2000–15

Porgera changed dramatically in the early 2000s. As 'an enclave of overdevelopment in a sea of (relative) poverty' (Banks 2005: 140), it attracted economic migrants who swelled the valley's population (Bainton and Banks 2018). While accurate figures are hard to come by, it may be that the valley's population increased from roughly 5,000 in 1980 to 40,000 in 2015 (Kemp and Owen 2015: 8). The greatest increase in population occurred after 2001 following the opening of Paiam town, a planned community built by the PJV. As a visitor to the valley myself, the change was striking. In 2001, the largest public space in Porgera was the disused airstrip at Yandiakale, which was largely empty. By 2016 it was the site of a bustling market which was thronged with people.

With this increase in population Porgera also became more violent. Jacka writes that 'in the eastern Porgera Valley … a series of tribal fights … raged from 2004 to 2012 … approximately one hundred people were killed, hundreds or more displaced, and nearly every structure in eastern Porgera was burned or razed to the ground' (Jacka 2019: 2). John Burton estimates that between 2002 and 2006 annually there were 103 homicides per 100,000 people. By contrast, the number for Iraq civilian casualties in 2006 was 101 and the number for Ciudad Juárez, Mexico, in 2009 (during a huge drug war), was greater than 132 (both per 100,000; Burton 2014: 44).

The PJV's internal assessments were equally dire. Unauthorised miners, many of them migrants, were increasingly breaching mine security and working gold in the open pit, leading to violent clashes with security guards (Bainton et al. 2020). By 2005, the mine's security staff was 380 people, including 10 expatriates, 100 Papua New Guineans from other areas of the country and 180 local hires (Gray 2005). Despite this large force, 'the prevailing view among senior management was that the security status quo at Porgera was no longer viable' (Rosenau et al. 2009: 28). The PJV was losing its ability to restrain the population. For instance, in early January 2007 'Police Mobile Squads from Mt Hagen, Mendi and Laiagam' were 'sent into the valley … following reports of continuous law and order problems in the area' (Tiptip 2007). In late April and early

May 2007, work was halted for a week when roads into the mine were blockaded by disgruntled landowners who felt a relocation program was not moving fast enough (Anon. 2007).

As a result of the worsening law and order situation in the valley, the PJV made the decision to move to a strategy not only of restraint, but control—that is, becoming more directly involved in the maintenance of law and order in the valley. As a result, in 2008 Philip Kikala (the MP for Porgera), Ila Temu (a Barrick executive) and Peter Ipatas (Governor of Enga) wrote a discussion paper outlining a public–private partnership between the mine, the Enga provincial government and the national government to restore order to the valley (Kikala et al. 2008). This was part of a public–private initiative undertaken between the mine and government called the 'restoring justice initiative' (RJI), whose steering committee was led by Sir Barry Holloway (Safihao 2009). The organisers of the RJI did not see it as a case of corporate repression of the community—as might be familiar from resource conflict in Africa—but rather as a situation in which the PJV was aligned with the 'community' against 'troublemakers', many of whom were categorised as 'squatters' or 'outsiders'.

The RJI had many components, including strengthening civil society. One of those components was a state of emergency declaration similar to those used in Enga in the past. In February 2009, Kikala petitioned PNG's National Executive Council (NEC) to 'take action against an increase in violent crime and lawlessness' in Porgera (BGC 2009: 2). On 26 February 2009 the NEC responded by announcing plans for 'the call-out of additional police resources' to 're-establish law and order in Enga province, including the Porgera region' (ibid.), and in April 'Operation Ipili', as the call-out was known, got underway. According to the *Post-Courier*, the operation involved 200 policemen, including a police air unit and an 'intelligent unit' [sic] from the Papua New Guinea Defence Force (PNGDF). It was designed to last six weeks (Timb 2009).

Newspaper coverage describes broad support for the operation (Timb 2009). Nixon Mangape, a prominent landowner and future parliamentarian, said that 'all leaders and the population of Porgera welcome the call-out' (ibid.). The governor of Enga, Peter Ipatas, also 'welcomed' the call-out after initial misgiving about not being consulted, and his concern that this was a temporary solution to a long-term problem (Kepson 2009). The most widely covered aspects of the call-out were the eviction of residents and the burning of houses on 27 and 28 April

2009 on Wuangima, a piece of land at the base of the open pit where police had previously razed a settlement in 1987. Mine management believed that this area was home to unauthorised miners. Mark Ekepa, the chairman of the Porgera Landowners Association (PLA), arrived in Port Moresby on 29 April to meet with the prime minister and withdraw PLA support for the operation, because the state of emergency was targeting 'not the homes of illegal settlers' but rather houses which 'belong to the second and third generation landowners who were not thought of by the National Government and Barrick Gold in their relocation plan in 1989' (Eroro 2009). Ultimately a team from Amnesty International found that 'at least 131' buildings had been destroyed by police forces during the three months of Operation Ipili (Amnesty International 2010: 3).

The RJI may at first appear to be a shift in PJV strategy from a focus on restraint to a focus on control. But it appears that once successful operations were over, the PJV returned to a focus on restraint. After the burnings, for instance, the mine built a fence around the open pit which was finished in late 2010 at the cost of USD30 million (Stirton 2010). One consultancy report noted that by 2011 the RJI 'had to a degree served its original purpose' even though 'the community had struggled to manage many of the impacts of the mine' (Robinson 2014: 7), suggesting that the original purpose was to restrain activities against the mine rather than control and order society.

Indeed, throughout this period the police were understaffed and under-resourced. From 2008 to 2012 there were 14 police in the valley to police perhaps as many as 73,000 people, 'equating to one (1) police officer to over 3529 people' (Robinson 2014: 8). In comparison, US cities with 50–100,000 people in them averaged 16.1 police officers per 10,000 of population (Anon. 2018). Police facilities at the station in Paiam were a shambles:

> Prisoner numbers were often above forty (40) housed in three excessively sub standard cells … There was an ageing fleet of three (3) police vehicles and no equipment such as firearms, computers, ammunition, basic office supplies and cleaning services … The Police Station did not maintain an up to date 'Occurrence Books' which is one of the most basic functions of any police station. (Robinson 2014: 9)

Conflict had not abated either. 'Tribal clashes in the Porgera District were a major contributor to a serious breakdown of law and order in the final quarter of 2011 and early 2012', Robinson notes, 'as violence between the Ano and Nomali clans resulted in the deaths of 15 clan members. There were violent attacks towards the mine and its employees, in particular the mine's security guards' (Robinson 2014: 5). She continues:

> In mid to late 2012 when half the police station was burnt down and tribal fighting continued a serious breakdown in law and order resulted, culminating in the forced shut down of numerous businesses including the local bank, post office and looting of local business IPI. The highlands highway was closed and the operations of the mine were seriously impacted in various ways including industrial action by employees who refused to travel to work. During this time, several employees were injured and one killed while travelling to and from work. (Robinson 2014: 7)

As a result, the RJI was 're-invigorated' (ibid.: 7). In October 2011 the PJV hired an expatriate, Julian Whayman, to manage the project, which he did until late 2014 (Whayman 2015). In 2012 the PJV built a new police barracks, a project managed and paid for through the tax credit scheme (see Bainton and Macintyre, Chapter 4, this volume). It also purchased furniture for the houses. As a result, the number of police in Porgera was increased to 50 (Robinson 2014: 6). It is hard not to notice this renewed interest in controlling Porgera came at the same time as the mine was losing the battle to restrain the valley's inhabitants.

Disorder in the valley continued to threaten both the mine and the community. In 2013 four people were killed on 2 December, allegedly by PJV security and police mobile squads who shot them at the Pongema bridge. The next day, a mob attempted to move towards Paiam but was blocked by police. It turned towards the mine site but did not enter it. As a result, work at the mine was suspended for 12 hours as the next shift of workers could not be safely transported to the site (Boyle 2013). Unauthorised miners, mostly young men who entered the open pit of the mine to work for gold, were also increasing in number. In December 2013, 1,000 illegal miners entered the mine site in a fortnight (Kepson 2013). By March and April 2014, the situation in the valley had gotten so bad that the government deployed two mobile squads and 40 members of the PNGDF, who had been made 'Special Constables with arrest powers', to the valley. The total force, counting the soldiers, was 140 people.

In May the *Post-Courier* reported that the total number of police in the valley was 200 (Anon. 2014). On 6 June 2014 police again burned houses in Wingima, with estimates varying from 20 to 200 (Poiya 2014).

At this point it appears that Porgera may have entered a cycle in which increasing disorder led to a state of emergency or call-out operation, followed by a period of calm before the pattern repeated itself. Although the rhetoric and perhaps the intention of the stakeholders involved in planning the call-outs was one of control and normalisation, in practice this appears to be a circular pattern of restraint and disorder, not a linear increase in control of valley life.

This may not be entirely true, however. In 2016 landslips and sinkholes challenged the mine's feasibility and caused the destruction of a large settlement in the valley. In 2017, elections created their own tensions similar to those seen in the discussion of central Enga above. The *Post-Courier* and other newspapers focused less attention on the valley, perhaps because beat reporters no longer visited there. By 2018 life in Porgera entered a new phase, as negotiations began to renew the mine's lease, which was set to expire in 2019. These negotiations continued up to the time of expiration. With these negotiations ongoing at the time of this writing, the mine has been put on care and maintenance and is not currently operating (see Afterword, this volume).

Conclusion

In this chapter I have provided a history of policing, violence and governance in Enga and Porgera, which reveals the shortcomings of several popular narratives about the province and the valley. Degenerationist narratives claim that government control of the country declined beginning in the independence period. This account has shown, to the contrary, that neither the Australian nor the Papua New Guinean government has controlled the population of central Enga. Rather, they have merely been able at times to restrain it. While this narrative is correct in that violence has been on the rise in the province, it incorrectly places this rise in the independence era and not in the pre-independence era. Insofar as Erskine's promise has never been realised, the government in Enga has always been an 'absent presence'. This raises the wider question of how representative Enga is of the country as whole. While PNG is a large and complex place with extremely diverse histories of governance, it may be that as historians turn their attention to the twentieth century, they provide a different history of colonialism and policing in PNG then that produced by the 'men on the spot' who rule the country.

Claims that the mine operates in a 'rough neighbourhood' clearly have some prima facie validity. However, this chapter demonstrates that Enga is 'rough' not only because of pre-existing cultural norms but because of its history. Banks astutely notes, for instance, that it is not merely the case that 'large resource developments transform the nature and focus of people's lives'. Rather, 'the mine … and the possibilities and problems it creates, become captured into the existing society, lives and ways of understanding of the local population' (Banks 2008: 30). Taking this point even further, I would emphasise the dialectic nature of this ongoing encounter: both governance and local norms are transformed in their encounter with one another, and at each further encounter they are confronted with a transformed partner whose identity and structure they are, in some part, responsible for (Sahlins 1985). The mine does not merely work in a rough neighbourhood, it has 'roughened up' the neighbourhood itself through its actions and indeed its mere presence in the valley. In particular, its strategy of restraint rather than control has had long-term consequences for both the community and the PJV itself. Banks et al. (2013) are correct to distinguish between 'immanent' and 'spontaneous' social change around Porgera, but it must be pointed out that the mine's response (or lack of response) to these changes becomes part of the causal chain which produce these unintended and unwilled events in the valley.

That said, we should not overstate the role of the PJV in the history of policing and governance in the valley. Illiberal use of force, states of emergency, house burnings and violent actions are not methods which are newly invented by the government and the company to keep Porgera secure. On the contrary, they are the historical norm in the province. Without historical perspective, it may appear that these forms of violence are a novel result of the mine's presence and thus part of a generalised 'resource curse' phenomenon which occurs across multiple resource extraction projects. This is incorrect. In fact, they are part of an ensemble of governance techniques which the PJV inherited.

The case of Porgera shows that more care must be taken in developing the concept of the 'resource curse'. There is a danger in using this term indiscriminately simply to mean anything bad that happens near a mining or petroleum project. Porgera is very different from the cases of Bougainville and West Papua described by McKenna (2016) in that the PJV has never lost its 'social license to operate'—the question has been who should benefit from the operation and how much. Disorder in Porgera is not an insurgency against the mine. Nor is Porgera like some resource conflicts in West Africa described by Le Billon (2014), in which profits from resource extraction fuel civil war. Each of these cases are different, and adequate comparison must avoid historical particularism (which treats each case as too unique to be comparable to any other) on the one hand and a reductive analysis which cannot discriminate between cases on the other. Historical anthropology is, I believe, well-suited to this purpose because it can explain structural transformation across time and can work to build mid-level comparisons across cases (Sahlins 1985; Tilly 2010).

Finally, this discussion has implications for discussions of policing and the 'absent present' state. It is a truism to say that extractive capitalism takes on the role of the state in and around project areas. On the surface, Porgera is a clear example of this. But this more detailed historical overview also complicates this picture: in Porgera, security guards are also Reserve Police officers. Police vehicles, housing and uniforms are provided by the mine. A specialised police unit has been created at the urging of business interests. To understand these complexities it may be worthwhile to abandon the commonly used narrative that there are two collective actors called the 'mine' and the 'government'. It may be more useful to ask instead: What are the networks of individuals that lie behind these supposedly solidary corporate actors? When and why is a use of force attributed to the PJV or the state? Why call political conflict associated with elections 'tribal'? Whose interests and what moral purpose are served when the agency behind violence is disambiguated one way and not the other? Studies of corporate capitalism have demonstrated how managing blame and attribution are central in managing the fallout of environmental pollution. It may be that fine-grained analyses of resource conflict could adopt this lens when they turn their attention to policing and governance.

In closing, I would like to note that there is something tragic about the history of Enga and Porgera. In 1884 Erskine promised that Melanesians would benefit from the blessings of government. Engans believed in this promise because they value wealth and prosperity, not warrior prowess. Yet Western-style government has too often participated in the amplification of the negative things in Enga culture rather than the positive ones. The result is a tragic situation in which structural forces encourage a situation which no one in the province wants, but which too few seem able to escape. I hope that this essay offers an understanding of this situation, and in some small way contributes to creating a better world for the province and the people in it.

Acknowledgements

I'd like to thank my editors Nick Bainton and Emilka Skrzypek for their close attention and patience, both of which greatly aided this chapter. I'd also like to thank the staff of the Pacific Collection at the University of Hawai'i at Mānoa for providing me access to the historical material I draw on. Special thanks to Cristela Garcia-Spitz at the Tuzin Archive for Melanesian Studies for providing me a copy of the very scarce Enga Law and Order Project report.

References

Akali Tange Association, 2005. *The Shooting Fields of Porgera Joint Venture: Now a Case to Compensate and Justice to Prevail*. Report.

Amnesty International, 2010. 'Undermining Rights: Forced Evictions and Police Brutality around the Porgera Gold Mine, Papua New Guinea.' London: Amnesty International.

Anon., 2007. 'Porgera Remains Shut.' *Post-Courier*, 1 May.

———, 2014. 'PNG Police Deploy Staff to Mining Town.' Radio New Zealand, 2 May. Viewed 17 September 2020 at: www.rnz.co.nz/international/pacific-news/243134/png-police-deploy-staff-to-mining-town

———, 2018. 'Police Employment, Officers Per Capita Rates for U.S. Cities.' Governing blog, 2 July. Viewed 18 September 2020 at: www.governing.com/gov-data/safety-justice/police-officers-per-capita-rates-employment-for-city-departments.html

Bainton, N.A. and G. Banks, 2018. 'Land and Access: A Framework for Analysing Mining, Migration and Development in Melanesia.' *Sustainable Development* 26: 450–460. doi.org/10.1002/sd.1890

Bainton, N., J.R. Owen, S. Kenema and J. Burton, 2020. 'Land, Labour and Capital: Small and Large-Scale Miners in Papua New Guinea.' *Resources Policy* 68: 101805. doi.org/10.1016/j.resourpol.2020.101805

Banks, G., 1997. Mountain of Desire: Mining Company and Indigenous Community at the Porgera Gold Mine, Papua New Guinea. Canberra: The Australian National University (PhD thesis).

———, 2000. 'Razor Wire and Riots: Violence and the Mining Industry in Papua New Guinea.' In S. Dinnen (ed.), *Reflections on Violence in Melanesia*. Canberra: Asia Pacific Press.

———, 2005. 'Globalization, Poverty, and Hyperdevelopment in Papua New Guinea's Mining Sector.' *Focaal* 46: 128–144. doi.org/10.3167/092012906780786799

———, 2008. 'Understanding "Resource" Conflicts in Papua New Guinea.' *Asia Pacific Viewpoint* 49: 23–34. doi.org/10.1111/j.1467-8373.2008.00358.x

Banks, G. and S. Bonnell, 1997. 'Porgera Social Monitoring Programme: Annual Report 1996.' Unpublished report to Porgera Joint Venture.

Banks, G., D. Kuir-Ayius, D. Kombako and B. Sagir, 2013. 'Conceptualizing Mining Impacts, Livelihoods and Corporate Community Development in Melanesia.' *Community Development Journal* 48: 484–500. doi.org/10.1093/cdj/bst025

Bashkow, I., 2006. *The Meaning of Whitemen: Race and Modernity in the Orokaiva Cultural World*. Chicago: University of Chicago Press. doi.org/10.7208/chicago/9780226530062.001.0001

BGC (Barrick Gold Corporation), 2009. 'Re: Urgent Appeal of MiningWatch Canada Relating to the Porgera Valley.' Letter to the UN Special Rapporteur on Human Rights, 2 June. Viewed 18 September 2020 at: media.business-humanrights.org/media/documents/files/reports-and-materials/Barrick-Gold-appendix-B-re-Porgera-mine-16-Jun-2009.pdf

Biskup, P., B. Jinks and H. Nelson, 1968. *A Short History of New Guinea*. Sydney: Angus and Robertson.

Bonnell, S., 1999. 'Social Change in the Porgera Valley.' In C. Filer (ed.), *Dilemmas of Development: The Social and Economic Impact of the Porgera Gold Mine, 1989–1994*. Canberra: Asia Pacific Press. doi.org/10.22459/dd.12.2012.02

Boyle, P., 2013. 'PNG: Deadly Clashes at Barrick Gold's Porgera Mine.' *Greenleft Weekly*, 6 December.

Burton, J., 2014. 'Agency and the "Avatar" Narrative at the Porgera Gold Mine, Papua New Guinea.' *Journal de la Société des Océanistes* 138/139: 37–51. doi.org/10.4000/jso.7118

Clifford, W., L. Morauta and B. Stuart, 1984. *Law and Order in Papua New Guinea* (2 volumes). Port Moresby: Institute of National Affairs (Discussion Paper 16).

Derkley, H., 1997. 'Enga Province, 1987–1991: The Transformation of the *Tee.*' In R.J. May, A.J. Regan and A. Ley (eds), *Political Decentralisation in a New State: The Experience of Provincial Government in Papua New Guinea.* Bathurst (NSW): Crawford House Publishing.

Dinnen, S., 2001. *Law and Order in a Weak State: Crime and Politics in Papua New Guinea.* Honolulu: University of Hawai'i Press. doi.org/10.1515/9780824863296

Dorney, S., 2000. *Papua New Guinea: People, Politics and History Since 1975.* Sydney: Australian Broadcasting Corporation Books.

Eroro, S., 2009. 'Porgera up in Flames.' *Post-Courier*, 30 April.

Ferns, N., 2020. *Australia in the Age of International Development, 1945–1975: Colonial and Foreign Aid Policy in Papua New Guinea and Southeast Asia.* New York: Palgrave Macmillan.

Gammage, B., 1998. *The Sky Travellers: Journeys in New Guinea 1938–1939.* Carlton (Vic.): Melbourne University Press.

Golub, A., 2001. *Gold Positive: A Short History of Porgera 1930–1997.* Madang: Kristen Press.

———, 2014. *Leviathans at the Gold Mine: Creating Corporate and Indigenous Actors in Papua New Guinea.* Durham (NC): Duke University Press.

———, 2018. 'Introduction: The Politics of Order in Contemporary Papua New Guinea.' *Anthropological Forum* 28: 331–341. doi.org/10.1080/00664677.2018.1545108

Gordon, R. and A. Kipalan, 1982. 'Law and Order.' In B. Carrad, D. Lea and K. Talyaga (eds), *Enga: Foundations for Development.* Armidale: University of New England, Department of Geography.

Gordon, R. and M.J. Meggitt, 1985. *Law and Order in the New Guinea Highlands: Encounters with Enga.* Hanover (VT): University Press of New England.

GPNG (Government of PNG), 1973. 'Report of Committee Investigating Tribal Fighting in the Highlands.' Port Moresby: GPNG.

———, 1987. 'Report of the Auditor-General for Papua New Guinea on Enga Provincial Government in Terms of Section 74 of the Organic Law on Provincial Government for the Fiscal Year Ended 31 December 1986.' Boroko: Auditor-General's Office.

———, 1994. 'Report of the Auditor-General for Papua New Guinea on Enga Provincial Government in Terms of Section 74 of the Organic Law on Provincial Government for the Fiscal Year Ended 31 December 1993.' Boroko: Auditor-General's Office.

Gray, B., 2005. 'Current Security Issues in the Oil and Mining Sectors.' Powerpoint presentation to the sixth community affairs conference hosted by the PNG Chamber of Mines and Petroleum.

Gulliver, P.H., 1971. *Neighbours and Networks: The Idiom of Kinship in Social Action among the Ndendeuli of Tanzania.* Berkeley: University of California Press.

Hasluck, P., 1976. *A Time for Building: Australian Administration in Papua and New Guinea, 1951–1963.* Carlton (Vic.): Melbourne University Press.

Hawksley, C., 2007. 'Constructing Hegemony: Colonial Rule and Colonial Legitimacy in the Eastern Highlands of Papua New Guinea.' *Rethinking Marxism* 19: 195–207. doi.org/10.1080/08935690701219025

ICRC (International Committee of the Red Cross), 2017. 'Tribal Conflict in Papua New Guinea | On the Frontline.' Comment made by 'j t', n.d. Viewed 7 October 2020 at: www.youtube.com/watch?v=wo4uf-fXsUk

Jacka, J., 2016. 'Development Conflicts and Changing Mortuary Practices in a New Guinea Mining Area.' *Journal of the Polynesian Society* 125: 133–147. doi.org/10.15286/jps.125.2.133-147

———, 2019. 'Resource Conflicts and the Anthropology of the Dark and the Good in Highlands Papua New Guinea.' *Australian Journal of Anthropology* 30: 35–52. doi.org/10.1111/taja.12302

Jungk, M., O. Chichester and C. Fletcher, 2018. *In Search of Justice: Pathways to Remedy at the Porgera Gold Mine.* San Francisco: The British School at Rome.

Kanaparo, P., n.d. 'Political Cold War: Negipi Nanengipi Yanda Mende Pilyo Lakapupa (Eaters and Non-Eaters are Fighting).' Unpublished paper.

Kemp, D. and J.R. Owen, 2015. 'A Third Party Review of the Barrick/Porgera Joint Venture Off-Lease Resettlement Pilot: Operating Context and Opinion on Suitability.' Brisbane: University of Queensland, Centre for Social Responsibility in Mining.

Kepson, P., 2009. 'Ipatas praises Porgera SoE.' *Post-Courier*, 23 April.

————, 2013. 'Illegal Miners Reported.' *Post-Courier*, 13 December.

Kikala, P., I. Temu and P. Ipatas, 2008. 'Restoring Justice: Law and Justice Sector Partnerships in Enga Province, Papua New Guinea.' Port Moresby: Department of National Planning, Barrick Gold (Australia Pacific) and Enga Provincial Government.

Kituai, A., 1998. *My Gun, My Brother: The World of the Papua New Guinea Police, 1920–1960.* Honolulu: University of Hawai'i Press. doi.org/10.1515/9780824863692

Kumbon, D., 2017. *I Can See My Country Clearly Now: Memoirs of a Papua New Guinean Traveller.* Scotts Valley (CA): CreateSpace Independent Publishing Platform.

Lakane, J. and P. Gibbs, 2003. 'Haves and Have-Nots: The 2002 Elections in the Enga Province, PNG.' *Catalyst* 33: 96–116.

Lakau, A., 1988. 'Compensation on Government Land: The Wabag Town Centre.' In P.J. Hughes (ed.), *The Ethics of Development: Choices in Development Planning.* Port Moresby: University of Papua New Guinea Press.

Le Billon, P., 2014. *Wars of Plunder: Conflicts, Profits and the Politics of Resources.* New York: Oxford University Press. doi.org/10.1093/acprof:oso/9780199333462.001.0001

LiPuma, E., 2000. *Encompassing Others: The Magic of Modernity in Melanesia.* Ann Arbor: University of Michigan Press.

Lutz, B. and R. Rivers, 1968. 'Enga Understanding of Government.' In *Anthropology Study Conference (Amapyaka) Report.* Wabag: New Guinea Lutheran Mission.

MacKellar, M.L., 1975. 'The Enga Syndrome.' *Melanesian Law Journal* 3: 213–266.

MacWilliam, S., 2013. *Securing Village Life: Development in Late Colonial Papua New Guinea.* Canberra: ANU E Press. doi.org/10.22459/SVL.05.2013

Mapusia, M., 1986. 'Police Policy towards Tribal Fighting in the Highlands.' In L. Morauta (ed.), *Law and Order in a Changing Society.* Canberra: The Australian National University, Research School of Pacific Studies, Department of Political and Social Change (Monograph 6).

Markham, E.A., 1985. 'Lambchops in Papua New Guinea.' *Bikmaus* 6: 81–122.

McKenna, K., 2016. *Corporate Social Responsibility and Natural Resource Conflict*. London: Routledge.

Poiya, J., 2014. 'Porgera Burns: 200 Houses Razed, Expat Attacked in Retaliation Over Raid on Illegal Miners.' *Post-Courier*, 6 June.

Robinson, R., 2014. 'Restoring Justice Initiative: Contributing to a Just, Safe, and Secure Society in the Porgera Distinct: Progress Report 2008–2014.' Unpublished report to Porgera Joint Venture.

Rosenau, W., P. Chalk, R. McPherson, M. Parker and A. Long, 2009. *Corporations and Counterinsurgency*. Santa Monica (CA): Rand Corporation.

Safihao, J., 2009. 'Porgera SoE a Success Story.' *Post-Courier*, 9 September.

Sahlins, M., 1985. *Islands of History*. Chicago: University of Chicago Press.

Scott, N. and K. Pitzz, 1982. 'Administration and the Department of Enga Province.' In B. Carrad, D. Lea and K. Talyaga (eds), *Enga: Foundations for Development*. Armidale (NSW): University of New England, Department of Geography.

Spruth, E., 1968. 'Enga Attitudes to Government and Law.' In *Anthropology Study Conference (Amapyaka) Report*. Wabag: New Guinea Lutheran Mission.

Stirton, B., 2010. 'Papua New Guinea: Gold's Costly Dividend.' *Getty Images*, 23 November.

Talyaga, K., 1982. 'Liquor Sale and Consumption in Enga Province: Some Personal Observations.' In M. Marshall (ed.), *Through a Glass Darkly: Beer and Modernization in Papua New Guinea*. Boroko: Institute of Applied Social and Economic Research (Monograph 18).

Tilly, C., 2010. 'Mechanisms of the Middle Range.' In C. Calhoun (ed.), *Robert K. Merton: Sociology of Science and Sociology as Science*. New York: Columbia University Press.

Timb, S., 2009. 'Police Move into Porgera.' *Post-Courier*, 20 April.

Tiptip, N., 2007. 'Porgera District Takes to Special Police Operation.' *Post-Courier*, 31 January.

UN (United Nations), 1951. 'Mission de Visite des Nations Unies dans les Territoires sous Tutelle de Pacifique: Rapport sur la Nouvelle-Guinée et Resolution y Afférente du Conseil de Tutelle.' [United Nations Visiting Mission to the Pacific Trust Territories: Report on New Guinea and Related Trusteeship Council Resolution]. New York: UN.

Whayman, J., 2015. 'A Public–Private Partnership Tackling Law and Order in PNG.' DevPolicy blog, 5 June. Viewed 18 September 2020 at: www.devpolicy. org/a-public-private-partnership-tackling-law-and-order-in-png-20150605/

Wiessner, P. and N. Pupu, 2012. 'Toward Peace: Foreign Arms and Indigenous Institutions in a Papua New Guinea Society.' *Science* 337: 1651–1654. doi.org/ 10.1126/science.1221685

Wiessner, P., A. Tumu and N. Pupu, 2016. *Enga Culture and Community: Wisdom from the Past*. Wabag: Enga Provincial Government and Tradition and Transition Fund.

Wolfers, E., 1975. *Race Relations and Colonial Rule in Papua New Guinea*. Sydney: Australia and New Zealand Book Company.

Wormsley, W.E and M. Toke, 1985. 'The Enga Law and Order Project: Final Report.' Manuscript in the Melanesian Collection, University of San Diego Library. Viewed 18 September 2020 at: roger.ucsd.edu/record=b4800822~S9

Young, D.W., 2004. *'Our Land is Green and Black': Conflict Resolution in Enga*. Goroka: Melanesian Institute.

4

Being Like a State: How Large-Scale Mining Companies Assume Government Roles in Papua New Guinea

Nicholas Bainton and Martha Macintyre

Introduction

It is now a common orthodoxy to regard the state as an absent institution, or missing 'stakeholder', around large-scale resource extraction projects. In Papua New Guinea (PNG), local-level actors and external commentators frequently bemoan the dysfunction within local-level and provincial governments; the ineptitude of the national government; the lack of basic service provision and governance in these project settings; the absence of accountability among elected members and public servants, and the expectation that resource developers will somehow fill the void. Media outlets and online forums are replete with accounts of enraged communities demanding that the state fulfil its role under the terms of the various agreements that have been signed for these extraction projects. At the same time, many people also see the state as a competitor for the benefits perceived to be available from resource extraction projects, rather than a neutral arbiter, which leads many people to conclude that the state is ineffectual and corrupt.

In this chapter, we move beyond these default descriptions of the state to consider how the state is instantiated and experienced in these extractive contexts, or 'Melanesian mining arenas' (Bainton and Owen 2019). While it is not our intention to disprove these everyday observations and criticisms of the state, we argue that the state exhibits a selective and strategic presence throughout the life cycle of resource extraction projects. At different times it will be more or less present for different actors and institutions, resulting in the prioritisation of some interests and the marginalisation of others. The state often exhibits a heightened presence in these settings through the local agreements to which it is a party, and the obligations that it habitually fails to enact. This 'absent presence' creates situations where mining companies are routinely manoeuvred into 'being like a state'. The failure of the state to fulfil its duties in regard to the regulation of the industry and the management and implementation of benefit-sharing agreements, for example, frequently forces companies to fill these gaps when they would rather concentrate on the task of extracting natural resources and accumulating capital.

Local perceptions of the government's failure to protect the interests of its people coincide with those of analysts who have characterised PNG as a 'weak', 'fragile' or 'failing' state (Dinnen 2007; Firth 2018). Here, failures in service delivery (especially health and education), infrastructure provision and maintenance of law and order, and incapacity to ensure the security of its territory and population, are definitive—the state being the politically defined entity that is responsible for such functions. The legislation and policies surrounding mining in PNG do give the appearance of sovereign state management of economic activity. But as we demonstrate, the government's contractual commitments are regularly ignored or fail to materialise, which is exacerbated by the tendency of national and provincial-level governments to reduce the amount of funding for delivery of services on the presumption that mining companies will fund the difference. As a result, mining companies frequently assume responsibility for state functions in the interests of maintaining stable operations, shareholder investment and profitability.

To illustrate this point we describe how these processes and the absent presence phenomena have unfolded around a large-scale gold mining project in the Lihir group of islands in the province of New Ireland. Even though the state often appears to be invisible in Lihir, because of its inability to fulfil its duties, the idea of the state has served as a point of reference for actions carried out by a local public administration, community members and the company. These manoeuvrings and manipulations are not simply the result of government absence and inaction or corporate acquiescence to community demands. Contemporary events and processes have historical roots and antecedents (in colonial administrations and Christian missions) that are unearthed, or brought to the surface, through the work of resource extraction. For this reason, the absent presence of the state must be understood within its historical context.

Ambivalence and Tension: Emergent State Formations in Lihir

In 1985 the Kennecott Niugini Joint Venture Company established its exploration camp on the site of the current mining operation in Lihir (Figure 4.1). The following year Colin Filer and Richard Jackson were contracted to undertake a socioeconomic impact assessment for the proposed mining project (Filer and Jackson 1986). Their observations confirmed a certain degree of state neglect. On the one hand, the level of basic services in Lihir was thought to be above the national average; on the other hand, it was somewhat below the average for New Ireland Province. Filer and Jackson concluded that it was the second of these facts, not the first, which was evident to the people of Lihir, and this no doubt encouraged the local belief that the government was giving them a 'raw deal'. While existing services were largely funded by the government, they were in practice dispensed by other agencies: health and education services, and economic activities like the marketing of copra, and the provision of shipping services, were primarily managed by the Catholic Church. Approximately 50 public sector workers were employed by the provincial government to undertake decentralised government functions in Lihir, 'serving' a resident population of roughly 5,900 people in 1985. These public sector employees included a single patrol officer and an agricultural officer, 11 health workers, and the rest were teachers, but there were no police. The distance to the islands from mainland New Ireland certainly made it harder to attract teachers and health workers from other parts of the province, and provided an excuse for these workers (and their administrators) to shirk all sorts of responsibilities. The airstrip had also fallen into a state of disrepair, which prevented more regular travel and communication between Lihir and the provincial capital in Kavieng at a time when the provincial government should have been taking a closer interest in the islands.

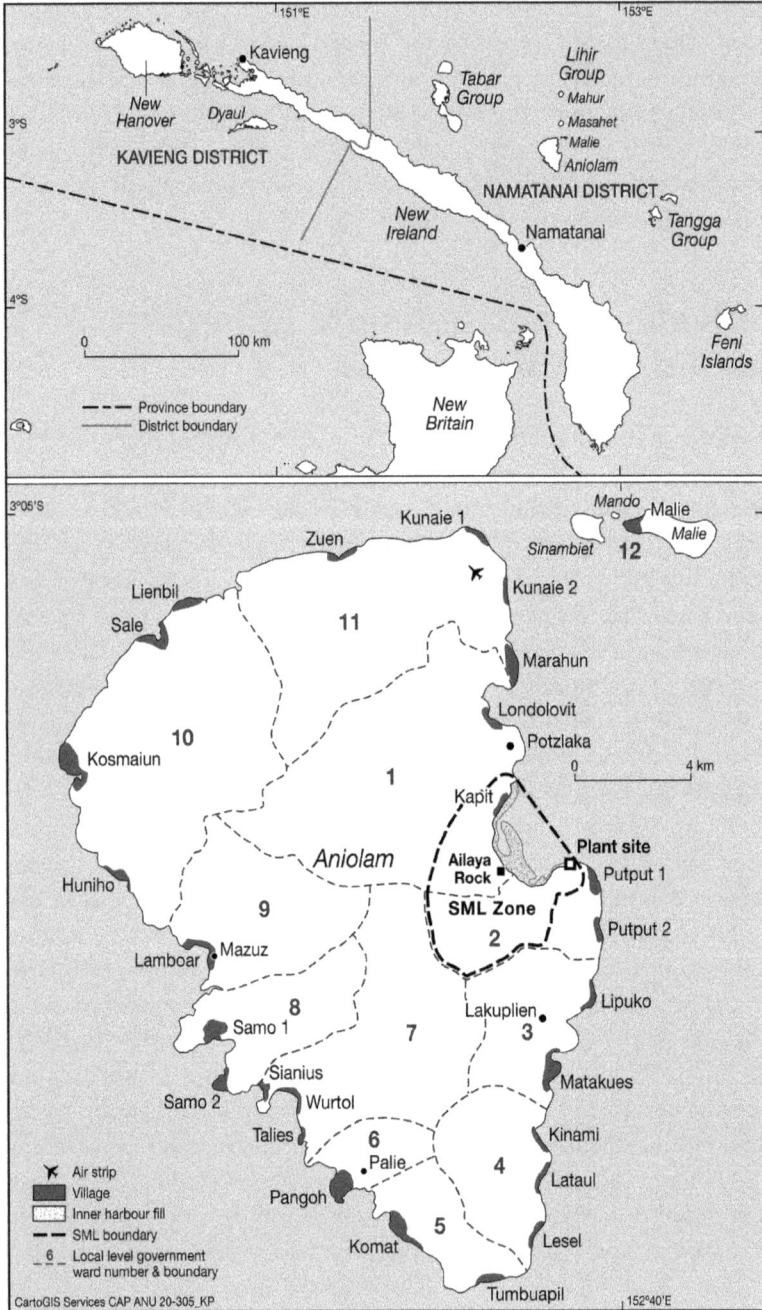

Figure 4.1 The Lihir group of islands, and key locations noted in this chapter

Source: ANU Cartography

At this stage, Lihirians were officially represented in the province by the Namatanai Local Government Council, Namatanai being the closest administrative centre on mainland New Ireland. Since the institution of the provincial government in 1977, Lihir had been represented on the provincial assembly by a succession of three members, the latest of whom, in 1986, was said to be a supporter of the People's Progress Party, while the national Member of Parliament was the leader of that party. The Lihirian member of the provincial government had not been seen in Lihir since he was elected, but few people were surprised, or even particularly resentful of this fact, since most constituents apparently did not expect their member to do anything but enjoy the temporary spoils of office (Filer and Jackson 1986: 121).

In the 1980s, the Lihirian representatives on the council maintained an uneasy relationship with the members of a local sociopolitical movement that had flourished in Lihir, known as the Tutavul Isakul Association or the 'TIA' (meaning stand together and work), which had its origins in the so-called Johnson cult that had emerged in northern New Ireland during the 1960s (Billings 2002; Bainton 2008). In 1984, the TIA broke away from the mainland movement and changed its name to Nimamar—an acronym derived from the names of the four islands in Lihir (Niolam, Malie, Masahet and Mahur)—which was intended to convey the idea of cooperation. The Nimamar Association held hostile views towards formal political institutions, which largely stemmed from Lihirian experiences with government and business and the belief that these institutions had been more of a hindrance than a help in unifying the community.

Nimamar Association leaders preached a form of millenarianism combined with singularly economic objectives. The desire for a radical reordering of society was explicit in the belief that 'Lihir will become a city', the state will be abolished, the Nimamar Association will become the government, schools will be abolished, there will be universal literacy, and that society will be divided between the 'faithful members' due to receive heavenly blessings (here on earth), and those condemned to a life of labour and toil (Bainton 2010: 55–69). Filer and Jackson eventually settled upon the term 'ritual communism' (1989: 116) to describe the mixture of religious commitment and aspirations for economic progress, and the desire for increased social and political unity.

We might therefore say that even before the mining project had begun, the state exerted an absent presence in Lihir. This was expressed through a combination of weak administrative functions, and the presence of formal political institutions which most people thought had limited capacity to represent their interests—especially in regards to the decisions that would have to be made in relation to the development of mining operations. Anti-government feelings were widespread in Lihir, and there was a growing desire for local independence. In other words, despite its material absence, or the lack of visible presence, the state loomed large in the collective consciousness of the residents of Lihir as an outside force to be replaced by a local organisation that could respond to the 'real' needs of the community.

In the 1986 provincial elections, a local Nimamar member, Ferdinand Samare, was elected to the provincial assembly, further obscuring the division between the Nimamar Association and 'official politics'. Lihirians were growing eager for their own government and by 1988, councillors and Nimamar members decided to break away from the Namatanai Local Government Council. It was not until 1988 that the New Ireland Provincial Government instituted their Community Government Act, largely to appease Lihirians after the feasibility of the mine had been established. Consequently the Nimamar Community Government was formed, representing a real transition towards more independent local governance. At this point Nimamar leaders stated their intention to withdraw from 'politics' and 'business', at least in the sense in which the government would regard these domains. This shift was compounded by the belief that the forthcoming mining project was the fulfilment of earlier prophesies for change. Many Lihirians were suspicious of the government's interest in the project, believing that the gold was theirs and that the state would ultimately benefit at their expense.

As mining negotiations continued into the 1990s Lihirians were determined that they would retain a greater portion of potential economic benefits. The New Ireland Provincial Government passed a bill in 1994 to transform the Nimamar Community Government into the Nimamar Development Authority (NDA), which was supposed to function as a type of interim local-level government until the Organic Law reform in 1995. This coincided with the signing of the mining agreements between Lihirians, the company and the state in 1995. It was not until 1997 that Lihirian leaders held the first elections to officially reconstitute the NDA into the Nimamar Rural Local Level Government (NRLLG).

The delay may well have been the result of internal divisions between supporters of the newly established government and older versions of Nimamar. It might have also been because the NDA primarily functioned like a technical administrative unit during the construction of the mine (to control the spending of money received from the mining operation on community development projects), rather than a political body (see Filer 2004: 3). And as we discuss below, the blurring of political interests and administrative functions has been a recurring theme in the history of the mining project and its agreements. In the end, the political inactivity of the NDA opened the door for the Lihir Mining Area Landowners Association (LMALA)—originally formed in 1989 to deal with the land disputes that were arising as exploration activities and negotiations got under way—to emerge as the major political force on the island, primarily due the intensifying heat between the company and the landowners during the construction of the mine (1995–97).

The new local-level government comprised 15 Ward Members elected by their ward constituents. The first Ward Members internally elected their chair, but in subsequent elections this has been a directly elected role. Only two members of the old Nimamar Association group were re-elected, representing a gradual shift in political stance. Together they formed the *Tumbawinlam* Assembly. The name derives from the local term for the larger of the two moieties that structure Lihirian clans—the other moiety being known as *Tumbawinmalkok* (small clan). While the original NDA remained as the administrative arm of the local-level government, Lihirians continued to refer to the local-level government as the NDA, which may have been a reflection of public confusion surrounding the role of the local-level government which was now in control of substantial amounts of money received through the mining operation, intended for community projects. Problems surrounding the legitimation of the NDA into the local-level government were finally solved in 2001 when the Nimamar Special Purposes Authority (NSPA) was established as a body separate from the political functioning of the local-level government that would administer government-funded projects (Filer 2004). However, this did little to resolve the underlying issues around the actual execution of their new responsibilities.

If the NRLLG functioned as the lower branch of the state that represented all Lihirians, it found itself in competition with LMALA, which represented the 'landowners' and the 'mine affected communities' but was not part of the state, but was already firmly entrenched in Lihir and recognised by the company. While both organisations shared a certain level of ambivalence, if not antipathy, towards the state and institutional governance in general, as we shall see, they were increasingly divided over their roles and the control of mining benefits in Lihir. This early representational division signalled the beginning of a much larger and more complicated set of issues that would split the members of these institutions, and enable the proliferation of other 'corporate bodies' purporting to represent different sections of the population of Lihir.

The Mine and its Agreements

In 1993 Lihirian leaders from LMALA and the NRLLG travelled to Port Moresby to participate in the state's newly established development forum process, which entailed a set of tripartite discussions between the national government, the provincial government and the project area landowners to secure their joint endorsement for the terms of the project and to spell out the distribution of the costs and benefits, and the roles and responsibilities arising from the development of the mine (Filer 1995, 2012). Throughout proceedings, Lihirian representatives insisted upon their primacy in the distribution of royalties and other benefits. On several occasions, they rejected the notion that the minerals below the ground were owned by the state and dismissed arguments by government representatives in terms that were redolent of secessionism and political autonomy. Their main representative, Mark Soipang, brushed aside provincial government claims stating that 'The developers are foreigners and the State is only a concept. It is us, the landowners who represent real life and people' (Soipang in Filer 1995: 68). Nevertheless, by 1995 Lihirian leaders had come to an agreement with the state and the Lihir Management Company, which was appointed as the developer and operator of the mine.[1] This new agreement highlighted the tension between the presence of the 'state-system', which required Lihirians to enter into formal arrangements with the government, and local resistance

---

1   The exploration program and feasibility studies for the Lihir project were conducted by a joint venture between Rio Tinto and Nuigini Mining Limited. Under the terms of the original project approvals, the Lihir project was transferred to a new Papua New Guinean company—Lihir Gold Limited (LGL). Rio Tinto established the Lihir Management Company (LMC) to develop and operate the mine on behalf of LGL. In 2005, Rio Tinto divested its interests in Lihir, allowing the Lihir operation to become owned and operated by LGL, until a merger occurred in 2010 with Newcrest Mining Limited.

to the 'state-idea' (Abrams 1988). Under the terms of the Mining Act, the developer and the state then signed a Mining Development Contract, and a Special Mining Lease (SML) was granted.[2]

In the period leading up to this agreement, the landowners association (LMALA) presented the developer with a 'position paper' that captured their ideas about compensation and development.[3] Compensation was presented as an 'all-encompassing issue' that would need to be addressed in a 'comprehensive' way to secure the approval of the landowners. The resulting set of agreements, known as the Integrated Benefits Package (or the 'IBP agreement'), was structured around the landowners' fourfold classification of their 'compensation principle': compensation *for* 'destruction'; compensation *as* 'development'; compensation *as* 'security'; and compensation *as* 'rehabilitation' (Figure 4.2). The underlying objective contained in this all-encompassing concept of compensation was to ensure that the landowners received a 'permanent substitute for the permanent losses and impacts associated with the mine' (LMALA 1994):

> Compensation if properly defined should be the state of equilibrium reached when the forces of destruction and impact must be equal to the forces of compensation. The end result of this process should provide for a balance where the Landowners are forever happy and accept the losses and impact they will suffer. Stability, peace and harmony should be the end result of the compensation payment. (LMALA 1994)

From the perspective of the landowners, compensation encompassed the full range of payments and provisions that would offset the losses they would suffer and their desire for greater control over the development of their resources. In other words, this was an expression of their expectation that mining would provide the springboard to propel them into a desired development future. If some Lihirians regarded the IBP agreement as the fulfilment of previous millennial prophesies, it could also be regarded as a manifestation of the state's so-called 'preferred area policy' that had

---

2   One particular point of contention which had delayed the completion of the agreement was the issue of equity in the mining project. LMALA had demanded 25 per cent equity—for each stakeholder (i.e. the state, the company, the provincial government and Lihirians). After five years of intensive negotiations, it was only when the provincial government opted to give up its share of equity in the project in favour of Lihirians that LMALA agreed sign off at the last moment. A former government liaison officer involved in these proceedings described this as 'an amazingly precocious coup' (Ron Brew, personal communication, August 2020).
3   The total value of one of the first 'position papers' delivered to the company was apparently calculated at some USD4 billion.

already become a feature of the nation's mineral policy landscape (Filer and Imbun 2007). In practice, this policy effectively creates a set of concentric rings or 'zones of entitlement' around each major mining project, starting with the innermost zone occupied by the customary landowners of the mining lease areas who typically experience the greatest burdens and impositions, followed by the project area community, the residents of the region and finally the nation as a whole. This policy is notionally meant to reflect the belief that those communities who experience the most impacts should be granted the most benefits.

1995 Lihir Integrated Benefits Package Agreement			
Chapter 1 *Destruction*	Chapter 2 *Development*	Chapter 3 *Security*	Chapter 4 *Rehabilitation*
Compensation and relocation agreements: • Putput and Ladolam relocation agreement • Kapit relocation agreement • Lands, crops, water and air compensation agreements	Commitments outlined in the three MOAs between the state, New Ireland Provincial Government, Nimamar Local Level Government and Lihir Mining Area Landowners Association	Trust funds: • Putput and Ladolam relocation trust fund • Kapit relocation trust fund • Putput plansite trust fund • Ladolam landowners trust fund	• Environmental Monitoring and Management Plan • Mine Closure Plan / and Long Term Investment Fund
Community agreements and company commitments: • Putput community agreement • Londolovit township agreement • Village development agreement	IBP commitments including: • Village Development Scheme (VDS) • Sealing of the ring road around Aniolam island (joint undertaking between the company and the government)	• Londolovit customary landowners trust fund • Lihir landowners trust fund	

Figure 4.2 The 1995 IBP agreement
Source: N. Bainton

At the time, the IBP agreement was considered in some circles to be the new benchmark for community benefit-sharing agreements across the nation. The *Mining Act 1992* made a distinction between development forum (benefit-sharing) agreements (or memoranda of agreement) between the state and affected area communities, and compensation/relocation agreements between the developer and the landowners. The real innovation of the IBP agreement was to bundle together the compensation and relocation agreements and three memoranda of agreement (MOAs), along with the company's environmental management plan (see Figure 4.2). The MOAs established the distribution of royalties and defined the responsibilities and undertakings of the parties in relation to the development of the mine and its benefits for the community. The MOAs and the environmental management plan were included in the IBP agreement so that the landowners could see the entire package of commitments and benefits they would receive from the mining operation in exchange for their land. This created a certain level of confusion between agreement conditions and operational plans, although the community have generally shown much less interest in the specific details of the environmental management plan, just as the state has shown limited interest in monitoring the company's compliance with its responsibilities under the IBP agreement, or its own commitments contained in the MOAs.

In practice, the only integrating factor in the IBP agreement is the blue hardbound cover that holds this set of commitments together. This observation stems from the lack of integration between various commitments and the roles and responsibilities of different parties, and inability of these parties to implement the IBP agreement in ways that matched the original vision of the landowners, which has been the source of considerable frustration and local political conflict. If the IBP agreement was once the high-water mark for such arrangements, within a matter of years it had become the low tide mark. It is likely that the parties would agree that this state of affairs arose as a result of a lack of governance around the agreements that would allow for effective coordination and monitoring of the various activities and responsibilities. It is less likely, however, that the parties would necessarily agree on who is to blame, which is partly due to the different ways that each party has understood the IBP agreement. As the self-proclaimed architects of the IBP agreement, the landowners (or more specifically the leaders of LMALA) have defended their design and pointed the finger at the company and the government as the principal agents and funders for implementation. On the other hand, the company has often argued that it has complied with most of its responsibilities and the real issue is the government's failure to enact those MOAs to which the company is not a party. While there may be some element of truth in these assertions, this overlooks the propensity of the government to treat the company as a 'contract administrator', sidestepping its responsibilities to monitor how well the company has upheld its obligations—or to actually test the substance of the development programs it claims to have delivered. This selective absence of the state has helped to reinforce the view among the community that Lihirians should control the funds and benefits derived from the operation of the mine to secure their independence from the state, and for that matter, the company.

Government Neglect

As we have seen, prior to the presence of the mining company, Lihirians were of marginal interest to the Government of Papua New Guinea. Like many other rural people, they were poor and services were minimal. The New Ireland Provincial Government had a presence at Potzlaka where the local-level government offices were located with two resident officers. But with the promise of various mine-related funds for developing services in Lihir, and the provision of housing, offices and other facilities for provincial government officers, the approval of the SML and the signing of the MOAs was accompanied by recognition that state service provision had to be expanded and improved in Lihir, and that responsibility for changes associated with the mining project would be subject to regular review. During the mine construction phase and the first two years of gold production there were quarterly meetings in Port Moresby, Lihir and Kavieng (on a cyclical basis). The quarterly meetings were chaired by the Department of Mining and attended by representatives of relevant departments in the national government (Labour and Employment, Commerce and Industry, and Environment, who all chaired subcommittees relevant to their MOA and statutory responsibilities), provincial government heads of department and occasionally others, such as the Police Commander and the chair of LMALA. These meetings were meant to provide timely assessment of emergent problems, to present interim reports of the company's compliance with environmental and social commitments and to identify appropriate responses from appropriate agencies.

From the outset it was clear that the government and Lihirian parties to these meetings viewed their purpose as occasions for lobbying the company representatives to provide funding for their respective departments and projects. Although the company supplied written information and assisted in drawing up an agenda prior to each meeting, these were rarely read, and attendees displayed little interest in their content. For both the provincial and the Lihirian contingents, the main concern was to ensure that the company financed the extension of government services for Lihirians. Whereas Lihir had formerly been an unattractive 'backwater' for government employees, the prospect of new offices and houses, efficient communication systems and life in a township suddenly made it a desirable location. The agreements had already indicated the need for these expanded services and people were eager to participate.

For many in provincial government the mine meant new employment opportunities and the chance to work effectively because the company, rather than the government, would be funding them. Meanwhile in Lihir, the infrastructure for all of these local administrative functions—roads, a 'town site', houses for police and other government employees, and new offices—was viewed by many as evidence of the dawn of the era when Lihir would become a city. This also marked the beginning of a process whereby the company would support government infrastructure and services.

Lihir was formally designated a sub-district within Namatanai District, and a sub-district administration was established which employed local public servants responsible for various services, including health, education, law and order, land administration, community welfare and agricultural development. A sub-district administrator was appointed to maintain oversight of these functions, each of which notionally reported back to its functional counterpart at the provincial or national level. Sub-district officers are public servants, part of the provincial or national bureaucracy, and as such are meant to be independent of the local-level government and the mining company. Their independence was compromised from the start as government funding was regularly delayed, often inadequate and occasionally failed to materialise.

Public servants therefore depended on the mine. The simplest examples are communications (fax machines and telephones) and transport (vehicles and fuel). The provincial government didn't pay bills so phones were cut off. The people in the sub-district office had vehicles, but often the 'fuel allowances' didn't come through—so the company supplied fuel to enable people to actually do their jobs.[4] There were also numerous instances where community pressure demanded a response from the company. The refrigerator at the Catholic Church's Palie Health Centre, which stored pharmaceuticals and vaccines broke down and—in the absence of government action—was replaced by the company. During the early days of mine construction, the mobile squad police were called in to deal with community conflict following a murder. The police officers got drunk

---

4   One former government liaison officer for the project noted that the National Department of Mining had provided a lot of funding in the first five years of the mine to support other departments, including funds to meet state infrastructure commitments under the MOA. The funds began to dry up as other national projects were prioritised (Ron Brew, personal communication, August 2020).

and crashed the police vehicle, damaging it beyond repair. After months without any means of travelling around the island, the mining company supplied a new vehicle in the interest of maintaining law and order.

Perhaps the most egregious failure initially was that of the national government to provide equipment that would have enabled the government environment officer to independently monitor environmental impacts of the mining operation. The officer arrived around 1997 to find he had a desk and a computer, but no equipped laboratory, no boat or motor vehicle and no storage for samples. In order to fulfil his designated duties, he appealed to the company for use of their equipment and facilities, which were provided. While he did in fact work entirely independently, the integrity of his routine assessments could readily be questioned by any external review.

This pattern has persisted, and little seemed to change when the regulator transitioned from a government department (the Department of Environment and Conservation) to a statutory authority (the Conservation Environment Protection Authority or CEPA) in 2014. Among other things, this change was meant to ensure that the regulator would be better resourced to carry out its functions in an independent manner. But if a certain kind of 'wilful dependency' upon the company has persisted, in some respects this also extends in the opposite direction. The company's environment department relies upon the local CEPA officer to maintain its relationship with senior CEPA management in Port Moresby, to facilitate any changes to permit conditions, and to help manage local community–environment issues. While company management have sought to limit direct assistance to the local CEPA office—to avoid accusations of 'buying off the government'—some managers have also come to the conclusion that some assistance will be necessary to ensure that the regulator can carry out those basic functions that serve the interests of the company. These challenges are by no means limited to Lihir and similar frustrations are frequently voiced across the industry.

The failure of government departments to support their staff is simply one of the ways that the 'absent state' problem manifests itself. To outside observers, the company's responses might appear as the illegitimate assumption of state functions. But Lihirian people viewed them as entirely appropriate, maintaining that the presence of the mine created the need for these expanded services and that the company should therefore pay for all services. In the words of one woman: 'Before the mine we didn't need

4. BEING LIKE A STATE

police to patrol. Now people buy beer, fight and spoil our places, so we have to have policemen come and take them to the station. It's because of the mine there is beer and trouble.' Others referred to the way that company agents had promised that the project would improve living standards, health services and education. While government officers presented these as indirect effects of economic progress, local people saw them as direct benefits promised by the company.

At the provincial level, the abrogation of responsibilities was justified differently. In discussions with public servants in Kavieng in 2000 about the lack of provincial support for staff stationed in Lihir, the somewhat cynical response to questions posed by Martha Macintyre was that 'Lihirians are rich now—the money should be spent on other New Irelanders'. From their perspective, the mining project had brought economic advantages that other New Irelanders lacked. Public servants in other parts of the province did not have the nice new houses, new offices or other facilities that their colleagues working in Lihir enjoyed. As word spread about the houses, new roads, water tanks, electricity supplies (usually greatly exaggerated as in fact only the relocated Lihirian villagers had such benefits), so the idea of relative deprivation flourished.

The complexities of state neglect and the mismatched expectations of the various parties involved in benefit distribution require closer examination. Under the terms of the MOAs, the national government was to provide the provincial government and the local-level government with an annual Special Support Grant (SSG) to compensate for the decline in their share of mineral royalties under the terms of the Mining Act. The SSG was calculated in the same way as royalties (a percentage of annual gold sales from the mine) but this was not an additional tax levied on the developer.[5] The SSG also functions like a redistribution mechanism, whereby the state allocates a portion of its revenue derived from resource extraction projects to the region or the province hosting the operation to compensate those who do not directly receive benefits from the project. It was intended for use on projects that are approved by the national government and to be administered by the provincial and the local-level governments. The payment of the annual SSG has been a continual source of political frustration, and the Governor of New Ireland, Sir Julius Chan, has often publicly proclaimed that the national government has

---

5   The 1995 MOA calculated this at 1 per cent of annual revenue. The revised 2007 MOA calculated this at 0.25 per cent of annual revenue.

deliberately withheld or delayed the payment of the SSG, which in turn has supposedly impacted the ability of the provincial government and the NRLLG to carry out their functions and deliver infrastructure services to the people of Lihir. Some observers have suggested that the state has continually failed to deliver the SSG because it is a conditional grant and provinces must apply for project funding in a designated format through the National Planning Office, whereas provincial governors would prefer to think that the SSG is unconditional funding which they can 'manage'. This demonstrates the political struggles around the payment of funds related to the mining operation. It also reveals the antagonism between the different levels of government over access to mining 'benefits', and the connection between Sir Julius Chan's well-known political ambitions for provincial autonomy and the desire to contain resource wealth to the province.

Providing Services and Assuming Responsibilities

With the advent of the mine vast amounts of physical infrastructure soon emerged in Lihir. In the newly defined 'affected area' villages, this entailed the provision of improved housing, sealed roads, reticulated water supplies and electrical connections. While these were originally catered for under the terms of the IBP agreement, there was much less clarity around who would be responsible for their maintenance and upkeep, and who would foot the bill. In the absence of the government or any other capable local institution, this has largely fallen to the company, as it provides the only source of power and manages the supply of water. The company also manages the Village Development Scheme (VDS), which was originally proposed as a benefit for communities who were not covered by the negotiations, but was later expanded and redesigned as a housing improvement program across affected and non-affected villages.[6] A similar situation has emerged in regards to the other civic services.

---

6   Annual VDS provisions are outstripped by the expectation that each household will receive a new house. This is compounded by the need to use some of these provisions to maintain the existing housing stock, which creates a further cycle of dependency and community frustration. See Bainton (2017: 323–34).

In the years immediately prior to the establishment of the mine, policing on Lihir was carried out by a couple of policemen stationed at Potzlaka. Their station comprised a single room, and a small shipping container served as the holding cell. During the construction phase of the mine, a new police station was constructed with several holding cells, the number of officers increased and several new police vehicles were supplied. The influx of returnees, fly-in-fly-out employees, and migrants hoping for employment meant that police suddenly had more to do. Crimes increased—or at least appeared to do so, as people reported crime more. There were more court cases and so the need for a courthouse arose. The mining company soon established a working relationship with the police as they came to rely on them for local law and order enforcement and to maintain a stable operating environment. Over time these provisions have expanded, and were later captured in a memorandum of understanding (MOU) between the company and the Royal PNG Constabulary that outlined annual support for police operations, housing maintenance, vehicles, meals, uniforms and travel. In many other contexts this type of support would be regarded with suspicion, and in those places where the police have meted out forms of extrajudicial punishment it would be considered as evidence of corporate-funded state violence (see Golub, Chapter 3, this volume). But for the most part, Lihirians continue to expect the company to support law and order services due to the prevailing expectation that mining will deliver social and economic development, the belief that the mine is responsible for increased law and order issues and the lack of faith in any government department to undertake this role.

The provision of improved health services is often regarded as one of the most significant benefits of resource extraction. Many resource companies seek to improve their social and political standing through conspicuous contributions to community health programs, and governments often encourage companies to help them fill basic service gaps as a condition of their operating licence. Prior to the mine, health services were limited. Malaria was endemic, and there were high rates of infant mortality, tuberculosis and yaws. The Lihir Medical Centre was established in 1996 when an MOU was put in place between the state, the provincial government, the Nimamar Development Authority, the Catholic Church health agency and the mining company for an integrated medical facility to be built at Londolovit to serve mine employees and their dependents and other residents of Lihir. The medical centre is managed by a contracted international health service provider, and operational costs are covered by the company. The medical centre is staffed by national and expatriate doctors and national nurses, and in addition to servicing the company it provides primary healthcare services to the local community, runs community health outreach programs and supports the Catholic Church and government-run health services in Lihir.

While many Lihirians make good use of the medical centre—and the centre has undoubtedly helped to improve overall health conditions in Lihir and perhaps represents the most significant benefit of the mine— from an early stage many Lihirians were frustrated by the fact that the medical centre primarily existed to serve the needs of the company. As a result there have been several ill-fated attempts to develop an 'independent' Lihirian-managed health program. These initiatives reflect local desires for autonomy—and the expectation to have the same services as white Australians—and the tension between simultaneously wanting the mining company to provide these benefits and services, and wanting to own or at least control specific institutions.

In the pre-mining era, education services were just as limited as health services. The IBP agreement contained provisions for the government to construct the first high school in Lihir (which has only recently been upgraded to cater for grades 11 and 12), and to appoint a school inspector who would be responsible for improving education standards across the schools. The company also built an international primary school in the residential town site which is staffed with expatriate teachers and delivers an Australian curriculum. The school is owned and operated by the company and was initially intended to service expatriate families residing in the company town. This situation has gradually changed as more Papua New Guineans are employed by the company in senior roles that come with housing entitlements, meaning that their children also attend the international school, and as wealthy Lihirians and local business people opt to send their children there. In addition to its technical training centre, the company provides annual scholarships for tertiary studies and runs a distance education school in partnership with the University of Papua New Guinea, which offers local community members an opportunity to complete their high school certificate. The company has a special interest in the local education system, since low education standards in Lihir directly affect the 'talent pool' for its local employment obligations. In order to address these issues, at one point the company's human resources function was forced to establish bridging programs to develop basic numeracy and literacy among some junior and apprentice-level employees from Lihir. Taken as a whole, the image that emerges is one of total dependency upon the company to keep essential services in place.

The Lihir Destiny: New Agreements for Old Issues

By the turn of the century, there was a growing realisation in Lihir that the IBP agreement had failed to deliver all the expected goods. The IBP agreement contained a provision for review every five years, and so in 2000 LMALA and NRLLG representatives set about preparing themselves for this task. The review process soon became a political process as a self-styled group of elite male leaders who called themselves the Joint Negotiating Committee (JNC), comprising leaders from LMALA and the NRLLG, along with several members of Lihir's 'educated elite', assumed an uncompromising stance towards the company and the state which prolonged the review process for multiple years as they refined their catalogue of requirements. The JNC members experimented with various strategies and formulas as they attempted to articulate their preferred 'road to development' that would provide the basis for a revised agreements package. By the early 2000s the JNC members had concluded the that first era of the mining operation had been 'wasted', and with each revised position paper their demands looked more like a development manifesto or a blueprint for an independent micro-nation than a standard compensation package that would be recognisable to the company, let alone the state.

During this time, the JNC members were introduced to the self-help philosophies of the Personal Viability (PV) movement that was taking off throughout the country, and they recruited the founder and director of this movement, 'Papa' Sam Tam, to implement his PV training program in Lihir and advise them throughout the renegotiation process. The PV course contained a mixture of lessons in self-help, financial management and entrepreneurialism. By completing various grades of training, individuals would become 'personally viable' and responsible for their own 'destiny'.[7] The emphasis on individual responsibility resonated with

---

7   Sam Tam coordinates this program from his Entrepreneurial Development Training Centre. The program has been adopted in various places in Papua New Guinea, and other Melanesian countries. See Bainton (2010, 2011); Bainton and Cox (2009); and Macintyre (2013).

a broader set of concerns about local dependency upon the mine, and for some Lihirians PV tapped into historical aspirations to 'unlock' the secret to the wealth held by white Australians and represented the kind of knowledge that they suspected they had previously been denied. The JNC members eventually arrived upon the idea of the 'Lihir Destiny' vision, which they explained as the goal of creating a 'healthy, wealthy, happy and wise society' based upon 'financial independence and self-reliance for all Lihirians', which sounded like a more consolidated version of the landowners' original compensation principle presented to the company in 1994.

The path that would lead to this future state of balance was articulated in a policy document called the Lihir Sustainable Development Plan (LSDP). The JNC had devised the LSDP as the road map for the Lihir Destiny vision. According to this vision, the mine would provide the financial base to enable a form of sustainable development that would be aided by their newly acquired commitment to the teachings of the PV movement. If this sounded like a vernacular form of neoliberal theory— that placed the onus on the individual and obfuscated other structural inequalities—then it partially reflected the extent to which some Lihirian leaders had also given up on the idea that the state would provide the development they desired, the realisation that Lihirians would have to harness the future for themselves, and the belief that this would also create the conditions for a greater level of Lihirian independence.

In April 2007 LMALA, NRLLG, the New Ireland Provincial Government and the company signed the revised IBP agreement, or what came to be known as 'IBP2'. The original four-chapter structure was expanded to a five-chapter structure to incorporate the LSDP, and the three existing MOAs were consolidated into a single MOA between the same parties.[8] The IBP2 agreement was underpinned by a commitment from the company for 100 million kina over five years (or PGK20 million per annum plus CPI) for the implementation of the LSDP, which represented

---

8   The revised MOA set out the terms for the royalty rate and the distribution of funds; the Special Support Grant from the state to the provincial and local level governments; landowner equity in the mining project; the Tax Credit Scheme; the New Ireland Trust Fund; the roles and responsibilities regarding the provision of community infrastructure, health and education services, law and order programs, environmental monitoring, training and localisation programs for the mine, and local business development; and endorsement of the LSDP as the long-term economic development plan for Lihir.

a substantial increase the annual funding commitment.[9] This figure was not calculated on the actual cost of rolling out the plan, but was instead offered an impressive sum to unlock the negotiations and bring the review to a close. As one former company manager put it, 'this was a "nuclear" option to end what had been six years of torturous and at times acrimonious and expensive negotiation'.

These funds were referred to as the 'IBP2 grant', which was split across the five chapters of the agreement that cover various projects and programs, and was inclusive of the compensation paid to Lihirians for the physical impacts of the mine (see Figure 4.3). The scope of the final chapter ('Rehabilitation') was revised to include the mine closure plan and the trust funds, and the company's environmental management plan was removed. Ongoing obligations from the original IBP agreement (including compensation payments, the VDS program and the trust funds) were carried over to the revised agreement, along with outstanding obligations like the sealing of the ring road around the main island, and were classified as 'fixed obligations'. The remaining funds were then allocated against the other projects and programs across the chapters.

2007 Revised Lihir Integrated Benefits Package Agreement				
Chapter 1 **Lihir Destiny**	**Chapter 2** **Destruction**	**Chapter 3** **Development**	**Chapter 4** **Security**	**Chapter 5** **Mine Closure**
Nimamar Local Level Government, Lihir Sustainable Development Office	*Lihir Mining Area Landowners Association*	*Nimamar Special Purposes Authority*	*Lihir Sustainable Development Limited, Lakaka, Mineral Resources Lihir*	*Mine Closure Committee*
Fixed obligations • VDS housing and assistance; ring road **LSDP Projects** • VDS maintenance • School transport • Medical transfers • Health & education • Law & order • Lihir culture • Women, youth & sports • LSDP administration budget	**Fixed obligations** • Compensation • Community grants • Cultural site maintenance **Other LSDP projects** • Specific Issues Committees projects	**LSDP Projects** • Capacity building • NSPA infrastructure projects for health, education and law & order • NSPA administration	**LSDP Projects** • Capacity building (i.e. Personal Viability) • Lihir Integrated Livestock • Lihir Micro Bank • Agriculture and other non-mining economic activities • Lihir Sustainable Development Limited administration	**Fixed obligations** • Trust funds: mine closure, Putput, & Ladolam Relocation, Kapit Relocation, Putput Plantsite, Ladolam Landowners, Londolovit Customary Landowners, Lihir Landowners **LSPD Projects** • Mine closure administration

Figure 4.3 IBP2 agreement and LSDP structure
Source: N. Bainton

9   The annual PGK20 million + CPI payment has continued beyond the original five-year commitment date, due to the failure to renegotiate this agreement at the five-year mark (as discussed further under 'Derailing the Destiny').

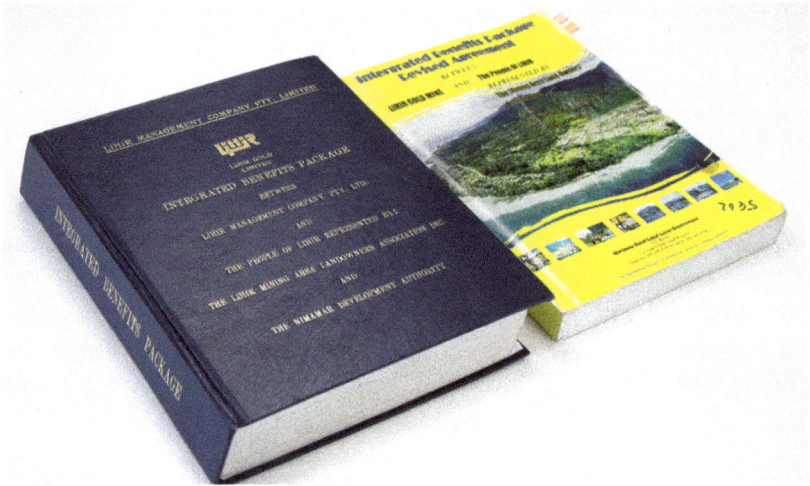

Figure 4.4 The Lihir IBP agreements (1995 and 2007)
Source: N. Bainton

The IBP2 agreement contained revisions to the terms of the original IBP agreement, combined with a set of policy statements on the LSDP as the blueprint for the future of Lihir, and a series of accompanying spreadsheets that detailed the various budget allocations for LSDP projects, along with the mine closure plan. The blue, hardbound cover of the 1995 IBP agreement was replaced with a yellow, soft-bound cover, replete with an aerial image of Putput village and the mine site that was accidently set in reverse on some copies of the agreement, which might be retrospectively read as a symbolic indication of the direction that the LSDP would eventually travel (see Figure 4.4).

The agreement required a novel set of institutional arrangements to govern and implement the new commitments. Some of the responsibilities contained in the different chapters were aligned with the functions of existing institutions like the NRLLG and LMALA, or the Nimamar Special Purposes Authority (NSPA). In other cases new 'offices' or corporate bodies were created to support the overall management of the LSDP, including a joint stakeholder 'LSDP Planning and Monitoring Committee', which was meant to steer Lihir towards its destiny. Figure 4.5 provides a simplified outline of the institutional landscape in Lihir. The number of state entities has evidently increased since the pre-mining era, just as the number of Lihirian or landowner entities has flourished to represent an expanding set of local interests. The creation of several 'joint entities' was meant to represent their different interests and convey their mutual commitment to the Lihir Destiny vision.

State Entities	Joint Entities	Lihirian / Landowner Entities
Lihir Community Government	Joint Negotiating Committee (JNC)	Tutuvul Isakul Association (TIA)
Nimamar Development Authority (NDA)		Nimamar Association
Nimamar Rural Local Level Government (NRLLG)	Lihir Sustainable Development Plan	Lihir Mining Area Landowners Association (LMALA)
Nimamar Special Purposes Authority (NSPA)		
Sub-District Administration (inc. Department of Mines, Environment, Health, Education)	Planning & Monitoring Committee (LSDP PMC)	Specific Issues Committees (SICs)
		Lihir Sustainable Development Limited (LSDL)
Mineral Resources Authority (MRA)		Mineral Resources Lihir (MRL)
Royal PNG Constabulary (RPNGC)	Lihir Sustainable Development Office (LSDO)	Specific Issues Companies (SICs)
New Ireland Provincial Government (NIPG)		Lakaka / Anitua Group of Companies

Figure 4.5 The institutional landscape in Lihir (past and present)
Source: N. Bainton

Under the terms of the IBP2 agreement, the NRLLG agreed to consolidate its budget with the IBP2 grant for the purposes of implementing the LSDP as the shared vision for Lihir.[10] The intention was to combine the benefits and compensation package from the company (the annual IBP2 grant of PGK20 million) with the funding received by the NRLLG (the royalties and SSG) to provide a larger consolidated resource base to support an integrated development plan for Lihir (the LSDP). The aim was to ensure that the landowners received their due entitlement, to supplement the government's resources to deliver services and infrastructure for all Lihirians, and to avoid duplication between projects and programs that are provided by the company (mainly for the affected area communities), and those that are provided by the government (mainly for the community as a whole).

Chapter 1 of the agreement therefore contained a raft of social development projects and programs that were meant to enhance those services that the government was already responsible for delivering under the terms of the MOA. This was to be coordinated through a newly established Lihir Sustainable Development Office (LSDO) in conjunction with the NRLLG. Chapter 2 set out the range of compensation payments and grants that would be received by the affected area communities, and

---

10   The IBP2 agreement also contained provisions for a capacity-building program to ensure that the primary stakeholder institutions would be able to implement the LSDP. Curiously, this did not include the company; it was incorrectly assumed that the company had sufficient capacity to play its role. The agreement also contained provisions related to Lihirian equity in the mine, and landowner participation in mining (i.e. business contracts).

commitments to a newly formed group of 'Specific Issues Committees' or 'SICs', that notionally represented the interests of specific landowner groups and their 'specific issues' in relation to the mining operation.[11] LMALA was responsible for the overall coordination of this chapter, which entailed oversight of these new SICs and their programs, and the delivery of various affected area community grants, while the company was responsible for directly managing royalty and compensation payments.

Chapter 3 outlined the social infrastructure projects (like bridges, roads and buildings) to be delivered by the NSPA, which served as the technical arm of the NRLLG. Funding provisions were included to cover the administration of the NSPA. Chapter 4 focused on providing 'economic security' for Lihirians through the delivery of the PV training program, and the development of 'non-mining dependent' economic activities, including a livestock farm and a micro bank to support those 'grassroots' entrepreneurs who would soon graduate from the PV course. The institutions notionally responsible for this chapter included a new local company called Lihir Sustainable Development Limited, the peak landowner company Lakaka (later rebranded as Anitua), and Mineral Resources Lihir, which managed Lihirian equity in the mine (which was later sold down to create the MRL investment and future funds[12]). The final chapter contained the annual commitments to the trust funds, and the conceptual mine closure plan to be administered by the company. As an entire package this was the basis for achieving long-held visions of 'modernity' and the transition towards an autonomous future.

Derailing the Destiny

Soon after the signing of the IBP2 agreement, the parties retreated to their respective corners of the island and the company was left with the task of working out how to handle these revised commitments. The new agreement, and the visionary plan that it underpinned, was meant to represent a new era of collaboration. But it was soon apparent that it would take a lot more to address the historical tensions between these groups and to align their competing interests. This situation was partly related

---

11   This included, for example, a 'Plantsite SIC', 'Putput and Ladolam SIC', 'Pit and Stockpile SIC', 'Kapit Relocation SIC', 'Airport SIC' and others. The list of SICs has since grown, as each aggrieved community group has sought formal recognition of their specific set of issues.

12   See MRL Capital: www.mrlcapital.com.pg/

to the structure of the new agreement, and the lack of detail around how these arrangements should be managed, especially the so-called 'LSDP consolidated funds' comprising the IBP2 grant from the company and the NRLLG's budget contribution.

The management and control of the LSDP consolidated funds has been at the centre of a long-term dispute between LMALA and the NRLLG, which started immediately after the signing of the revised agreement. The dispute arose around the implementation of the LSDP and because governance processes for the management of government royalties and company grants were not established as provided for under the terms of the revised MOA and new agreement. The consistent inability of the parties to work together has resulted in numerous revised and interim arrangements for the management of projects supported by the IBP2 grant.[13]

Up to 2012, IBP2 grant projects were budgeted by the joint stakeholder LSDP Planning and Monitoring Committee with the NRLLG. In June 2012, the NRLLG and LMALA signed a joint resolution that ended this short period of troubled collaboration. The LSDP Planning and Monitoring Committee was effectively folded into LMALA, who then assumed responsibility for implementing the IBP2 grant and the overall LSDP, while the NRLLG 'independently managed' the expenditure of its royalty funds. The parties justified the split by appealing to Biblical law: let the government (or Caesar) claim what was 'rightfully' theirs, which in turn confirmed the autonomy of LMALA over all land-related benefits and compensation (the IBP2 grant from the company). This immediately undermined the philosophical intent of the LSDP. Since 2012, the company and LMALA (and its LSDP Planning and Monitoring Committee) have jointly agreed upon annual budget allocations for LSDP projects and programs across the five chapters. The company has managed 'fixed obligations'—which largely amounts to the direct payment of compensation monies under Chapter 2 and the VDS program under Chapter 1—and LMALA has assumed responsibility for implementation of all other LSDP projects and programs.

---

13   There is insufficient space in this chapter to document the details of the disputes between these parties, the subsequent legal investigations, the various mechanics for the governance and implementation of LSDP programs, or the outcomes of these arrangements.

This arrangement partly reflects LMALA's efforts to control all funds and programs under the LSDP, and the commitment of LMALA leaders to the Lihir Destiny vision and their sense of frustration over the capacity of the government to provide services for the Lihirian community. It also reflects the belief within LMALA that the IBP agreement is a compensation agreement primarily between the developer and the affected area communities, which therefore entitles them to control and manage all funds derived from this agreement, along with an overarching desire to minimise any interference by the state. Both parties have attempted to claim the moral ground, with LMALA asserting that if it was not for the sacrifice of their members there would be no benefits or royalties, while the NRLLG has claimed that it existed well before the mine and will continue to exist once mining operations end, at which point there will be no further need for landowners to incorporate their interests. This has resulted in a deepening twist, as LMALA has deliberately assumed a quasi-government role as it has attempted to implement the LSDP, which has demanded even greater efforts by the company to manage the local fallout from this arrangement, leading it further down the path towards behaving more like a quasi-state than a developer.

These struggles and contradictions have been exacerbated by a further review of the compensation and benefit-sharing agreement that has run in parallel to the implementation of the existing arrangements. In 2012, the company commenced the second review of the IBP agreement, which came to be known as the Lihir Agreements Review process. This has been a stop-start affair that has divided Lihirians and come at great cost to the company, LMALA and the NRLLG. At the time of writing this chapter, the review process was still in progress because the parties have been unable to agree on the size, content and structure of a revised agreements package—or what may eventually become the 'IBP3'. The changes and issues we have described are set against the background of an unresolved and politicised review process and a local crisis of representation. This has meant, among other things, that implementation of the IBP2 agreement is thoroughly entangled with the politics of the Lihir Agreements Review, as the frustrations with the current arrangements are intensified by the delays to the review process, which in turn places even greater pressure upon the company to keep basic community services running, and increases the unrealistic expectation that a new agreement will somehow solve all of these problems. Under these conditions, it is unlikely that any new agreements package will be considered acceptable by the residents of Lihir.

The Tax Credit Scheme

In an attempt to address the gaps created by state absence, and to reconcile the wider expectation of service and infrastructure provision, in 2011 the company started implementing the tax credit scheme, which is perhaps the most blatant manifestation of the 'absent state' phenomenon and the blurring of the boundaries between state taxes and corporate-funded community development. This scheme was developed when it became apparent that neither national government agencies nor provincial governments are able to fulfil their commitments in the various benefit-sharing agreements to which they are a party. The tax credit scheme originated in 1992 at the Porgera Gold Mine, after the management of the mine came to the conclusion that the benefit-sharing agreements, to which they were not a party, presented a major risk to stable operations.[14] The national and provincial levels of government had failed to uphold their commitments under these agreements and as a result local community representatives were justifiably annoyed, inflaming an already volatile social landscape in the vicinity of the mine (see Golub, Chapter 3, this volume). The company understood the importance of maintaining local political support for the mine, so they persuaded the national government to introduce an 'infrastructure tax credit scheme' in the 1992 budget (Filer and Imbun 2007: 89). The scheme entailed an amendment to the Income Tax Act that then allowed the developers of large-scale resource projects to expend up to 0.75 per cent of their gross revenues on social and economic infrastructure projects. This expenditure would then be offset as a credit against the annual tax on profits ordinarily payable by these companies to the national government. In practice, this meant that companies could use their own engineering and project delivery capacities to satisfy the demand for improved infrastructure within local project areas and in adjoining areas that would otherwise not receive any project-related benefits. In theory, this would increase the level of local and regional support for individual mining projects at no extra cost to these developers.

---

14 A second key driver was the failure of the national government to maintain the Highlands Highway. The Porgera Joint Venture wanted to ensure that it remained open, as a primary lifeline for the operation, but it did not want to take on that particular state responsibility without some fiscal compensation.

Although the tax credit scheme already existed when the original benefit-sharing agreements for the Lihir gold mine were signed in 1995, the company did not start implementing the scheme until 2011, which was partly because the company had previously claimed it was not in a position to pay corporate tax to the national government, and was therefore not eligible for any credits. The program got off to a bumpy start that sparked a public relations debacle as the provincial governor, Sir Julius Chan, launched an offensive against the company for its failure to complete agreed tax credit infrastructure projects on mainland New Ireland. In 2013, the price of gold fell substantially, although not as substantially as the share price of the company, but by that stage the company had already made commitments to the provincial government to fund numerous infrastructure projects over the coming years. The governor insisted that these funding commitments must be upheld regardless of corporate profitability and so, as a show of compromise, the company agreed to fund tax credit scheme projects at a set annual level from its own internal cash flows even though the company still maintained that it was no longer in a tax-paying position.

The tax credit scheme originally arose as a company-led solution to the lack of capacity within provincial and national governments, and as a kind of strategy to enhance local political support for a specific mining project. In the case of Lihir, the primary source of pressure to deliver upon this scheme has come from the provincial government rather than the project area communities. This is partly due to the local expectation that the company will directly fund development projects as a form of compensation or as a benefit under the IBP agreement rather than an additional corporate tax offset. Management of the tax credit scheme in Lihir has been via a joint committee (comprising provincial, sub-district and NRLLG representatives), and has targeted projects that are not covered under the IBP2 agreement, which is Lihir-specific, whereas the tax credit scheme is notionally province-wide.

If multinational mining companies have their own reasons for implementing the tax credit scheme, or being seen to do so, this obviously lightens the load on national government agencies and provincial governments and delivers tangible benefits that might otherwise not be seen. But this should not blind us to the real decline in the capacity of relevant government agencies to accomplish their commitments and uphold their obligations to project area communities. Although the operator of the Lihir mine claims to have expended almost PGK100 million between 2011 and 2019 on various education, health and road infrastructure projects across New Ireland Province, including in Lihir (NML 2019), this has barely lifted the capability of the provincial government, much less the NRLLG, to provide comparable contributions and fulfil their commitments under the IBP agreement and the MOA. So while the tax credit scheme might represent a 'good news story' for individual companies and local members of parliament who are invited to participate in the ceremonial opening of these projects, in practice it has done little to alleviate the pressures arising from the lack of governance around the core benefit-sharing agreements at each resource extraction project. And in Lihir it more or less serves as a parallel or supplementary development program managed by the company to fill some of the spaces created by the failure of the IBP2 agreement and the selective absence of the state (or at the very least to divert some attention away from this development fiasco).

Conclusion: Being in a State of Compromise

In his account of the original planning and negotiation processes for the Lihir gold mine, Colin Filer suggested that the government may consider the geographical isolation of the project as a pretext for treating Lihir as an enclave that the company could then administer from its own plentiful resources, while the company may find that it is too engrossed in its engineering and accounting problems to pay serious attention to the needs and fears of the local community (Filer 1995: 73). A variation of this dilemma has ultimately come to pass, and equally so at every other major mining project throughout the country.

While the state has maintained a limited presence in Lihir relative to the expectations of some Lihirians and most managers of the company, the state clearly has a far greater presence now than it had in 1985, when it was almost completely absent. If the presence of the state is measured by the number of residents who are now on the government payroll, it does not appear to have any less of a presence than it has in other rural local-level government areas in New Ireland Province. The size or scale of its presence should therefore be distinguished from the source of the funds used to maintain it and people's perceptions of its capacity to keep its promises. And it could certainly be argued that its presence has been dwarfed by the expansion of the mining company, and further diminished by the company's assumption of governmental functions. Moreover, the selective absence of the state has cornered the company into a far more intensive relationship with the local Lihirian community, enhancing the perception of the absence of the state.

The inability of the government to fulfil its basic duties under the terms of the MOA is only one part of the problem. The other part of the problem is that the company has been lumbered with responsibility for the implementation of agreements that were substantially devised by other stakeholders. The original development forum agreements (or MOAs) were mainly based on a template devised by public servants in Port Moresby, which was then modified in light of pressure from local community (especially LMALA) representatives. The IBP2 agreement, and the LSDP, were predominantly the product of the JNC and reflect the thinking of its members who saw themselves as the vanguards for a cultural revolution in Lihir. This partly explains why the company has been unable to set in place robust governing controls around the agreements. This situation has been compounded by the emergence of a complex institutional landscape comprised of numerous competing entities or corporate bodies. The net result is a convoluted and constantly shifting set of governance arrangements for managing benefits and compensation payments and maintaining the delivery of basic services.

Several interrelated processes now begin to emerge. Historical experiences of neglect and local engagement with external forces—manifest in a combination of ambivalence, tension and enthusiasm for institutional governance and order—have culminated in various schemes to attain a form of independence, which have most recently taken the shape of the Lihir Destiny vision, captured in the LSDP. But the inability to fully enact these policies and programs (funded by the mining company) has reinforced local desires to maintain total control over all mine-derived funds and programs as a way of dealing with their real and perceived issues with the company and the state, and as a way of satisfying their needs and aspirations. At this point, we can ask whether local community and landowner representatives actually want the state, in the shape of the national or provincial government, to have a greater presence in the islands. In some respects, it seems that they do not.

The concept of fungibility also helps to explain the government's abrogation of its responsibilities in Lihir (just as it does in the case of the global aid industry). Mine-related funding intended for social development projects has frequently been diverted, directly or indirectly, to other programs or uses, including administrative expenses. This has been a constant concern for the company—in relation to both the NRLLG and its use of royalties and the SSG, and LMALA's use of the IBP2 grant. In order to protect its interests and keep the operation open, the company is therefore routinely required to assume the functions of the government, sometimes arbitrating between disputing parties or acting as a corporate community development agency, which ultimately creates conditions of total dependency upon the mining operation. The very structure of the agreements is partly to blame for this situation, and has exacerbated this state of corporate and community compromise.

From this perspective, the state is perhaps better described as an inattentive parent who periodically meddles in community affairs, or an absent landlord who occasionally comes to inspect their industrial estates, regularly collecting 'rental payments' (in the form of royalties and taxes), yet pays no attention to the harm their corporate tenants are inflicting upon their residential neighbours. However, these analogies only go so far, and from the perspective of the community, the government has no moral right to benefit from the mine and 'their' resources so long as it fails to enact its state responsibilities.

References

Abrams, P., 1988. 'Notes on the Difficulty of Studying the State.' *Journal of Historical Sociology* 1: 58–89. doi.org/10.1111/j.1467-6443.1988.tb00004.x

Bainton, N.A., 2008. 'The Genesis and the Escalation of Desire and Antipathy in the Lihir Islands, Papua New Guinea.' *Journal of Pacific History* 43: 289–312. doi.org/10.1080/00223340802499609

———, 2010. *The Lihir Destiny: Cultural Responses to Mining in Melanesia*. Canberra: ANU E Press (Asia-Pacific Environment Monograph 5). doi.org/10.22459/ld.10.2010

———, 2011. 'Are You Viable? Personal Avarice, Collective Antagonism and Grassroots Development in Melanesia.' In M. Patterson and M. Macintyre (eds), *Managing Modernity in the Western Pacific*. St Lucia: University of Queensland Press.

————, 2017. 'Migrants, Labourers and Landowners at the Lihir Gold Mine, Papua New Guinea.' In C. Filer and P.-Y. Le Meur (eds), *Large-scale Mines and Local-level Politics: Between New Caledonia and Papua New Guinea*. Canberra: ANU Press (Asia-Pacific Environment Monograph 12). doi.org/10.22459/lmlp.10.2017.11

Bainton, N. and J. Cox, 2009. 'Parallel States, Parallel Economies: Legitimacy and Prosperity in Papua New Guinea.' Research School of Pacific and Asian Studies, The Australian National University (State Society and Governance in Melanesia Discussion Paper 2009/5).

Bainton, N. and J.R. Owen, 2019. 'Zones of Entanglement: Researching Mining Arenas in Melanesia and Beyond.' *Extractive Industries and Society* 6: 767–774. doi.org/10.1016/j.exis.2018.08.012

Billings, D.K., 2002. *Cargo Cult as Theatre: Political Performance in the Pacific*. Lanham (MD): Lexington Books.

Dinnen, S., 2007. *Law and Order in a Weak State: Crime and Politics in Papua New Guinea*. Honolulu: University of Hawai'i Press. doi.org/10.1515/9780824863296

Filer, C., 1995. 'Participation, Governance and Social Impact: The Planning of the Lihir Gold Mine.' In D. Denoon (ed.), *Mining and Mineral Resource Policy Issues in Asia-Pacific: Prospects for the 21st Century*. Canberra: The Australian National University, Research School of Pacific and Asian Studies, Division of Asian and Pacific History.

————, 2004. 'Horses for Courses: Special Purposes Authorities and Local Level Governance in Papua New Guinea.' Canberra: The Australian National University, Research School of Pacific and Asian Studies, State, Society and Governance in Melanesia Project (Discussion Paper 6/2004).

————, 2012. 'The Development Forum in Papua New Guinea: Evaluating Outcomes for Local Communities.' In M. Langton and J. Longbottom (eds), *Community Futures, Legal Architecture: Foundations for Indigenous Peoples in the Global Mining Boom*. Abingdon: Routledge.

Filer, C. and B.Y. Imbun, 2007. 'A Short History of Mineral Development Policies in Papua New Guinea, 1972–2002.' In R. May (ed.), *Policy Making and Implementation: Studies from Papua New Guinea*. Canberra: ANU E Press. doi.org/10.22459/pmi.09.2009.06

Filer, C. and R. Jackson, 1986. *The Social and Economic Impact of a Gold Mine in Lihir*. Konedobu: Lihir Liaison Committee.

————, 1989. *The Social and Economic Impact of a Gold Mine in Lihir: Revised and Expanded (2 volumes)*. Konedobu: Lihir Liaison Committee.

Firth, S., 2018. 'Instability in the Pacific Islands: A Status Report.' Sydney: Lowy Institute.

LMALA (Lihir Mining Area Landowners Association), 1994. 'Memo to PNG Law Reform Commission.' Unpublished paper.

Macintyre, M., 2013. 'Instant Wealth: Visions of the Future on Lihir, New Ireland, Papua New Guinea.' In M. Tabani and M. Abong (eds), *Kago, Kastom and Kalja: The Study of Indigenous Movements in Melanesia Today*. Marseilles: Pacific-CREDO Publications. doi.org/10.4000/books.pacific.166

NML (Newcrest Mining Limited), 2019. 'Newcrest Mining Tax Credit Scheme Expenditure Details for the Quarter Ending December-2019.' Unpublished report.

5

Absence as Immoral Act: The PNG LNG Project and the Impact of an Absent State

Michael Main

Inside Gigira burns Laitebo, which when kindled, the flames will illuminate Hela and the land beyond.

'The Hela Prophecy' (Cuthbert 2009)

Introduction

In May 2014, Papua New Guinea (PNG) exported the first shipment of liquefied natural gas from its newly completed Papua New Guinea Liquefied Natural Gas (PNG LNG) project. The PNG LNG project had been promoted as a nation-defining project that was predicted to 'transform the economy of Papua New Guinea' creating benefits that would 'improve the lives of Papua New Guineans by providing essential services' and declaring that the project's landowners 'stand to benefit from direct payment of royalties on production of gas and associated petroleum products, as well as improved social and economic infrastructure' (ACIL Tasman 2008: v). The project has since become synonymous with the struggle of aggrieved landowners against the social and economic failures of resource development, and the corruption and neglect of a largely absent Papua New Guinean state. Analysis has shown that in reality 'the PNG economy is in a worse state than it would have been if it stayed on the underlying growth path of the 2000s and had no PNG LNG project' (Flanagan and Fletcher 2018: 12).

The PNG LNG project began construction in the mountainous terrain of what is now PNG's Hela Province in 2010. The project took around four years to construct at a reported cost of USD19 billion, making it the largest and most capital-intensive resource project in the history of the region. The project is vast in scale and includes a purpose-built international airport at Komo (see Figure 5.1), a gas conditioning plant at Hides where the gas extraction wells are located (the Hides Gas Conditioning Plant (HGCP)), a 290 km onshore pipeline to the Omati River and a 407 km offshore pipeline beneath the Gulf of Papua to the LNG plant at Caution Bay. The large footprint of the project means that it affects a number of landowner groups. The HGCP and extraction wells are located along a mountain ridge, known as Hides Ridge, the Huli name for which is Gigira. The overarching agreement signed between the landowners for the project and the state is known as the Umbrella Benefits Sharing Agreement (UBSA), under which landowners in specific Petroleum Development License areas signed License Benefits Sharing Agreements (LBSA). The landowners of the gas resource are the Huli-speaking clans of Hela Province who have genealogical ties to land associated with the PNG LNG project, and whose representatives signed up to a series of LBSAs prior to the project's construction. These LBSAs also included development promises that pertain to the entire province, which is expected to benefit as a whole from the project. These agreements were signed between the landowners and the state of PNG, and included promises of direct royalty payments to landowners in addition to a range of infrastructure projects including the construction of roads and schools, power supply and medical centres. Landowners from other parts of the project such as the pipeline and the LNG plant at Caution Bay in Central Province also signed various agreements. This chapter focuses on the Huli population of Hela Province, who are by far the largest landowning population and the customary landowners of the gas resource.

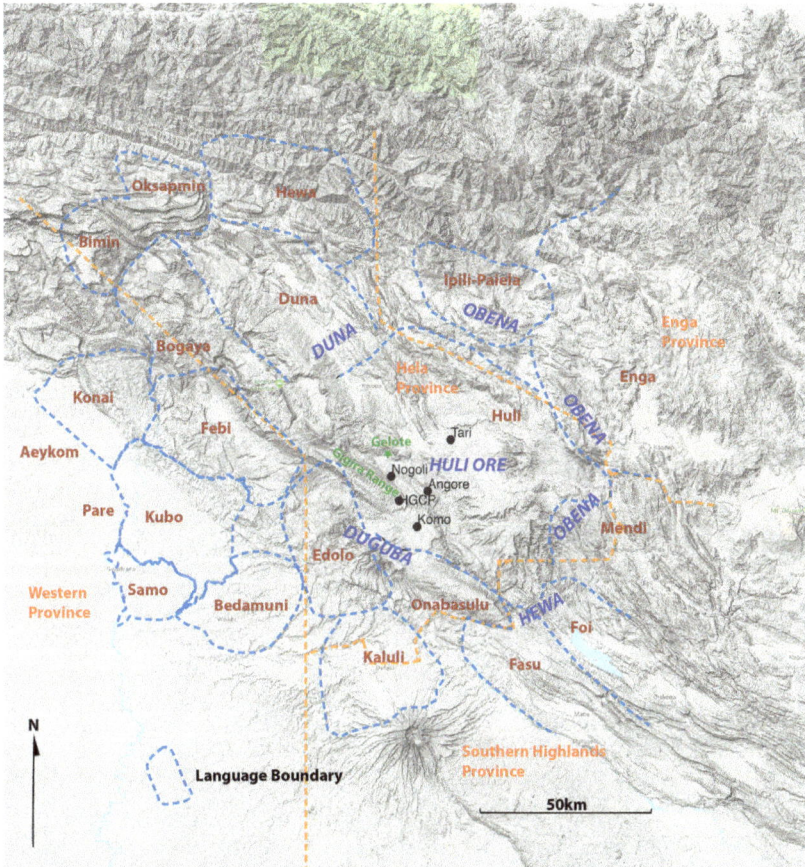

Figure 5.1 Hela geography
Source: Modified from Ballard (1995): Vol. 2, Figure B11

During fieldwork conducted in 2016, two years after the project began operating, it was immediately clear to me that Hela Province and its landowners had benefited little from the project, and it seemed that the majority of the population were arguably worse off as a result of the project. Frustration seethed among landowners at the absence of promised infrastructure and the failure to pay royalties because the landowners were yet to be officially identified by the state through the 'landowner identification' processes required under the terms of the Oil and Gas Act (see Filer 2019). Health and education services that existed prior to the project were in decline, and blame for the situation was largely attributed to the corrupt behaviour of politicians and various public servants. Yet the

PNG LNG project remains a venture that defines Hela as a province, and even, as this chapter will argue, as an Indigenous nation for its majority Huli-speaking population.

The concept of the state itself has been a highly contested subject for social theorists. The state of PNG exists firmly as an idea in the minds of its citizens, despite the extraordinary diversity of its cultural terrain, hundreds of language groups, immense poverty, lack of state infrastructure and the inability of the state to project power to its own borders. For many Papua New Guineans, the state of PNG therefore exists largely in spite of its apparent absence. This chapter reverses the focus from James Scott's concern with how the state sees its citizens (Scott 1998), to an interest in how citizens view the state, and by extension, how it is that the state can exist at all when it refuses to see its citizens. This poses the obvious question: how can the state be described as something that sees or doesn't see? The state is not an object but the subject of itself, consisting of its own citizens, some of whom come to represent the state and occupy positions of state power. The state is not other to itself, and when it sees, it is doing the seeing of itself. This has important implications, as highlighted in this chapter, for when state officials are caught between their simultaneous roles as state representatives and state subjects. I take the view proposed by Chernilo 'that social theory's ambivalent attempts at conceptualizing the nation-state reflect the actual ambivalence of the position of the nation-state in modernity' (2006: 6). Out of this embrace of ambivalence that might liberate social theory from the need to define the state, I propose that the state be viewed simply as a value proposition held by a collective. The value propositions that form modern nation states are encoded and symbolised in ways that are common to all of these states: in foundational text, in national song, in flags and in the arts. This value proposition invests the state with a great deal of power, including the power to dispense justice. I would go as far to suggest that it is unreasonable to ask what it is that defines the state, because it is the undefined nature of the state that makes it possible for the state to exist. A better question to ask is: why does the state exist? The state is a desired and idealised collectivity with strongly defined borders. The state exists because it consists of subjects who collectively value its existence. The difference between a state and a nation is a legal one; both are value propositions that are collectively made. The question as to why humans have come to form collectives that value the existence of states is beyond the scope of this chapter. This understanding of the state creates the space for the state to be materially absent, yet still exist firmly as a set of values according to which the state should behave even if it does not. An analogy might be made with Lao Tzu's famous observation that the wheel is defined by the hole in the middle (Tzu 1963: 67). The state is an enabling constraint encircling the desire that the state should exist.

This chapter explores the role of natural resources in the value proposition that is the state of PNG. The country's Constitution includes the declaration of its fourth goal 'for Papua New Guinea's natural resources and environment to be conserved and used for the collective benefit of us all, and be replenished for the benefit of future generations' (GPNG 1975: 4). Former prime minister Sir Julius Chan's famous statement that PNG 'is a mountain of gold floating on a sea of oil' seeks to legitimise the state by establishing a natural resource base upon which PNG's founding values can be proposed. In the eyes of the state and the Huli landowners, these values ideally underpin the establishment of the PNG LNG project.

Once Were Hegemonic

At the beginning of the PNG LNG project's construction Hela did not exist as a province, and the Hela landscape was part of the larger Southern Highlands Province. Yet Hela did exist as a concept and calls for the creation of Hela as a separate province date to the 1960s. During a sitting of the third House of Assembly of Papua and New Guinea in 1974, Andrew Wabiria, member for the Koroba-Kopiago electorate of PNG's Southern Highlands District, moved a motion that called for the establishment of a new Hela Province to be carved out of the existing Southern Highlands District.[1] It was not until the full potential of Hela's vast gas reserves began to be exploited by the PNG LNG project that the realisation of Hela as a province started to become a reality. By the time the province of Hela, covering an area of 10,498 km² with an estimated population that in 2011 exceeded 250,000 (GPNG 2013), was finally established in August 2012, the struggle for its creation had become inextricably linked with the development of the vast natural gas reserves that exist beneath Hela's mountainous terrain.

---

1   Districts became Provinces when PNG became independent in 1975.

The historical underpinnings of the contemporary notion of Hela need to be described in order to understand the contemporary relationship between Hela, the PNG LNG project, ExxonMobil and the state. The population of Hela Province is dominated by Huli-speaking clans that identify a geographical and cultural core of Huliness which includes the main township of Tari (Ballard 2002). The mythology at the basis for the concept of a Hela people involves an apical ancestor named Hela who was the progenitor of the five groups that occupy the Tari basin and surrounds. These five groups are derived from the offspring of Hela's four sons and one daughter: Huli, Duna, Obena, Duguba and Wana Hewa.[2] This is the contemporary rendition of the Hela story, which has solidified into a neatly packaged concept that is widely accepted and understood across the population of Hela, and that has been disseminated through PNG's media to become part of PNG's national story. For Huli, Obena include Ipili and Enga groups to the northeast and east; Duna lie to the northwest; Duguba include Papuan Plateau groups to the south; and Wana Hewa to the southeast (see Figure 5.1).

Many different versions of Huli and Hela origin stories were in existence prior to the influence of the Australian colonial administration, which began in earnest during the 1950s. Following the exploration and/or exploitation of nearby gold, oil and gas reserves during the 1980s, Hela as a political project took on a much greater urgency, solidifying into a singular Hela narrative and vision of development for the province and its people.[3] However, the concept of Hela itself pre-existed both the exploitation of natural resources, and the earlier influence of the Australian administration, as did a Huli hegemonic tendency towards the material and cosmological control of their entire known world. After the first Patrol Post in Huli territory was established at Tari in 1951, the early patrol officers soon realised that the people of the Tari area were part of a much broader understanding of a Hela people: 'The word Hela Huri[4] is often heard and seems to embrace all true Huri people' (Esdale 1955: 21). Esdale's interpretation of the word 'Hela' as 'altogether' or 'one kind', while semantically incorrect, does describe the sentiment

---

2   Wana Hewa is usually contracted to Wan Hewa in common parlance. Other versions have Wana Hewa as a son, rather than a daughter, however the version that includes a daughter has emerged as the dominant mythology.
3   These include the Mt Kare gold rush in neighbouring Enga Province, and gas exploration undertaken by BP in what is now Hela Province.
4   Huli was initially spelled as 'Huri' by the Australian administration during the 1950s.

of a people united by a singular, apical ancestor named Hela. Hela is simply the name of the apical ancestor described above. The idea of Hela, as it is held by its Huli-speaking majority, may be reasonably described in terms of an Indigenous nationhood. If we take Benedict Anderson's (1983: 49) model for nationhood as a community where most people are unknown to each other and 'yet in the minds of each lives the image of their communion', then the size and extent of the shared historical, cosmological and genealogical understanding of Hela, linked by extensive trade routes and ritual connections, is a snug fit. This understanding of nationhood is incorporated by my definition of the state, and there is no reason why the two cannot coexist, as one value proposition encompassed by another. The state of PNG is able to incorporate a diverse range of Indigenous groups who all subscribe to the value proposition that is the state, which is something that Huli have done enthusiastically since PNG gained independence.

Huli view themselves as having a central location in relation to surrounding groups, and have historically imposed their own hegemonic understanding of themselves on their neighbours. This understanding is realised simultaneously in terms of their trading relationships, mythical genealogical recall and cosmology. At this point it should be stated that much of Huli traditional knowledge has either vanished or been radically transformed over the past few decades; however, what is being described here does provide the foundation for contemporary Huli understandings of themselves and of their relationship to both the PNG LNG project and the PNG state. In cosmological terms the most important Huli ritual site was a place called Gelote, which roughly translates as 'the support of the universe' (Goldman 1981: 75). This site was at the centre of a vast cosmological system known as *dindi pongo* ('root of the earth') that linked together several other ritual sites that are dotted across the extent of Hela. Gelote was also referred to idiomatically as *dindi hanuni*, or 'the middle land'.[5] In material terms, Huli are situated centrally in relation to their trading partners. Prior to colonial intervention, Huli not only identified themselves as being at the cosmological centre of the universe, but also played a central role in the control of regional trade (Ballard 1994). Here I differ from Aletta Biersack when she writes that Huli, Duna and Ipili people 'were "ex-centric," centred not on themselves as geographical isolates but on the culturally diverse field in which their mythology,

5   Quite literally, 'middle earth'.

trade routes, and marriage practices embedded them' (1995: 7). Huli did not consider themselves to be 'geographical isolates', but their geographical connectivity was certainly centred upon themselves, which was also recognised by their neighbours as Huli functioned as a regional trading language. Indeed, their centrist and hegemonic understanding of themselves depended upon their connections throughout the diverse field that surrounded them.

Much has been written on the cosmological underpinnings of this expansive Huli world-view (Goldman 1983; Frankel 1986; Ballard 1994, 1995), but economic factors in the form of trade routes and centres of production peripheral to a centralised Huli-controlled trading zone are at least equally important to providing an understanding of the Huli-centric notion of Hela, and it is these material concerns that have transformed to become the most important aspect of people's lives. As Tom Ernst writes, 'The degree of freedom of movement by Huli and to Huli territory is more likely to have been "the peace of the trade" than a commonly subscribed to Huli ritual or cosmological hegemony' (2008: 31). Today the primacy of these material concerns has emerged as the strongest motivator of contemporary events. The PNG LNG project has returned Huli to the centre of economic activity, which is a position to which they feel both cosmologically and historically entitled. The major difference between the present situation and the past is that power is now located external to Huli territory, and the material destiny of Hela is largely in the hands of the state rather than Huli themselves.

Huli have retained an understanding of their historically superior status at the centre of economic activity, and have incorporated the existence of gas reserves into their trading cosmology. A Huli friend in Tari explained to me his understanding of the history of Huli trading relationships in the following way:

> Engans get salt, they cut small trees and throw them into a spring and stay for months in the water. Salt crystals grow on the wood. Then they dry and burn the wood and collect the salt to make bundles of salt. The Engans (Obena) want to trade with Komo people, but not the Tari Huli. Men from Hulia (Tari Huli) stopped the Obena from travelling down to Komo and Bosavi by telling the Obena that the Duguba people will eat them. Obena are fatter, stockier people and the Duguba will like to eat them. It is safe for the skinnier Huli to trade with the Duguba, so they take the salt from the frightened Obena and trade with Duguba for cassowary

feathers and bones, bird of paradise feathers, tree oil, and wood that Huli use to make bows. The Huli give some of these things to Obena people but keep the best for themselves. Huli people must keep the Obena and Duguba people apart. The prophecy states that one day they will see each other and then an eternal fire that exists beneath the Gigira Range, the Lai Tebo fire, will light all the way from Gigira to Obena territory.

This is now interpreted to be the power line that runs from Nogoli to power the Porgera Gold Mine. Nogoli is located about 20 km southwest of Tari and is the site of the gas turbine that generates power for Porgera in neighbouring Enga Province.

This cosmological aspect of the Huli trading relationship, expressed as a desire to prevent the meeting of the Duguba and Obena people lest the land be destroyed, has been described by Ballard (1995).[6] The economic and cosmological drivers for this concern are not easily separated, and my opinion is that the movement of people across distances amounting to tens of kilometres into neighbouring territories should be viewed through the lens of the primacy of material interests. Huli cosmology more broadly, I would argue, needs to be understood in material terms, and it is these material desires that are driving the reconfiguration of Huli cosmology to incorporate the presence and extraction of natural resources. This view is supported by Huli themselves who consistently describe the purpose of their cosmological rituals in terms of the maintenance and resurrection of the fertility 'of the entire universe' (Ballard 1994: 142). With the development of the PNG LNG project, the material underpinnings of Huli cosmology are well and truly dominant, but the power to realise material desires is in the hands of the state.

A Prophecy of Abundance and the Immorality of an Absent State

The PNG LNG project is so vast in scale and conception that its gravitational pull has reduced the development aspirations of Hela to a singularity. Every hope and dream of every Hela citizen was expected

---

6   Ballard reports 'a somewhat cynical young Huli businessman' who posited that 'this stricture was formulated by his ancestors precisely to maintain the monopoly enjoyed by Huli on the lucrative trade between Obena and Duguba groups of such wealth items as salt, tree oil and blackpalm bow staves' (Ballard 1995: 57).

to be realised via the LNG wealth that had been imagined without end. PNG LNG was imagined by the project's main protagonists, ExxonMobil and Oil Search, and the PNG state, as well as the Huli population, as a paradigm shift in the material landscape of the people and of the future. To an extent, Huli had been here before when in the 1660s the Long Island volcano erupted and changed their world utterly and instantly. A Plinian eruption of equivalent magnitude to that of Krakatau in 1883 deposited a layer of volcanic ash across the PNG highlands (Blong 1982). Huli stories of a time of darkness, known as *mbingi* in Huli, were first recognised as being related to this eruption by Blong (1979). The details of the stories varied, and were recorded by the first patrol officers assigned to Huli territory during the 1950s. During a patrol into the Komo basin from Lake Kutubu, Charles Terrell (1953: 18) recorded a story about a time of 'white rain' when the sky became dark and 'white stuff that was like the ground' fell for seven days. After that event the people found that their sweet potato crops grew faster than they had before. The volcanic ash brought such enhanced fertility to the soil that ritual practices emerged that were dedicated to bringing on the event again. Crucially, the line between disaster and renewal was finely balanced and depended on correct moral behaviour and observance of strict moral codes, breaches of which are known in Huli as *ilili* (forbidden act). Improper behaviour could result in the failure of *mbingi* and the destruction of the entire known Huli universe.

A preoccupation with *mbingi* combined with the rapid embrace of Christianity from the mid-1950s onwards was shown by Ballard (2000) to have intersected neatly with Christian notions of the apocalypse and, in particular, the role of human moral agency in bringing about the destruction of the land. One of the most enduring features of a Huli world-view has been the perception that the world is in a constant state of decline. This 'entropic' perception of the Huli universe has been influenced by recurring natural events such as droughts and floods and much of Huli ritual was devoted to reversing this downward trajectory of material and social change (Ballard 2000: 207). By the time of the PNG LNG project, the cultural memory of *mbingi* was in retreat and the time of darkness was no longer evoked in Huli expectations of the future. But an expectation of apocalyptic transformation remained as a habitus of anticipation that was ready and open to the promises of PNG LNG to which Huli believed themselves to be ancestrally entitled. PNG LNG became the new *mbingi* and the prophecy known as Gigira Lai Tebo (see below) was elevated into the realm of truth and demonstrable fact, evidenced unmistakably by the giant fire that burns day and night from the top of ExxonMobil's flaring tower. As a technical-political project, the PNG LNG project is also finely balanced between success and development failure. Notwithstanding the technical success of the PNG LNG project, and its ability to generate revenue for the companies and their investors, from the point of view of its Huli landowners, the immoral behaviour that has tipped the project towards development disaster is found in the corruption and neglect of an absent state. The immorality of state absence is a material threat to the future aspirations of Huli and the prophetic entitlement of Hela to benefit from its gas.

Gigira Lai Tebo

As stated above, the gas resource for the PNG LNG project lies beneath a northwest–southeast limestone mountain ridge commonly known as Hides Ridge. The Huli name for this mountain is *Hari* Gigira (Mt Gigira).[7] ExxonMobil drilled its gas wells along this ridge and constructed its vast Hides Gas Conditioning Plant (HGCP) towards the southeastern end of the ridge between 2010 and 2014 (see Figure 5.1). The topography of the Gigira Range results in soft cloud formations that settle blanket-like and fall down the sides of the range. The effect is similar in appearance to the white smoke that oozes from the kunai-grass roofs of Huli houses when a fire is burning within (see Figures 5.2 and 5.3). This visual effect is the source of the belief that there is a fire that burns beneath the Gigira Range. Huli houses are built with a central fireplace that is fed by long pieces of split firewood extending towards the doorway. The hardest and hottest-burning firewood comes from the *lai* tree (*Dodonaea viscosa*), commonly known in Australia as sticky hop bush, of which an unusually tall variety grows in the PNG highlands. The Huli word 'tebo' refers to the glowing coals of a fire. Running beneath the Gigira Range is a giant, burning piece of lai wood—the Gigira Lai Tebo. The Gigira Lai Tebo story is one of many Huli mythological tales that relate to land formations across Hela. One of the LNG ships dedicated to the PNG LNG project even bears the name 'Gigira Laitebo' (see Figure 5.4).

---

7   *Hari* means 'mountain' in Huli. Gigira is a range or ridge formation rather than a mountain peak, however, *hari* is a more generalised descriptor for a mountain form rather than a geographically specific mountain peak.

Figure 5.2 Huli house with smoke oozing through the roof
Source: Photograph by author

Figure 5.3 Gigira Range with cloud
Source: Photograph by author

Figure 5.4 Gigira Laitebo ship
Source: Geraldo Pietragala on VesselFinder (www.vesselfinder.com)

Figure 5.5 ExxonMobil's HGCP on Gigira Range
Source: Photograph by author

When gas extraction wells were first drilled on Gigira, people living on the slopes of the Gigira Range fled their homes in fear that they would be engulfed by the Gigira fire. The fact that ExxonMobil built its conditioning plant and gas flare on the mountain further cements the legend. Figure 5.5 shows the visual impact of the PNG LNG project on the landscape. The prophecy associated with the project is widely known and was often repeated to me whenever I asked anyone about the PNG LNG project: One day a man with red legs will come to take the Gigira fire. You may share with him some of the fire, but do not give him the whole fire lest the world will end. Everyone who told me this story concluded with the observation that the Huli have given away the whole fire.

The Gigira Lai Tebo prophecy has not featured in previous academic literature, although it has featured occasionally in PNG media reports since the PNG LNG project began, and little was known about its detail prior to the PNG LNG project. In 2009, just prior to the construction phase, the prophecy was described in a blog post that was published by the son of former schoolteachers at Lake Kutubu (Cuthbert 2009). Cuthbert wrote 'Inside Gigira burns Laitebo which when kindled, the flames will illuminate Hela and the land beyond'. He went on to say: 'When the man with the orange legs comes to take the fire from you, give him an axe and tell him to go get his own firewood for his own fire.' In 2016 I visited areas on the slopes of Gigira Range that I was told were now uninhabited because the people had moved away in fear of the fire spilling out. Warnings that the world will come to an end as a result of prohibited human behaviour is a Huli trope that is common to the majority of Huli prophetic understandings.

A Nation-Defining Resource within an Absent State

Historically, Huli have observed deterioration in productivity of the land, epidemics of disease, increase in social disorder and warfare, floods, drought and earthquakes as representative signs of the decline of the Huli universe (Ballard 2000). Humans had a role in controlling the fate of the universe by observing *ilili* codes, and implicit in the decline of the universe was failure to observe these moral codes (Ballard 2000). Giving away the Lai Tebo fire was one such act encoded in prophecy. During my time in Hela, signs of the Lai Tebo prophecy playing out were expressed in observations about the general malaise of Hela Province, increasing conflict, corruption, a decline in basic services and a general increase in poverty. Hela was rich in resources and therefore entitled to benefit from those resources. In moral terms, Hela's resources belonged to Hela itself, and to its people, and the state had a duty to facilitate Hela's development. The understanding that PNG's customary landowners are also the owners of the natural resources that exist beneath their customary lands is a commonly held view that Filer (2014: 79) describes as part of 'the founding fiction of the independent state' (see also Bainton and Skrzypek, Chapter 1, this volume). I would argue that this is one of the founding values of the independent state, which persists regardless of its legal standing. Huli have a cosmological entitlement to Hela's resources, and the historical and cultural depth to this entitlement precedes the colonially introduced legislative tradition that grants ownership of resources to the state. Moral failing therefore resided partly with Huli themselves, who had clearly given away their fire for nothing. But most of the blame was laid upon the state for failing to deliver on its promises, and therefore failing to live up to itself as a value proposition. The LBSAs had been negotiated and signed between landowners and the state, and it was the state, therefore, that landowners expected to make good on its promises. The nonappearance of promised development infrastructure represented the material absence of the state. The breaking of promises and demonstration of neglect represented the moral failing of the state. The failure of state officials to respond to the concerns of landowners was a demonstration of the absence of the state in moral terms.

Crucially, references to the Lai Tebo prophecy were based on observations of material decline primarily witnessed through a lack of government services and decaying infrastructure, rather than natural phenomena. The 2015–16 drought that severely impacted much of the highlands was coming to an end, and its impact was noted in terms of the lack of government response and the fact that the majority of people still had to eke out an existence as subsistence farmers. One reported example was the failure to maintain downpipes and water tanks at schools, which resulted in the forced closure of many schools due to lack of water (Bourke et al. 2016). The same report documented an increase in the crude death rate as a result of the drought. Development was expected to change the dependent relationship that people have with the land. The potential for development to save lives in the context of the drought and the failure of the state to realise that potential is a classic case of structural violence (Galtung 1969). It is also a failure of the value proposition that constitutes the state's existence. Drought was therefore a phenomenon that served to highlight the lack of development and the extent of state neglect in the province and their continuing dependence upon the land. The same can be said for the magnitude 7.5 earthquake that struck Hela Province in February 2018, as described later in this chapter. The PNG LNG project was widely regarded as a tool of the state to enable it to provide development in the form of infrastructure, health and education services, business development, employment, and cash for the landowners. Some landowners even expressed a certain level of sympathy for ExxonMobil because, like them, it had no choice but to deal with a corrupt state.

Attacking the State by Proxy

This sympathy was expressed in a letter to ExxonMobil that was drafted by a group of landowner leaders who had gathered near the Hides conditioning plant to organise a blockade in August 2016. These leaders had taken their case to the Supreme Court in Port Moresby to have PNG LNG project royalty payments, that were supposedly being held in abeyance, released to them and their clans. Two leaders in particular, who had previously been rivals for the leadership of the Petroleum Development License 7 (PDL7) landowner's group, had united in their efforts against the state.[8] When they finally got their day in court, the court turned down their application on the basis that they had not demonstrated themselves to be legitimate landowners. As a result, they took matters into their own hands and came to Gigira and the epicentre of the PNG LNG project— the conditioning plant compound. I was invited to witness the drafting of the letter to ExxonMobil. The letter was addressed to Andrew Barry, managing director of ExxonMobil PNG, and it politely and apologetically explained the reason that the project had to be shut down and made promises that no ExxonMobil property or personnel would be harmed. The authors specifically requested that ExxonMobil turn off the gas flare that burns within the compound. They drew an explicit distinction between the gas agreement that had been signed between the state and the developer in 2008, and the UBSA that had been signed between the

---

8   The Hides gas field and HGCP are located within the boundaries of PDL7.

state and the landowners in Kokopo in 2009. The letter explained in some detail the grievances between the landowners and the state and finished with the following statement:

> Thus we do not have hope and trust in this government as they continue to dishonour the agreements entered in to in Kokopo and in Hides respectively.
>
> Therefore we PDL 01, 07 and 09 landowners are calling on Exxon and its partners to voluntarily shut down the HGCP and well pads B and C. We also want Exxon to put off the flare in the HGCP so that the whole operation can be shut for indefinite period until the state responds to the petition positively.
>
> Thank you so much for your cooperation.

The landowner leaders' grievances were quite clearly directed towards the state rather than the company. In establishing a 'social license to operate' for the PNG LNG project, ExonMobil had been an outspoken advocate of the benefits that the project was expected to bring to the landowners. Prior to the project's construction, ExxonMobil had promoted numerous expected benefits from the project via its community affairs officers as they engaged with local communities. Yet when it came to confrontation over the failure of the project to deliver on its promises the landowner leaders directed their concerns to the state rather than the company. Why did these leaders focus almost exclusively on the state and not consider the responsibility of ExxonMobil and the potential for it to engage directly with the development of Hela Province, as Oil Search has done via its investments in the Hela Provincial health system?[9] I argue that the landowner leaders at PNG LNG were representatives of Hela, while the state of PNG is legitimised by its valued role in facilitating the viability of Hela and its transformation into a wealthy and prosperous example of a desired Papua New Guinean modernity. ExxonMobil is regarded as a tool, as an exclusively material entity being used by the state for its own benefit. The landowner leaders viewed ExxonMobil as a pawn in a larger game of corruption that was being played out by various levels of government and the judiciary. During the course of my fieldwork several people made the comment to me that ExxonMobil was just a company doing its job. ExxonMobil, in their eyes, was doing exactly what it was

---

9   Oil Search, via the Oil Search Development Foundation, has taken over the Hela provincial health system, including significant upgrades to the Tari hospital. A common complaint made to me was that ExxonMobil should behave more like Oil Search.

supposed to be doing: extracting gas, selling the gas for a profit, and paying royalties and taxes to the state.[10] It was the state, therefore, that had the responsibility of passing on agreed royalties and benefits to the people. Poorly understood by the landowners, however, was the aggressive tax-minimisation regime that the state had signed up to with the PNG LNG project. Details of this regime were reported by a World Bank investigation that attributed the lack of state revenue from the project to 'a complex web of exemptions and allowances that effectively mean that little revenue is received by government and landowners, either through taxes, royalties or development levies' (World Bank 2017: 27).

The landowner leaders were playing a dangerous game and, after two days, having received no response from ExxonMobil, they organised their clansmen to lock the main gate to the HGCP and blockade the project. My companions would not take me back up to the HGCP to witness these events as they were concerned for my safety and were worried about the potential for a shootout between the landowners and the police mobile squad that was stationed to protect the project. When I did return to the HGCP the next day, I was told by one of the landowner leaders who was the main organiser of the blockade that he had directed all his people to get hold of any weaponry they had and to enter the HGCP and lock the gate. The landowners responded by arming themselves and approaching the gates of the HGCP, compelling security personnel to step aside and allow them in. The organiser of the blockade then ushered the landowners back out and the gate was locked. Landowners also attempted, albeit unsuccessfully, to shut down the gas wells by breaking through the security fences that surrounded them and manually turning off the taps. When I later spoke with members of the police mobile squad, they told me that they had no intention of risking their lives by going against the landowners in defence of ExxonMobil, and they could see for themselves that the landowners were getting nothing from the project. The landowner leaders stated that the project would remain in lockdown until the government came to listen to their concerns.

---

10 This contrasts with the situation in Lihir, where landowners have maintained very strong expectations that the mining company will deliver development (see Bainton 2021; Bainton and Macintyre, Chapter 4, this volume).

Four days later the state responded by materialising at Gigira on the playground of the local school at Para to face questions from landowner leaders.[11] This was a crucial moment of state presence that stood in stark contrast to the experience of state absence that had thus far defined the relationship between the Huli landowners and the state. The landowner leaders had a very good grasp of the complexities of the gas agreement, and were able to hold their own when arguing about various forms of equity, well head production values, the value of LNG shipments and the amount of money that the government was receiving. The majority of landowners in the crowd, or people that I met at the market also spoke of royalty percentages, the number of LNG shipments to date and the structure of the state's Alternative Dispute Resolution (ADR) process that was then underway, which had failed to resolve any of their issues over landowner identification and the distribution of benefits.[12] The questioning was aggressive and accompanied in equal part by a torrent of abuse and threats that rained down upon the delegation for several hours in the baking sun. The highlight of the event was the moment the finance minister and Huli member for Tari-Pori, James Marape, was challenged to partake in a ritual practice known as *dindi nabaya*, which translates as 'we eat the ground'. The challenge was made by a representative of the Juha landowners who had questioned Marape about the fate of certain development funds, in response to Marape's answer, which was considered to be a lie by the majority of the crowd. In situations where people have competing claims over the historical truth of land ownership, the practice of *dindi nabaya*, which entails each person eating a handful of the disputed land, serves as a way of divining the truth. The land itself, being the subject of the argument, is the material objectification of a singular truth being disputed. The person who is lying about the truth of land ownership, specifically about the historical truth claims made in relation to the land, will become sick and die as a result of eating the land. Death is not instant but may take some months. This was perhaps a novel application of *dindi*

11   The government representatives included the Huli finance minister James Marape (who later became the prime minister of the country in 2019), the prime minister's departmental secretary, the managing director of the Mineral Resources Development Corporation (MRDC) Augusten Mano, the mining and petroleum minister Nixon Duma, the managing director of Kumul Petroleum Holdings, the secretary for the Department of Petroleum and Energy David Manan, the member for Koroba-Lake Kopiago Philip Undialu, and the district administrator for Hela William Bando.
12   Many landowner leaders were in dispute over the identification of landowners and the failure to undertake a proper process of landowner identification. Once these disputes were taken to court, Justice Ambeng Kandakasi ordered that landowner benefit distribution be halted until the landowner issues could be resolved through the ADR process (Filer 2019: 41).

nabaya, but the intent was clear. Marape was invited to come down off the stage and the crowd parted to create an open space for the two men to perform the ritual of *dindi nabaya*. His response to this ritual challenge was stunned silence. His evident fear and non-response was damning. And for many people in the crowd it represented the moral failure of the state.

Apart from the appeal to Huli tradition in the invoking of *dindi nabaya*, there was no invoking of the Gigira Lai Tebo prophecy or any other mythological understanding. The arguments were well-informed, pragmatic, and based on the material concerns of the landowners. The appeal to *dindi nabaya* was a body blow to the finance minister, as it was personal and a way of weaponising a shared ancestral and spiritual connection to land. The invoking of *dindi nabaya* was an extension of its original application in land disputes. Marape and the landowners were not arguing over land ownership, but over the truthfulness of Marape's statement about the fate of money that had been generated by the PNG LNG project. The landowner leaders decided that Marape, as a son of Hela, had a moral duty to tell the truth about Hela's resources; his duty to the state was secondary. Marape's silent response was taken by the crowd as evidence that he understood that his primary moral duty was to Hela. In the nation of Hela, state trumps company, but land trumps them both, and as a Huli it was generally expected that the Papua New Guinean finance minister James Marape's first duty was therefore to Hela.

Hela Trumps State

Prior to independence, Huli eagerly subscribed to the value proposition that was the colonial state. For Huli, the colonial state was symbolically represented by the crown worn on the policemen's headwear. When I asked John Hocknull, the last patrol officer to be stationed at Komo until independence in 1975, how Huli at Komo reacted to independence, he responded that the strong reaction was not to independence, but earlier when the national emblem of PNG replaced the crown that was worn on the policemen's uniform. The crown had been recognised as the head of power, out of which all social order emanated, and its removal, according to Hocknull, was like the sky had fallen in. Huli desire a benevolent power that is invested in the state, but if the state itself is absent then decay and disorder is the feared result. This is crucial to understanding Huli perceptions of the independent state as a value proposition. Today the absence of the PNG state in Hela, and especially in the remoter areas of the PNG LNG project operations at Komo and Hides, is palpable. When the landowners attacked the PNG LNG project, they were not attacking ExxonMobil but the state of PNG. Hela, much like the imagined state of PNG, or the imagined status of the crown, is constituted as an idea, and has more salience in its idealised form than does the material presence of the world's largest oil and gas company, which was ideally understood as a tool of the state. The landowners did not attack the tool, but the entity that wielded it. The state, constituted as an absence, and therefore as a value proposition that has collapsed, only loomed larger as its absence became the defining issue of the conflict.

Crucial to understanding Huli perceptions of ExxonMobil and its material presence in Hela is the perceived relationship between the PNG LNG project and the land. Unlike other extractive projects in PNG, or some large-scale mines in Australia (see Everingham et al., Chapter 9; and Lewis, Chapter 8, both this volume), the environmental footprint of the PNG LNG project is relatively benign. What impact the PNG LNG project has on the landscape consists mainly of infrastructure: the Komo airfield, the gas conditioning plant, even the waste incinerator, are all viewed as orderly and technologically advanced examples of material development. Unlike the Ok Tedi and Porgera mines, there is no production of waste material or tailings being disposed in the river system (see Golub, Chapter 3, this volume). Land ownership in Huli, or more accurately custodianship, is as much about the historical use of the land as it is about genealogical connections to land. It is beyond the scope of this chapter to detail the complexities of Huli land ownership. However, one feature that is relevant here is the fact that it is possible to lose rights over land through neglect. If your garden is allowed to become overgrown and go to waste then you are at risk of someone else coming along to work and improve the land and thereby gain rights to its occupation. Neglect of land is regarded as a moral failing. If you allow the land to go bad by allowing it to become overgrown with weeds and fail to put the land to some productive use then someone else has a moral right to occupy and improve the land. In this sense, ExxonMobil is doing work to the land and has a degree of user rights as a result. Had the PNG LNG project resulted in pollution to large tracts of land and waterways it is likely that ExxonMobil would have been imagined as an active agent in the destruction of Hela and it is arguable that the PNG LNG project would not have survived for as long as it has.

A Seismic Presence

On 26 February 2018 Hela was hit by a magnitude 7.5 earthquake. The earthquake caused devastation to the landscape and people and caused significant damage to the PNG LNG project. Several days later, on 2 March, ExxonMobil declared force majeure on exports of LNG and the project was shut down for six weeks. The earthquake induced a radical shift in the way the PNG LNG project is perceived by its Huli landowners. Almost overnight the perception of state absence went from one of moral failing to active agent in the destruction of Hela. ExxonMobil went from misappropriated tool to malevolent partner in Hela's demise. The initial response to the earthquake was a widespread belief that it had been caused by the operations of the PNG LNG project. This belief had both a mythological and scientific basis and the two are not easily separated. That an earthquake should occur of such magnitude that it caused severe damage to project infrastructure as well as large areas of the PNG highlands, killing many and leaving many more in desperate need of assistance, was perfectly in line with the doomsday scenario that accompanied the Lai Tebo prophecy, as well as a more generalised entropic perspective of a world in constant decline. Many landowners as well as people across PNG and in the global activist community believed that the act of gas extraction had caused the tectonic activity that resulted in the earthquake. This view was supported by strong evidence that the process of fracking (a process not used by the PNG LNG project) has caused earthquakes elsewhere, particularly in the United States, and communicated via popular publications such as *National Geographic* (Gibbens 2017). The suspicion that the earthquake had been caused by the PNG LNG project was so strong and so widely held among business and political elites that the PNG government was moved to ask the Australian government to conduct a review into the cause of the earthquake, which it agreed to do via the services of Geoscience Australia (Anon. 2018).

An earthquake of such magnitude, occurring at that location, came as no surprise to anyone with an understanding of the history of tectonic activity in the PNG highlands. With the PNG LNG project shut down, the belief that the disaster had a metaphysical component was far from universal, and intense debate occurred on Huli social media networks where these views were often ridiculed. None of my Huli friends spoke to me in terms of traditional cosmology or prophecy. Rather, their concerns were framed around the inadequate response from the state and the plight of many who had been cut off from essential services. The aid post that serviced the Komo region was built on a low-lying area that sank into the ground due to liquefaction caused by the earthquake. Roads became impassable and desperately needed medical assistance was eventually supplied by the Australian military and Oil Search helicopters. It was these immediate concerns that were in the forefront of people's reaction to, and perception of, the earthquake.

The argument that prevailed more broadly across Huli society was a debate between two material causes: was it a natural or anthropogenic phenomenon? But at the same time the disaster was imbued with the sense that there was a moral component, not attached to its cause, but to its happening. For Huli, natural phenomena that happen without any particular cause, or those with a direct human cause albeit lacking in mindful intent, may still be classed within a family of events that are indicative of moral decline and a general trajectory of social and physical decay. The Oil Search managing director Peter Botten, was quoted as saying that the perception of an anthropogenic cause was 'a communications issue' (Barrett and Gloystein 2018). With this he missed a crucial point: the perception of the earthquake as having an anthropogenic cause was not a communications issue, but a development issue (Main 2018). Dozens of people were killed by landslides or collapsed buildings, however the majority of the suffering that resulted from the earthquake could be attributed to failing or non-existent infrastructure, a lack of government capacity at the local level, an absence of government response at the national level and the failure of the PNG LNG project to have contributed to a more developed Hela Province that would have otherwise been better placed to respond to natural disasters of any kind. In the eyes of the Huli, the earthquake also imposed a heavy moral status on the company, which had previously enjoyed a more morally neutral position. Lack of development was a moral failing of both the state and the project developer, and in the context of the earthquake, this lack, born of absence, became in itself a powerful agent of suffering.

The Rise of ATALA

Several weeks after the earthquake in June 2018, part of the PNG LNG project was attacked with new a level of ferocity and organisation by a group calling themselves the Angore Tiddy Apa Landowners Association, or 'ATALA'.[13] Their attack was a direct response the earthquake and the impact of the absent state. ATALA was formed in 2014, declaring itself as the umbrella association for the Angore PDL8 landowners. According to ATALA's founding document the UBSA that was signed in May 2009 was done so before Petroleum Retention License (PRL) areas were converted into smaller Petroleum Development License (PDL) areas. ATALA were claiming that the social mapping was undertaken while Angore was still part of the old PRL11 and the results of this process were therefore null and void. It is beyond the scope of this chapter to undertake an analysis of the legal framework around petroleum development in PNG and I make no comment about the legitimacy or otherwise of ATALA's claims. Rather my intention is to provide an outline of the point of view of ATALA, who claim the support of 150 clans spread across the former PRL11 area, totalling some 100,000 people (Woods 2018a).

The gas resource tapped by the Angore wellhead is a later addition to the main resource at Hides and was planned to connect to the existing HGCP via an 11 km pipeline. The ATALA leadership is quite separate from the group that came together to blockade the HGCP in August 2016. But during that time ATALA gave their support for that action via their Facebook page, which has been the main forum for ATALA to communicate its objectives, which included asking 'the Member for Tari Pori, Minister for Finance and Government Business Leader, Honorable James Marape to acquit the Angore PDL8 Project Area Funds which he has been paying to his political cronies'. Up until February 2018, ATALA was primarily a political voice that was generally well-informed and articulate and engaged in online debate around PNG LNG land ownership issues and royalty payments. The 26 February earthquake induced a seismic shift in the attitude and approach taken by ATALA. The last post published by the association prior to the earthquake was a lengthy and considered statement related to the continuing issue of landowner identification and included the observation that 'Failures of successive attempts by the key State Agency stakeholders of the project is embarrassing for such an important project that underpins the economy of PNG for the next 30 years'.[14] The first post published after the

---

13   It is more common to spell 'tiddy Apa' as 'dindi apa', which means 'land father'. The Huli 'd' is pronounced closer to the apical consonant sound of 't' and is often interchangeable with 'd'. It is unclear why the 'n' has been left out.
14   See: www.facebook.com/pg/angorePDL8/posts/?ref=page_internal (10 November 2017).

earthquake simply stated: 'Our struggle is a struggle that has no definition to the Government and Stakeholders of the PNGLNG Project. One day, Hell will break loose!'[15]

On 21 June 2018, heavily armed ATALA representatives 'unleashed hell' on ExxonMobil's Angore camp, destroying buildings and machinery and shutting down the Angore tie-in project for the foreseeable future. Sources based in Tari told me that during this event 35 Papua New Guinea Defence Force (PNGDF) soldiers were flown by helicopter to secure the camp. The soldiers positioned themselves to mount an attack against the landowners but withdrew at the last moment when they realised that they were vastly outnumbered and outgunned. When the soldiers retreated, their helicopter was shot at. ATALA published an account of the attack on their Facebook page that included the statement: 'Exxon, we missed that chopper but its gonna come down. Look across and look at the smoke and flames. When you say so we talk.' ExxonMobil supplied the helicopter to the PNGDF for use in their action against the landowners. At the same time landowners at Komo under the umbrella leadership of ATALA placed a log across the Komo airfield to symbolise that they had shut down the facility for the foreseeable future.

Their actions were a clear violation of their own mission statement, which reads:

> Work with honesty and integrity, and equitably share all benefits with neither fear nor favour.
>
> Work hard and transform the wisdom of the laitebo prophesy into socioeconomic and political achievements for the Angore people now and into the future.
>
> Negotiate without violence, and work honestly and diligently with the key stakeholders of the PNGLNG Project.

The principle of negotiation without violence was buried with the earthquake, and this change in approach is documented in a statement released on 21 May 2018 that asserts: 'Since the EQ Disaster in the Hela, Southern Highlands, Gulf and Western Province; the shift of the negotiating power is now with the people.' This statement gets to the core of the fundamental shift in power relations between people and

---

15   Ibid. (28 April 2018).

the state that was created by the earthquake. The earthquake provided ATALA with the moral authority to assert themselves as a military power and to bypass the practice of negotiation and deploy violence to achieve their ends. In this action, ATALA has made a concerted effort to gain the support of a large number of clans across a greater part of Hela. State neglect in the context of the earthquake and its aftermath has become a form of structural violence, as the inability and unwillingness of the state to provide adequate disaster relief has resulted in increased suffering for a great many people. For ATALA and many Huli landowners the earthquake amounted to the total collapse of the value proposition that is the state of PNG. This opened up the space for ATALA to reinvent itself as a morally superior alternative to the state. Furthermore, the failure of the state to respond to a natural disaster of such magnitude removes the logic that accompanied previous efforts at negotiation and the pursuit of state-based mechanisms such as court actions and political directions. For ATALA and its supporters, violent action has become both a morally justifiable and rational method of advancing their cause. Hela must assert itself as a nation without the need for a state, since the value proposition that is the state barely exists. State neglect has become more than just a source of frustration and a perceived moral failing—it has emerged as a form of negative agency that must be combatted.

The boldness of ATALA was expressed not just through violent action, but in its demand for a stand-alone gas stripping facility for the Angore resource, and its offer to buy ExxonMobil out of its gas resource at Angore. A statement released by ATALA on 20 June 2018 asserts a sovereign claim to the gas resource, something that had previously been ceded to the state in the expectation that the state and developer would act in the best interests of the landowners:

> In Angore, ATALA Incorporated is the landlord, not the government of Papua New Guinea. ATALA Incorporated has taken over Angore PDL8 Operations including the 5.4 Trillion Cubic Feet of LNG.

> This project, the PNGLNG Project, has brought curse and suffering to the people of Angore and we will need that now. ATALA Inc as the landlord is taking over the Angore PDL8 and will develop separately from the PNGLNG Project.

The document then outlines in some detail the economic mechanisms by which the Angore gas resource is to be utilised specifically for the benefit of the landowners within PDL8. It goes as far as demanding 'a standalone gas stripping facility so as to determine and quantify how much LNG and Condensate are produced from our reserve which is an independent service to other gas-fields within and outside of the Hela Province'. The political impact of the earthquake was to negate the state and its development apparatus in ExxonMobil as legitimate entities. Instead of desiring the absent state to become materially present, the material consequences of state absence in the context of the earthquake reversed this desire: the absent state bereft of value was to be banished altogether. Emerging from this negation is an organisation within the nation of Hela that believes it has no choice but to go it alone, backed by a sense of legitimacy based on a concept of Gigira Lai Tebo, which provides both the entitlement to the benefits of the gas resource and the moral authority to control it.

Conclusion: The State of Nothing

The PNG LNG project has profoundly shaped the PNG economy, and has helped to define PNG as a state that has absented itself from its basic obligations to its own people. The PNG LNG project has simultaneously transformed and helped to define an understanding of Hela as an Indigenous nation endowed with resource wealth and entitled to a parent state that can fulfil its development aspirations. The experience of an absent state in Hela Province in the context of greatly increased expectations of the state resulting from the PNG LNG project reveals the extent to which the state itself is a concept that exists as a shared idea that may increase in salience as the material presence of the state disappears. The PNG LNG project was expected to enhance the value proposition of the state by enabling it to provide a bonanza of promised development projects as well as direct cash payments to landowners. The state exists materially in the form of the built environment that is constitutive of its value: the schools and hospitals, roads and services that the state is expected to provide. The state also exists materially in the form of its representatives who hold positions of power and authority who are valued for their ability to deliver development outcomes. At Hides in August 2016, when this absence became intolerable to a group of Huli elites, they were able to exercise power over the state using the PNG LNG project as a proxy, and force the state to become present and to take notice of those that it had been refusing to acknowledge. One Huli elite, the minister for finance James Marape, was caught between his identity as a son of Hela, and his role as a representative of the state. As part of the state Marape could retreat into absence, but Hela was part of Marape and that was a presence that he was unable to deny.

This chapter has demonstrated how the concept of state absence took on a special significance in the context of the state's failure to respond to natural disaster. The state, as an externalised, parental authority, derives its legitimacy and is defined as a value proposition through its ability to service the viability of the nation of Hela. The destructive power of the earthquake of February 2018 was the ultimate exposure of the state to the abrogation of its most fundamental responsibilities. Where corruption and neglect had only made the state loom larger for its absence, the trauma of the shaking of Hela has turned this absence into a form of active malevolence. In the process, ExxonMobil has become a weapon rather than a tool of the state, and has endured the consequences at Angore. On 26 June 2018 the PNG government offered ATALA PGK20 million in return for ending their protest (Woods 2018b). The landowners rejected this initial offer, asking that the state also provide their organisation with official recognition as a legitimate organisation, to which the state duly complied. With this request the landowners were keeping a foot in both camps, as they were simultaneously trying to bypass the state and asking to be recognised by the state and extracting as much value from the state as they could. The state continues to exist, even as its existence is repelled, as its value has not entirely collapsed. It is unclear how the future will unfold for Hela and the PNG LNG project, however the granting of cash and negotiating power to ATALA as an official organisation has thus far prevented further attacks against the project, as no further threats have emerged from this organisation. In both August 2016 and June 2018, landowners demanded that the state 'present itself', and on both occasions the state complied. But these occasions have only had the effect of satisfying the demands of particular landowner leaders and their supporters, many of whom had borrowed substantial amounts of money in the expectation that a development windfall was coming, and needed the state to bail them out when the debt collectors started calling. These protests, although made possible by the absence of the state, have done nothing to create the substantial reform that is desired and expected by the broader Huli population. As the gas continues to flow and be shipped out to be sold offshore, it remains to be seen whether Hela landowners will continue to tolerate the impact of the state's absence into the future.

Postscript

On 12 April 2019, James Marape resigned as finance minister, citing disagreements with the prime minister over the signing of the agreements for the new Papua LNG project in neighbouring Gulf Province, which appears destined for a similar fate to that of the PNG LNG project. The next day Marape returned to Tari, the capital of Hela Province, to a hero's welcome, stating that the government had brought nothing tangible to Hela, only promises. After seven years as finance minister of an absent state, Marape chose to be present as the representative of the Nation of Hela as his self-perceived value as a representative of the state was finally exhausted. On 7 May 2019 Marape was elected 'alternative prime minister' by PNG's opposition party, and, after a complex succession struggle, was sworn in as prime minister on 29 May. Marape's rise has done much to dissolve the boundary between the nation of Hela and the state of PNG. Marape based his power struggle on the failure of PNG to benefit from its considerable extractable resources, and, following his swearing in, immediately called for economic independence for PNG, and derided the systemic impairment of the translation of PNG's 'opportunities and blessings' to its people (Auka-Salmang 2019). Since then Marape's 'take back PNG' rhetoric has been consistent, bold and has translated into robust policy positions (Kama 2019). Marape has thus far maintained a consistent position in negotiations between the state and ExxonMobil over the P'nyang gas field, and, in February 2020, suspended negotiations describing ExxonMobil's position as 'unacceptable', stating that PNG's resources must 'allow us to improve our people's way of life. We don't believe the last offer made by ExxonMobil would have done that' (Wilkinson 2020).

By the time of writing, in July 2020, Marape has thus far avoided the pitfalls of his predecessor, has the appearance of being genuine in his motives and engages much more intimately with PNG's extraordinarily diverse grassroots population. Marape communicates directly to the nation via Facebook, making himself present in households across the country, and was initially signing off with the Huli term 'Hulukuyama'. This is a praise term for a Huli man, and will only be recognisable to Huli speakers. Huli could not be more legible to the state, and yet their struggle to obtain state royalty payments for the PNG LNG project continues. The reason for this is that their struggle is about making themselves legible internally, rather than to the state. Dispute over land ownership is part of a centuries-old argument over the control and interpretation of Huli land ownership history, and this has intersected with disputes over royalty payments. But royalty payments are not what the promise of the PNG LNG project was ever about. It is about development, and that means infrastructure, it means services, it means meaningful employment, improved standards of living—all of which are only possible via the accumulation of capital, and the effective use of capital. Royalty payments to thousands of landowners are about the thinning of capital, they are not a pathway to development, and they have never been the sole objective of the Huli population. Royalty payments are at least an indication that the state is paying attention, and that perhaps real development might be possible. It is arguable that the PNG LNG project, and its stark reveal of an absent state, are responsible for Marape's rise to power. Marape's term might finally reveal whether the presence of the extractive industries, and the activities of multinational extractive corporations, should even be considered as viable pathways towards the development of the nation.

References

ACIL Tasman, 2008. 'PNG LNG Economic Impact Study: An Assessment of the Direct and Indirect Impacts of the Proposed PNG LNG Project on the Economy of Papua New Guinea.' Melbourne: ACIL Tasman Pty Ltd for ExxonMobil.

Anderson, B., 1983. *Imagined Communities: Reflections on the Origin and Spread of Nationalism*. London: Verso.

Anon., 2018. 'PNG Earthquake: Calls for Formal Inquiry into Cause as More Aid Arrives, ADF Sends Globemaster Packed with Supplies.' Australian Broadcasting Corporation, 10 March. Viewed 29 September 2020 at: www.abc.net.au/news/2018-03-10/calls-for-inquiry-into-cause-on-png-quake/9534898

Auka-Salmang, G., 2019. 'Marape Sworn in as Country's 8th PM.' *Post-Courier*, 30 May.

Bainton, N., 2021. 'Menacing the Mine: Double Asymmetry and Mutual Incomprehension in the Lihir Islands.' In N. Bainton, D. McDougall, K. Alexeyeff and J. Cox (eds), *Unequal Lives: Gender, Race and Class in the Western Pacific*. Canberra: ANU Press. doi.org/10.22459/UE.2020.14

Ballard, C., 1994. 'The Centre Cannot Hold: Trade Networks and Sacred Geography in the Papua New Guinea Highlands.' *Archaeology in Oceania* 29(3): 130–148. doi.org/10.1002/arco.1994.29.3.130

———, 1995. The Death of a Great Land: Ritual, History and Subsistence Revolution in the Southern Highlands of Papua New Guinea. Canberra: The Australian National University (PhD thesis).

———, 2000. 'The Fire Next Time: The Conversion of the Huli Apocalypse.' Ethnohistory 47: 205–225. doi.org/10.1215/00141801-47-1-205

———, 2002. 'A History of Huli Society and Settlement in the Tari Region.' Papua New Guinea Medical Journal 45(1/2): 8–14.

Barrett, J. and H. Gloystein, 2018. 'Shakes and Superstition: Exxon Faces Backlash in Papua New Guinea.' Reuters, 8 March.

Biersack, A., 1995. 'Introduction.' In A. Biersack (ed.), Papuan Borderlands: Huli, Duna, and Ipili Perspectives on the Papua New Guinea Highlands. Ann Arbor: University of Michigan Press. doi.org/10.3998/mpub.14775

Blong, R.J., 1979. 'Huli Legends and Volcanic Eruptions, Papua New Guinea.' Search 10(3): 93–94.

———, 1982. The Time of Darkness: Local Legends and Volcanic Reality in Papua New Guinea. Canberra: Australian National University Press.

Bourke, R.M., B. Allen and M. Lowe, 2016. 'Estimated Impact of Drought and Frost on Food Supply in Rural PNG in 2015.' Canberra: The Australian National University, Crawford School of Public Policy, Development Policy Centre (Policy Brief 11).

Chernilo, D., 2006. 'Social Theory's Methodological Nationalism: Myth and Reality.' European Journal of Social Theory 9: 5–22. doi.org/10.1177/1368431006060460

Cuthbert, N., 2009. 'The Hela Prophesy.' Malum Nalu blog, 22 November. Viewed 29 September 2020 at: malumnalu.blogspot.com.au/2009/11/hela-prophesy.html

Ernst, T., 2008. 'Full-Scale Social Mapping and Landowner Identification Study of PRL02.' Unpublished report to ExxonMobil Ltd.

Esdale, F.V., 1955. 'Tari Patrol Report No. 6 of 1954/55.' Patrol Reports. Southern Highlands District, Tari, 1954–1955. National Archives of Papua New Guinea, Accession 496.

Filer, C., 2014. 'The Double Movement of Immovable Property Rights in Papua New Guinea.' The Journal of Pacific History 49: 76–94. doi.org/10.1080/00223344.2013.876158

————, 2019. 'Methods in the Madness: The "Landowner Problem" in the PNG LNG Project.' Canberra: The Australian National University, Crawford School of Public Policy, Development Policy Centre (Discussion Paper 76). doi.org/10.2139/ssrn.3332826

Flanagan, P. and L. Fletcher, 2018. 'Double or Nothing: The Broken Economic Promises of PNG LNG.' Sydney: Jubilee Australia.

Frankel, S., 1986. *The Huli Response to Illness.* Cambridge: Cambridge University Press. doi.org/10.1017/CBO9780511521072

Galtung, J., 1969. 'Violence, Peace, and Peace Research.' *Journal of Peace Research* 6(3): 167–191. doi.org/10.1177/002234336900600301

Gibbens, S., 2017. 'How Humans Are Causing Deadly Earthquakes.' *National Geographic*, 2 October.

Goldman, L., 1981. Talk Never Dies: An Analysis of Disputes Among the Huli. London: University College London (PhD thesis).

————, 1983. *Talk Never Dies: The Language of Huli Disputes.* London: Tavistock.

GPNG (Government of PNG), 1975. *Constitution of the Independent State of Papua New Guinea.* Port Moresby: National Parliament.

————, 2013. 'National Population and Housing Census 2011: Final Figures.' Port Moresby: National Statistical Office.

Kama, B., 2019. '"Take Back PNG": Prime Minister Marape and his Audacious Vision for PNG.' DevPolicy blog, 8 August. Viewed 23 September 2020 at: devpolicy.org/take-back-png-prime-minister-marape-and-his-audacious-vision-for-png-20190808/

Main, M., 2018. 'How PNG LNG Is Shaking Up the Earthquake.' Envirosociety blog, 28 March. Viewed 23 September 2020 at: www.envirosociety.org/2018/03/michael-main-how-png-lng-is-shaking-up-the-earthquake

Patrol Reports. Southern Highlands District, Lake Kutubu, 1953 - 1954. National Archives of Papua New Guinea, Accession 496.

Scott, J., 1998. *Seeing Like a State: How Certain Schemes to Improve the Human Condition Have Failed.* New Haven (CT): Yale University Press. doi.org/10.2307/j.ctvxkn7ds

Terrell, C.E.T., 1953. 'Lake Kutubu Patrol Report No. 3 of 1953/54.'

Tzu, L., 1963. *Tao Te Ching* (trans. D.C. Lau). London: Penguin Classics.

Wilkinson, R., 2020. 'PNG stops P'nyang Gas Agreement Negotiations with ExxonMobil.' *Oil and Gas Journal*, 3 February.

Woods, L., 2018a. 'Papua New Guinea Landowners Take Up Arms Against Natural Gas Project.' Mongabay blog, 26 June. Viewed 23 September 2020 at: news.mongabay.com/2018/06/papua-new-guinea-landowners-take-up-arms-against-natural-gas-project/

———, 2018b. 'Deal in Sight for PNG Landowners Protesting Exxon-Led Gas Project.' Mongabay blog, 31 August. Viewed 23 September 2020 at: news.mongabay.com/2018/08/deal-in-sight-for-png-landowners-protesting-exxon-led-gas-project/

World Bank, 2017. 'Papua New Guinea Economic Update, December 2017: Reinforcing Resilience.' Viewed 23 September 2020 at: documents.world bank.org/curated/en/150591512370709162/Papua-New-Guinea-Economic-Update-Reinforcing-Resilience.pdf

6

In Between Presence and Absence: Ambiguous Encounters of the State in Unconventional Gas Developments in Queensland, Australia

Martin Espig

Introduction

Australia's natural gas industry has grown rapidly over the last two decades. Sourced from conventional and unconventional resources, a majority of this new capacity is supplying a liquefied natural gas industry that exports to overseas energy markets. In the country's northeastern state of Queensland, the extraction of gas located within the cleats of underground coal seams has become technologically and economically viable since the turn of the millennium. This unconventional coal seam gas (also called coal bed methane) is not contained within pressurised reservoirs like conventional oil and gas but dispersed throughout coal-bearing strata of large geological basins. Accessing coal seam gas (CSG) therefore requires numerous wells to be drilled in close grids across broad areas (see Figure 6.1). Groundwater is removed to lower the hydrostatic pressure that holds the gas in place, and both then flow to the surface where they are separated. Although less than in other unconventional gas sources, this process also includes controversial extraction techniques such as hydraulic fracturing ('fracking'). Following the industry's swift growth and over AUD70 billion invested over the last decade and a half, around 11,000 producing CSG wells are currently operational across Queensland. The total number of wells is expected to double once the industry is fully established. The processing and transport of gas and water involve additional infrastructure such as gathering pipelines, well pad access tracks, compressor stations and water treatment facilities.

Establishing these extractive networks and accounting for the various impacts of their collective footprint has so far resulted in some 5,700 Conduct and Compensation Agreements negotiated between gas companies and landholders (GCQ 2018). However, CSG developments and associated regional transformations have not occurred without disputes over potential risks associated with the dynamically evolving industry (de Rijke 2013a: 46–8; Espig 2018: 11). Particularly salient points of contestation within host communities and among the wider public revolved around impacts to groundwater and negative health effects for those living in proximity to CSG infrastructure. These debates were exacerbated by perceived absences in state regulatory and oversight function and the fact that, considering the pace and complexity of the industry's developments, many decisions involved scientific uncertainties regarding its short and long-term impacts.

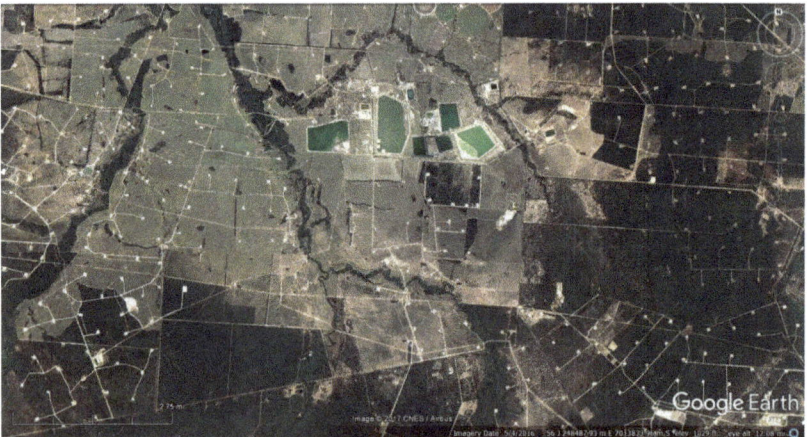

Figure 6.1 A coal seam gas field in the Western Downs region
Source: Google Earth 2020

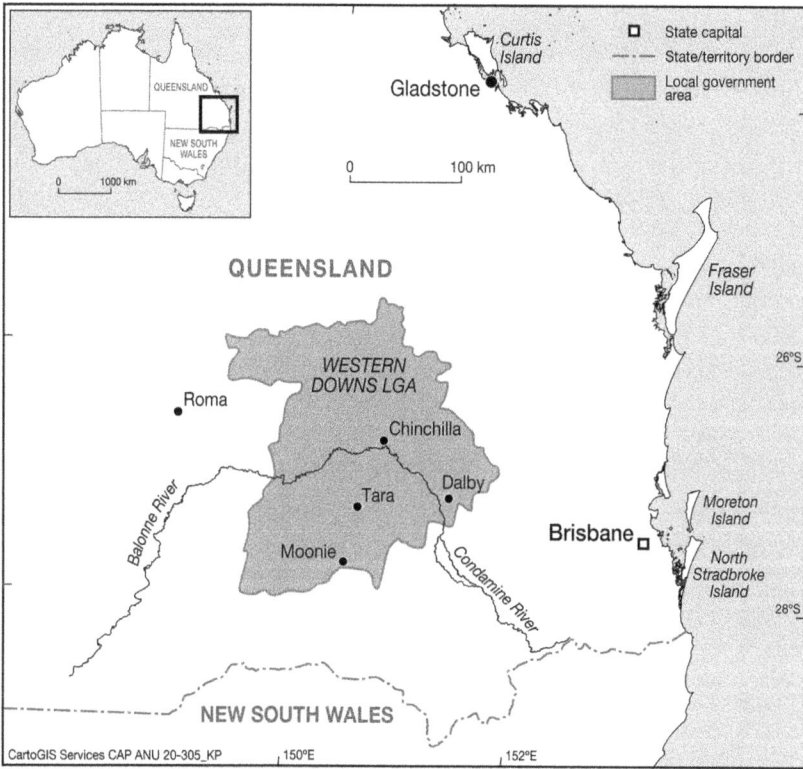

Figure 6.2 The Western Downs local government area and Curtis Island, the location of LNG processing and export facilities some 500 km to the north
Source: ANU Cartography

This chapter focuses on the risk controversy surrounding CSG developments in Queensland and how they pertain to the role of the state in natural resource developments. My analysis draws on research of CSG-related topics since 2011, including multi-sited ethnographic fieldwork conducted between late 2014 and 2016 in one of the CSG hotspots in Queensland, the Western Downs local government area (see Figure 6.2). This predominantly agricultural region covers a land area of almost 38,000 km² and is home to some 33,000 people. With thousands of operational wells, the Western Downs is the most intensively developed area in Australia in terms of unconventional gas projects. Its gas fields supply three major liquefied natural gas (LNG) plants that are located some 500 km to the north on Curtis Island near the town of Gladstone (Eriksen 2018).

Many irrigating farmers and graziers in the Western Downs, who rely on historically already overused aquifers, voiced concerns about uncertainties regarding the quantity of groundwater extracted over the CSG industry's multiple decades-long lifespan. Especially early on, these concerns were well-founded, as levels of uncertainty were high even among scientific researchers, industry professionals and regulators (Espig and de Rijke 2016). Debates also emerged around the potential impacts of chemicals used during extraction. Some residents were further opposed to possible human health risks from environmental pollution associated with, for instance, leaking wells and infrastructure, which prompted them to emphasise potentially unknown side effects of CSG extraction and perceived shortcomings in state departments' regulatory oversight. In this sense, uncertainties not only existed around the scientific aspects of CSG developments and potential environmental impacts. For many concerned residents, uncertainties also manifested in a normative sense regarding the roles of state departments and their regulatory capacities, as well as in moral terms concerning the trustworthiness of state and CSG industry actors. This chapter is developed around interlocutors' accounts regarding these aspects.

While my research also involved Queensland state politicians, regulators and urban-based scientists, the chapter is primarily based on ethnographic work with rural landholders and regional residents in the Western Downs. Among those community members, only small minorities either unequivocally supported or outright opposed CSG developments. I more often encountered mixed feelings towards the industry, with many interlocutors describing positive as well as negative aspects of CSG projects. This is in line with surveys conducted at the time of my fieldwork, which show that a majority of participating Western Downs residents either tolerated or accepted the CSG industry (Espig 2018: 160–6). In this chapter, I focus specifically on interlocutors who expressed concerns over CSG projects' risks and uncertainties, regardless of their overall attitude towards the industry.[1] Those interlocutors dealt with a variety of Queensland state departments, including the Department of Natural Resources, Mines and Energy and the Department of Environment and Heritage Protection, as well as associated entities such as the Office of

---

1   By focusing on one segment of my ethnographic data, I do not intend to oversimplify community members' differences nor to 'sanitise' the local politics of representation and conflict (Trigger 2000: 203; Ballard and Banks 2003: 306; see Espig and de Rijke 2018 for a corresponding discussion). For a relevant debate on the 'problem of ethnographic refusal' (Ortner 1995) in the context of mining, see Bainton and Owen (2018) and Kirsch (2018: 9–12 and 28–9).

Groundwater Impact Assessment or GasFields Commission Queensland. It would therefore be misleading to regard the Queensland state as a single monolithic entity and community members' encounters with the state as uniform. The following account should be read accordingly.

Through the lens of CSG risk debates, it is possible to understand community members' experiences of the state within natural resource developments, especially the ambiguous encounters and frictions that projects can create between the state and concerned members of host communities. My primary intention behind developing this perspective is not to analyse the Queensland state as a government apparatus of specific institutions, regulatory instruments and bureaucratic procedures. Instead, I follow Sharma and Gupta (2005: 27) in regarding states 'as culturally embedded and discursively constructed ensembles', and Trouillot (2003: 89) in presenting the state as 'a set of practices and processes and the effects they produce'. My aim is to contribute towards an anthropological understanding of 'how states are culturally constituted, how they are substantiated in people's lives, and … the sociopolitical and everyday consequences of these constructions', which makes it necessary to focus on 'everyday practices and representations as modes through which the state comes into being' (Sharma and Gupta 2005: 27). Rather than attending to its mere institutional contours, I focus on the community–state interface and examine how concerned host community members encountered 'the state' in CSG risk debates.[2]

The following section briefly describes the regulatory foundations of CSG developments and highlights the Queensland state's presence in terms of granting and guaranteeing land access rights. I outline residents' reactions to this presence, and how the Queensland state's risk governance policies account for uncertainties. The two subsequent sections paint a contrasting picture of a partially absent state, with CSG projects outpacing regulating departments in some cases and pushing them beyond their capacity, especially once developments fully commenced. Concomitantly, concerned members of affected communities and the wider public struggled to receive clear answers while navigating what I described as the resulting organised irresponsibility around CSG developments. This leads to a discussion on the displacement, or delegation, of state functions and a responsibilisation of community members. Exploring the

---

2   See Bainton and Skrzypek (Chapter 1, this volume) for a corresponding discussion of Philip Abrams' distinction between the 'state-system' and the 'state-idea'.

theme of 'absent presence', I argue that a dialectic of state presence and absence reconfigures the relationships between the Queensland state, its subjects, and multinational gas companies. In these relational contexts, gas companies cannot merely rely on the state for their operational and 'social licence'. Conversely, material presented here also demonstrates that a privatisation of risk emerged around CSG developments, which can create significant uncertainties and (re)produce social inequalities among members of host communities. The ambiguous encounters of actors in the spaces in between state presence and absence thus produce a set of interwoven sociocultural and political challenges. The chapter concludes with a critical question regarding the responsibility and accountability of the various actors involved in contemporary resource developments.

The State within CSG Risk Debates

Australia has an identified 257 trillion cubic feet of conventional and unconventional natural gas resources, of which CSG accounts for about one third (AFG 2018). The legal and regulatory framework for the oil and gas sector is broadly divided into onshore and offshore activities. Under Australian law, ownership of onshore gas resources is vested in the Crown, namely the states or Commonwealth. This means that exploration and production titles are granted by respective state and federal governments (Hepburn 2015).[3] Throughout this chapter, I therefore refer to the institutional, legal and regulatory aspects of the Queensland State and, to a lesser extent, the Western Downs local government when discussing 'the state'. One of the main legal instruments regulating the CSG industry's activities is the *Petroleum and Gas (Production and Safety) Act 2004* (Queensland).[4] The Act's main purpose is, among others:

> to facilitate and regulate the carrying out of responsible petroleum activities and the development of a safe, efficient and viable petroleum and fuel gas industry, in a way that (a) manages the State's petroleum resources (i) in a way that has regard to the need for ecologically sustainable development; and (ii) for the benefit of all Queenslanders; and ... (c) creates an effective and efficient regulatory system for the carrying out of petroleum activities and

3   See QSG (2020: 62) for a process map that details the responsibilities and entities throughout the coal seam gas process, from tendering, exploring, producing to decommissioning.
4   See www.legislation.qld.gov.au/view/pdf/current/act-2004-025

the use of petroleum and fuel gas; and ... (f) ensures petroleum
activities are carried on in a way that minimises conflict with other
land uses.

Two important notions, which are commonly found in similar regulatory
instruments, are that publicly owned natural resources are to be developed
for the 'greater good' and that the state assumes the entrusted responsibility
to regulate these developments on behalf of its citizens, including those
with interests in other land uses (Hepburn 2015; Taylor and Hunter 2019;
also Bainton and Skrzypek, Chapter 1, this volume). For landholders in
the Western Downs, as in most other jurisdictions around the world,
this means that subterranean hydrocarbon resources underneath their
properties are the domain of the state and not of their private ownership.

More than any other Australian state, Queensland has followed a pro-
development agenda in support of the CSG industry (Cronshaw and
Quentin Grafton 2016; Espig 2018: 79–84 and 206–7). With resource
extraction being part of a wider Australian 'culture of development
ideology' (Trigger 1997; also Cleary 2011) and a 'growth-first' political
economy (Mercer et al. 2014), the prospect of substantial CSG project
investments was embraced by successive Queensland governments.
As some observers have argued, this included approaching uncertainties
over environmental impacts and risk concerns through 'an attitude
of "technological optimism" ... a confidence that the engineers and
geologists (and markets) will solve whatever disposal or environmental
problems arise' (Edwards 2006: 97). With resource ownership vested in
the Queensland state, landholders had limited authority over whether and
how CSG reserves were developed.

Gas companies that have been granted exploration or production titles are
within their rights to demand access to private properties and can request
the state to enforce this right, subject to a number of legally prescribed
procedures and compensation provisions. For example, access disputes
can be referred to the Queensland Land Court that specifically deals with
matters relating to land and resources. For some landholders, encounters
with the state in the context of CSG developments resemble Ballard and
Banks' (2003: 299) observation of the mining sector:

> Disputes over access to and ownership of mineral resources have
> generated protracted confrontations between the legal apparatus
> of the state and the precepts of local communities, many of which
> first encounter the full power of the state's sovereign claims to
> resources only through this process of dispute.

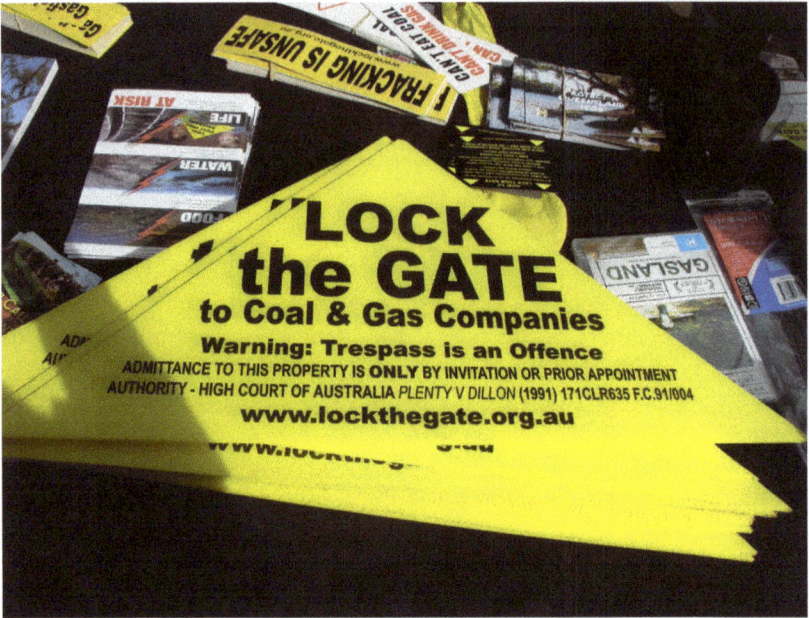

Figure 6.3 Lock the Gate Alliance paraphernalia at a demonstration in Brisbane, Australia, 2015
Source: Photograph by author

While only a relatively small number of cases actually proceeded to Land Court, the mere possibility already asserted enough coercive force in many instances. Acknowledging companies' state-guaranteed rights, several landholders who negotiated access, primarily through Conduct and Compensation Agreements, likened the process to a 'shotgun wedding' rather than voluntary negotiations. As the facilitator of the industry, the state is, in this sense, all too present for regional landholders, whether its sovereign powers are actually enforced or not (see also Holcombe, Chapter 7; and Lewis, Chapter 8, both this volume).[5]

Other members of host communities and the wider Australian public have instead chosen to resist and challenge gas companies and the state. The Lock the Gate Alliance, for instance, is an organisation that promotes a path of civil disobedience by advising landholders to refuse gas and mining companies access to their properties. They provide warning signs

5   Notions of coexistence with agricultural land uses or the industry's social licence to operate—predominantly invoked by staff in CSG companies and some state departments—might therefore first and foremost be about successful land access rather than the establishment of mutually beneficial relationships (Owen and Kemp 2017: 24).

that can be attached to gates and fences, bright yellow triangles that are well known in some parts of Australia—although less so in the Western Downs (see Figure 6.3). The alliance and associated individuals also make regular submissions to CSG and mining-related government bills and appear in regional and national media. Lock the Gate initially formed in the Western Downs with the first wave of opposition to local CSG developments. Since then, key protagonists have left the region and the alliance has shifted its focus and head office into the adjacent state of New South Wales, where proposed CSG projects are still at exploration and planning stages. According to some alliance members, part of this transition has been due to the industry progressing too rapidly in Queensland. This meant that organised resistance only formed once projects were in full development and many landholders had already entered into agreements. However, Makki (2015) investigated the identity politics within the Western Downs and put forward an alternative explanation. He noted that the perceived 'protester' identities of alliance members and associated community groups was incompatible with the predominantly conservative agrarian identities of the region's other residents (also Makki and van Vuuren 2016). Throughout my ethnographic research, I similarly observed that residents' critical engagements with CSG often occurred through forms of 'conforming resistance' instead of active protest; that is, by focusing on (scientific) evidence-making as a means of producing legitimate claims that can influence the decision making and regulation around CSG activities. This has, at least partly, been an attempt to avoid delegitimising 'protester' labels (Espig 2018: 196–204). These forms of resistance emerged in response to what was perceived as unacceptable uncertainties over CSG projects' various impacts. In turn, highlighting these uncertainties became an important aspect of their challenges of the state and gas companies.

Uncertainties are also an essential part of the 'adaptive management' regulatory regime applied to Queensland's CSG developments. This 'learning by doing' framework is a reaction to the pro-development ambitions of Queensland governments to establish a substantial gas export industry while pivotal questions regarding potential impacts were still uncertain. As Taylor and Hunter (2019: 51, footnote omitted) emphasise:

> The dilemmas in creating effective petroleum regulation are the estimation of risks and benefits, which are often based on limited scientific knowledge, particularly in relation to hydraulic fracturing technology. Thus, the regulatory conundrum presents

itself as follows, 'regulating too soon may stifle innovation, but regulating too late may result in significant or irreversible damage'. Such an approach is very much reactive, as the State responds to regulatory issues as they occur, rather than trying to anticipate and legislate for problems prior to the activity taking place. For example in Queensland, regulating unconventional gas activities 'too soon' has led to the search for a regulatory mechanism to manage coexistence.

Swayne (2012: 184) thus noted during the initial rush of CSG activities that 'only time will tell whether the current adaptive approach will be able to protect the Queensland environment from what the Queensland Government acknowledges are the "unknown and unintended impacts" of CSG production'. Uncertainties therefore remained not just with residents in host communities but also within the Queensland state's regulation of CSG projects—albeit in different forms and to varying levels of acceptability.

By focusing on these uncertainties, community members' conforming resistance brought forward more nuanced challenges wherein the epistemic and, ultimately, regulatory authority of gas companies and the Queensland state in relation to CSG developments' associated risks is at stake. Such challenges and their counter-claims are often articulated through debates over the status of scientific research in relation to environmental and health risks—discourses that tie into appealing and powerful tropes of environmental degradation and human rights violations. However, these scientific debates can mask wider conflicts around crucial sociocultural aspects of resource extraction, including envisioned development futures and the relationship between the state, its citizens and multinational companies (Ballard and Banks 2003: 299; Espig 2018: 222–6). The political dimension and critical role of the Queensland state's 'go grow' attitude in this regard is apparent in comparison to other Australian states that followed a 'go slow' approach or have banned onshore unconventional gas developments altogether (Cronshaw and Quentin Grafton 2016; de Rijke 2018). From this vantage point, the next section outlines Western Downs residents' experiences of CSG risk debates with the aim of exploring the wider sociocultural aspects of resource developments and the role of the state within them.

An Arms Race of Sorts: Outpaced and Beyond Capacity

The scale and pace of the CSG boom in the Western Downs played a major role in the narratives of my interlocutors. The authors of an industry review note that companies got 'swept up in a groundswell of enthusiasm and a "get it done at any cost" mentality', resulting in 'an "arms race" of sorts in assuring access to scarce resources' (Reid and Cann 2016: 4, 10). The pace and manner of development created a range of social impacts, from labour shortages to housing pressure, and gave rise to questions around CSG developments' potential impacts and associated uncertainties. These were not merely issues for local community members. Similar views were also expressed by urban-based industry professionals, consultants and researchers who, for instance, discussed the inexperience of some government departments and companies with large-scale unconventional gas developments. Many of these interlocutors were critical of CSG projects' initially fast pace:

> When this whole tsunami started in 2010 nobody, nobody, not even the lawyers, had any experience in what was going to happen and what the impacts were. Yet we are supposed to work out this agreement [with CSG companies] right at the start ... it has been a real problem. (Female landholder and CSG-specific regional NGO representative, Senate hearing, Brisbane, 27 July 2015)

> That's the nightmare about it, it's so fast and that's why they're pushing so hard. (Female landholder with nearby CSG infrastructure, interview, June 2015)

> It's the first time it's been done on this scale. ... It was a bit of a race. Everyone wanted to be first. I speak for the CEOs of the companies, they all wanted to get in before the rules changed. Because it was very favourable early on ... We described it as the big steam roller ... If we don't move fast enough to stay ahead of it, it's just going to get you ... Once they had invested the money, there was no stopping the project. (Male industry-based senior hydrologist, interview, April 2015)

The responsibility for this rapid development lies largely with pro-development Queensland state governments that approved CSG projects and the gas companies that pushed them. The local council in the Western Downs was frequently bypassed (de Rijke 2013b: 15), which is unsurprising given that legislative barriers usually prevent

Australia's local governments from taking more active roles in planning for major resource projects (Barclay et al. 2012).[6] Around the time of the CSG resource boom, Barclay and co-authors (2012: 19) also noted 'a widespread perception that [Australian] state governments are failing to provide adequate resources to assist local governments in meeting the challenges created by rapid expansion in the resources sector'. However, state governments themselves may lack these resources when confronted with multiple large-scale natural resource developments. Some observers have therefore linked the rapid expansion of Queensland's CSG projects to a then already decade-old Australian resource boom that saw governments approve mineral and hydrocarbon developments at a scale and pace that challenged—and possibly exceeded—their own regulatory capacity and policies (Cleary 2011; Taylor and Hunter 2019; cf. Eriksen's 2016, 2018 notion of 'runaway growth processes' more broadly).

Following this rush, many questioned whether the industry and Queensland state departments were able to understand and manage CSG's potential impacts. According to some observers, this was partly due to a long decline of departments' internal capacities and the transitioning of expert public servants into the CSG industry. Such transitions were prompted by departmental staff reductions and financial incentives from better paying industry positions. Concerns thus often involved departments lacking staff and time to adequately assess projects' Environmental Impact Statements (EISs) and an imbalance of scientific research capacities skewed towards CSG companies (see also Holcombe, Chapter 7, this volume). A female public servant in charge of the assessment of one CSG project noted in an informal conversation how she and colleagues often had to 'take things at face value from proponents', which was 'difficult, especially when you don't believe them'. They thus 'felt constantly behind' in their assessment (February 2016). The public servant's comments signal a disparity between the department's entrusted responsibilities of regulatory oversight and actual ability to fulfil this responsibility, perhaps primarily in terms of the

6   Barclay and co-authors (2012: 15) highlight that Australian local governments primarily perform two functions: a service/management role (delivery of local services and collection of taxes) and a political/representative role (represent the interests of their constituents at the local level). In terms of statutory authority over mineral and hydrocarbon developments, the Department of Infrastructure, Local Government and Planning (QSG 2017: 3) clarifies that 'resource development proposals are not assessed against local government planning schemes or by local governments'.

department's capacity rather than its willingness to perform these tasks (see also Skrzypek, Chapter 2, this volume). This perspective resonates with the concerns of other industry and government professionals:

> It outpaced organisations that were far better equipped to deal with that sort of thing than any individual in the community … We [the regional mayor and I] were talking about how many EISs they were having to deal with … He was talking about five or six at any one time, not just five or six in total … And they were worried at the time that they didn't have the capability or the capacity to do it full justice. We got one delivered here … It was 12,000 pages. It was ridiculous. Now tell me of anybody around, particularly regional areas, outside of the highly specialised knowledge of a university or government agency that can even begin to interpret that. (Male regional natural resource manager, interview, April 2015)

> Everybody from our side, from the environmental professional side, were saying 'there's just not enough people to do this job properly for all of these sites' … The three companies would have had the most water experts in any one room at any one time for any government or industry … Government had eight technical people and two managers in all of Queensland … In perspective, we had 32 technical people … So, we were essentially educating the government on what we were going to do with that water. (Male industry-based senior hydrologist, interview, April 2015)

Such limitations are not unique to the context of CSG projects in the Western Downs. State and provincial governments frequently struggle to effectively assess and manage impacts of large-scale resource developments (Measham et al. 2016: 107–8). However, a discrepancy between the state's sovereign power to approve extractive projects and its ability to oversee and responsibly regulate the sector can be problematic, particularly for local residents who are caught up in this disparity.

Legitimate doubts were therefore expressed regarding the initially low levels of scientific certainty of the CSG industry's risks and the ability of various departments to adequately monitor developments, including potentially unknown side effects. Scientific uncertainty was accompanied by normative uncertainties regarding the efficacy of regulatory processes and the state's capacity to enforce regulation. Against a background of rushed projects, these uncertainties became a point of contestation, with more prior scientific research or a slower pace of development being viable alternative pathways. The rapid progress of CSG projects and associated unknowns were not accepted by a number of actors who argued that many uncertainties could, and should, have been avoided (see more broadly Wehling 2006: 127–31; Beck and Wehling 2012: 40). Citizens' trust in the Queensland state responsibly facilitating CSG developments was therefore, among other things, contingent upon its perceived regulatory capacities. Unsurprisingly, some local community members who were confronted with uncertainties around potential impacts on their future agricultural livelihoods or health demanded answers to open questions. However, the responses they received were often unsatisfactory.

No Straight Answer: The Encounter of Organised Irresponsibility

In searching for certainty, a number of interlocutors noted that the responses they received to their questions from government or company staff were often limited to their area of expertise and assigned responsibility. Some were advised to instead contact another department or unit. As a result, these community members felt that they were sent around in circles and were not satisfied with what they regarded as specific questions remaining essentially unanswered. A sense of frustration was frequently palpable:

> The response that we got to a seven-page letter detailing all of the issues ... was embarrassing, particularly with regard to things like very specific information we've given them about actual harm ... It was three pages versus my seven pages. It's this: at arm's length, no actual depth, no investigation ... Please [government department] tell us what happened, please advise us? They have told us nothing ... And it's in free fall, hands off the wheel, choose your own adventure. (Female landholder with nearby CSG infrastructure, property tour, April 2015)

> And you ask them [a government department] questions about health impacts of CSG and they ask you a question for the answer. (Male regional town resident and church minister, interview May 2015)

> Who is taking responsibility for collating all this stuff? ... Somebody needs to be responsible ... So, there are four or five government departments in there conflicting and

belting each other up … Half the government agencies have skin in the game in this space and no one is ultimately saying or doing anything about the problems … What he [local grazier] is concerned about is the way the department has written that has made it so ambiguous, so farcical that it doesn't make any sense. Hence, he gets cranky because he doesn't understand it—none of us understand it because the letter is a pile of rubbish … And how poles apart are Natural Resources and Mines and Queensland Health? … So, then you throw in the middle of that Environment and Heritage Protection, then Agriculture … to muddy the waters. (Male state politician, interview, April 2015)

For citizens seeking clear answers to questions regarding risks associated with CSG developments, these perceptions of Queensland state departments' lack of responsibility can increase uncertainties, lead to mistrust—itself a form of moral uncertainty—and alienate them from government structures (Wynne 1996: 42; Gill 2011).[7] The above observations were echoed in a subsequent review of two Queensland departments with CSG-specific regulatory responsibilities and one entity with a legislated oversight role to the state's regulatory framework (QSG 2020).

From the perspective of concerned local community members, I interpret these findings by borrowing from sociologist Ulrich Beck who argues that in contemporary ecological risk debates '[r]esponsibilities can indeed be assigned but they are spread out over several social subsystems' (Beck 2009: 193). This, he contends, leads to 'responsibility *as* impunity or: organized irresponsibility' wherein 'no individual or institution seems to be held specifically accountable for anything' (ibid.: 194, original emphasis; also Douglas and Wildavsky 1982: 79–82; Dwyer and Minnegal 2006: 10–2 and 20). This organised irresponsibility stands in contrast to the Queensland state's authority to approve and regulate resource developments. As a result, concerned members of host communities and the wider public who submitted complaints were left searching for definite answers and advice. They questioned the perceived lack of government oversight, which caused problems of unclear responsibilities and ultimately accountability. This led to further normative uncertainties regarding the role of state departments and other entities with legislated oversight (e.g. the GasFields Commission Queensland). It is in these

---

7   Cf. Auyero and Swistun's (2008: 369–71) notion of the social production of toxic uncertainty through, among other aspects, state officials' 'labor of confusion and state (mis)interventions'. See also Stilgoe (2007).

liminal spaces, in between state presence and absence, where concerned community members who experience the rapid environmental and social transformations of a natural resource boom can encounter what Bainton and Skrzypek (Chapter 1, this volume) describe as the absent presence of the state. Here, everyday perceptions and practices, such as moving in circles between state departments, are important modes through which the state comes into being for concerned community members, mainly in the form of a heightened sense of its perceived failure to fulfil regulatory responsibilities (Sharma and Gupta 2005).

Considering the difficulties in forecasting impacts and projects' rushed progress, it is debatable whether definite answers could have been given to some of the questions raised by community members.[8] Several industry professionals also emphasised the indeterminacies of CSG developments; for example, exact infrastructure locations or the ultimate numbers of wells in a field can often not be determined at the onset but are influenced by outcomes of ongoing extraction. Other interlocutors appeared somewhat satisfied with the information and conflict resolution processes offered by, for instance, the Queensland Government's CSG Compliance Unit. However, for many concerned community members, perceived irresponsibilities have resulted in a diminishing trust in the relevant governing structures (Lloyd et al. 2013; Gillespie et al. 2016). Some addressed the perceived absence of adequate oversight by either individually or collectively trying to scrutinise the industry and government. One landholder argued at a community meeting that CSG companies 'need to be made accountable. No one else out here is going to do that. It's up to us, the community, to do that, because there simply is no other overseer. There should be, but there isn't' (April 2015). A grazier and representative of a regional CSG-focused NGO similarly criticised the Queensland Government's Mineral and Energy Resources (Common Provision) Bill 2014 that temporarily restricted public objection rights to extractive projects by noting, 'We are the watchdogs for our democracy and don't let any politician tell you otherwise' (group discussion April 2015; see Espig and de Rijke 2018: 220–1).

---

8   In contexts of uncertainties and unknowns one might also not be willing to accept responsibility and refuse to give a straight answer. Povinelli's (1993: 685) analysis of the language of indeterminacy in an Australian Aboriginal community is insightful, particularly 'the relationship among knowledge-claims, responsibility-culpability, and authority-status'. See also Gluckman (1972).

Such a felt absence of government oversight can, in tandem with scientific uncertainties and a heightened sense of mistrust, lead to insecurities and frustration among local residents and landholders (Checker 2007; Beck 2009: 45–6). Some citizens and groups find themselves compelled to scrutinise the CSG industry themselves, which can create privatised forms of risk management (Jasanoff 2002: 375; Rose 2005: 158; Pyysiäinen et al. 2017). However, whether scrutiny can be exercised effectively in light of the techno-scientific and ecological complexities of CSG developments, and by whom, must be critically evaluated.

Displacement, Responsibilisation and the Absent State

The two preceding sections portray instances of an absent presence of the state, within which securing investments and economic benefits might be perceived by some as taking priority over community members' risk concerns. A variety of local and 'outside' interlocutors (cf. Espig and de Rijke 2018: 220–1) attempted to assume regulatory responsibilities by challenging gas companies and the Queensland state's adaptive management regime through the means available to them, from conforming resistance to outright civil disobedience. Many concerned residents were not satisfied with a circular transfer of responsibility between companies and impotent state departments through, for instance, reference to regulation-compliant behaviour. As a female attendee at a company-organised gas facility tour emphasised in her response to a CSG company representative, '[p]eople want answers and details. It's not going to help to say, "We comply"' (July 2015). Such 'discourses of compliance' (Stilgoe 2005) can turn questions concerning unknowns into issues of measurement and compliance to set guidelines. These guidelines represent what is known about particular hazards and politically acceptable regulation at a given moment in time. Discourses that stress compliance with such guidelines might then create cognitive and social authority, but they can potentially obscure uncertainties and demarcate unknowns from further consideration. Reference to compliance also allows regulatory agencies to manage contested issues solely in scientific terms, which excludes non-scientific forms of knowledge from decision-making processes. Actors whose concerns remain unanswered, or are not even acknowledged, might consequently feel prompted to challenge the foundations of regulatory processes. The efficacy of discourses of compliance can thus be limited, especially when the underlying adequacy of guidelines is questioned (Stilgoe 2005, 2007).

Similar problems emerge when Queensland state departments passed on complaints about misconduct or impacts to companies for investigation. As a regional town resident explained:

> It's not just a blind mistrust. The mistrust is because we've seen it over and over again: here's my complaint and I'm giving it to [government departments]. They take it and go to bloody [company name] and say, 'Here's this complaint'. So, they hand it over and [company name] write their report, hand it back to them and then they take it back to the landowner saying, 'There's the report, everything is fine' … Self-regulation, self-reporting— it doesn't work. (Opposing female town resident, interview, May 2015)

Such instances constitute what Trouillot (2003: 90) describes as a displacement of state functions from governments to, in this case, private transnational companies (also Sharma and Gupta 2005: 21–2). In the context of the extractive industries, examples of such practices include the delegation of regulatory responsibilities to resource companies and their self-regulation through corporate social responsibility or voluntary industry initiatives (Brereton 2003; Dashwood 2012; Owen and Kemp 2017; also Bainton and Skrzypek, Chapter 1, this volume). One implication is that, within host communities, residents' uncertainties and distrust can increase when compliance is associated with irresponsibility among state departments as well as adaptive management regimes being reliant on company expertise and self-regulation. Unsurprisingly, concerned Western Downs residents and their allies in the wider public often contested the scientific foundation of state regulations and company conduct.

The resources they have at their disposal to put pressure on CSG companies and the government vary, though. The relatively wealthy and politically well-represented farmers in the Western Downs' eastern cotton growing area mounted a formidable challenge by investing considerable amounts of money into independent hydrological studies that demonstrated the limited understanding of interconnectivity between agriculturally used aquifers and those targeted by CSG companies. This kept proposed developments off prime agricultural land and prompted lengthy research trials. Small-scale residential landholders further west, on the other hand, found themselves in the midst of multiple gas fields. Among those living on so-called 'lifestyle blocks', located on agriculturally less valuable land, are members of vulnerable and lower income groups who could hardly afford a few hundred dollars to have water in their tanks or dams tested (Makki 2015; Makki and van Vuuren 2016).[9] The effectiveness of their challenges has therefore generally been lower than that of other groups in the Western Downs. I interpret this as the result of an interplay between multiple, unequally distributed financial, social and political 'capitals' (Bourdieu 1986; Everingham et al. 2014, 2015; QSG 2020). Residents' ability to mobilise such capitals influences experiences of regional change and the uncertainties associated with CSG developments by allowing some to develop a sense of agency and beneficial engagement with the industry. It seems that actors and groups with greater levels of pre-existing capitals tend to have benefited to a higher degree from CSG projects and were able to negotiate better compensation agreements, have a stronger say in where infrastructure is located on their property, or could avoid certain impacts (Measham et al. 2016: 107–8). So, while concerned Western Downs residents from a range of socioeconomic and cultural backgrounds challenged companies and the Queensland state, the outcomes they were able to achieve varied. This led to different experiences of CSG developments and of their encounters with the state.

The observation that individual community members become self-reliant on their own capabilities and capitals to secure their welfare is consistent with a broader neoliberal shift that has occurred in Australia's agricultural regions since the 1970s. This policy shift includes the private responsibilisation of farmers as individual, economically independent entrepreneurs and responsibility-taking agents (Pyysiäinen et al. 2017). In the context of Pennsylvania's shale gas projects (another form of unconventional natural gas), Malin (2014) critiqued similar developments as imposing a form of neoliberal individualisation and logic onto farmers. In debates over the complex environmental and health risks associated with CSG extraction, such an individualisation creates new demands and responsibilities for concerned community members. For some, profound limitations emerge if, for example, they are unable to conduct

---

9   There are some 2,100 rural subdivisions, or 'lifestyle blocks', between Tara and Chinchilla, typically ranging from 13–40ha. Established since the 1980s on poor agricultural land, these subdivisions are primarily affordable residential properties. Here, the civil disobedience movement against CSG developments began (see Makki 2015).

or afford scientific testing for suspected air pollution from nearby CSG activities. These residents and interested outsiders often face higher levels of uncertainty and what risk scholars describe as a decline in ontological security. As Beck (2009: 45–6 and 195, original emphasis) emphasises within his risk society thesis:

> a dramatic decline of ontological security now confronts lifeworlds, even in the peaceful corners of the earth. The three pillars of security are crumbling—the state, science and the economy … and are naming the 'self-conscious citizen' as their legal heir. But how are individuals supposed to accomplish what state, sciences and economic enterprises are unable to achieve? … The brutal fact of ontological insecurity always has an ultimate addressee: … the *individual*. Whatever propels risk and makes it incalculable … shifts the ultimate decision-making responsibility onto the individuals, who are ultimately left to their own devices with their partial and biased knowledge, with undecidability and multiple layers of uncertainty.

Similarly, Giddens (1991: 131, original emphasis) notes 'the *inevitability* of living with dangers which are *remote* from the control not only of individuals, but also of large organisations, including states'. This uncontrollability can lead to a 'sense of dread which is the antithesis of basic trust' (ibid.: 133). Notwithstanding the problematic generalising tendencies of these grand theoretical claims (see Dwyer and Minnegal 2006), Beck and Giddens highlight an often underappreciated aspect of what it might mean to live with the insecurities of complex 'manufactured risks'; that is, those predominantly created by the increasing scale of humans' application of techno-scientific capabilities. Negative psychological, sociocultural and political ramifications of risk and uncertainty can be observed especially within contexts where citizens perceive a disparity between the regulatory mandate of the state and its capacity to fulfil those responsibilities.

However, a retreat of the state and corresponding privatisation of risk management might also create opportunities for the reconfiguration of the power relationships between the state and its subjects. Rose (2005: 159) contends that the active citizens of 'advanced' liberal democracies can create new demands and challenges towards those in positions of authority:

Here we can witness the 'reversibility' of relations of authority—what starts off as a norm to be implanted into citizens can be repossessed as a demand which citizens can make of authorities. Individuals are to become 'experts of themselves', to adopt an educated and knowledgeable relation of self-care in respect of their bodies, their minds, their forms of conduct and that of the members of their own families.

Such challenges cannot only be established in relation to the private self but also in terms of wider societal debates, such as within environmental risk controversies. As indicated above, engaged landholders who could mobilise their various capitals effectively were able to create significant obstacles for gas projects. In the case of CSG developments in the adjacent state of New South Wales, the combination of local resistance and a national anti-gas movement was partly responsible for prompting the state government to even declare a moratorium on unconventional gas developments (Cronshaw and Quentin Grafton 2016; de Rijke 2018).[10]

These cases indicate that driven individuals, well-organised groups and international networks of interest can indeed challenge states and extractive companies in response to what they experience as an absence of adequate government regulation and a subsequent privatisation of risk management. However, this does not answer the questions of whether those who can bring forth such challenges should have to assume this responsibility, what eventually constitutes successful challenges, and whether their claims can be articulated in time before rapidly progressing projects create realities on the ground that foreclose alternative development pathways. It is thus possible that one element of conforming resistance is a 'neoliberal "unloading" of public services onto empowered and "responsibilized" selves and communities who ... are thereby made complicit in the contemporary workings of power and governance'

---

10   Similar challenges can also be found in other extractive sectors. Stuart Kirsch (2014, 2018), for instance, describes how Yonggom community members and a network of international NGOs and sympathisers were able to successfully sue BHP in Australia over the environmental damages caused by riverine tailings disposal at their Ok Tedi copper and gold mine in Papua New Guinea.

(Sharma and Gupta 2005: 21).[11] Local community members who are already vulnerable, marginalised or in socioeconomically weaker positions all too often find themselves on the downside of the accompanying individualisation of risk. The responsibilisation of the self can, in this sense, lead to the inconspicuous (re)production of social inequalities and unequal risk positions. As Jasanoff (2015: 17) notes, '[i]f networks diffuse responsibility, they can also depoliticize power by making its actions opaque or invisible'. With limited means to respond to what community members regard as the transfer of responsibility for their environments and health, their options are restricted to either moving away—as a number of interlocutors did—or to protest with their physical bodies against CSG development as a last resort. In any way, challenging regulatory institutions and those in positions of power and authority never occurs without significant personal investment, effort and sacrifice. Many individuals and groups can therefore only temporarily resist the pressures that arise from a perceived absence of the state.

Conclusion

In this chapter, I discussed concerned host community members' experiences of the state regarding CSG developments in the Western Downs. This ethnographic account should not be read as a thorough analysis of the Queensland state's regulatory instruments and agencies. Neither does it address the experiences of residents supporting CSG development or those who were not concerned about associated risks (see Espig 2018: 141–66). Here, I focused specifically on community members for whom CSG projects posed a clear and present threat. While I do not suggest that the Queensland state should be considered as 'failed' in this regard—especially in comparison to other jurisdictions discussed in this volume and elsewhere—my aim was to demonstrate that marginal

---

11 Pyysiäinen and co-authors (2017) present a critical analysis of Australian farmers' reactions to neoliberal 'responsibilisation' that provides complementary insights into individuals' experiences of the state. Contrary to neoliberal theory, most farmers they surveyed did see the causal explanations for their situation in external structural factors, such as government policies or macroeconomic conditions. As neoliberal theorists argue, however, a majority reacted with an internally-oriented response, or 'psychological reactance', wherein they took on the responsibility of actively regaining control and changing their circumstances, while only a smaller group demanded structural changes. These observations are insightful for the presented ethnographic findings: one might interpret individuals' or groups' attempts to regain control and 'hold the industry accountable' as an internalisation of responsibility in reaction to a perceived absence of satisfactory state regulation.

spaces of partial state presence do not only emerge in relatively recent postcolonial, new or 'weak' states (Das and Poole 2004: 4). Instead, a dialectic of presence and absence might also manifest within highly industrialised and, at first sight, well-regulated democratic states. The experiences of concerned community members in the Western Downs highlight that a perceived lack of regulatory oversight can become all too present in their daily encounters with the state, which shows that 'phenomena may have a powerful presence in people's lives precisely *because* of their absence' (Bille et al. 2010: 4, original emphasis). This dialectic of presence and absence is part of an ongoing reconfiguration of the relationships between the Queensland state, its subjects and private, multinational gas companies. In this sense, CSG risk debates do not just concern ecological hazards in a narrow sense but constitute a broader commentary on the social world (Espig 2018: 222–6) and the sociocultural transformations associated with large-scale natural resource developments.

While Ballard and Banks (2003) emphasise that, as part of these reconfigurations, local communities encounter the full power of the state's sovereign claims to natural resources, they also recognise the countervailing powers communities can mobilise. The outlined cases of resistance to CSG developments detail how states and private companies can face formidable challenges from local community members. This has critical implications for the relationship between citizens and the extractive industries. Writing about mining contexts, Kirsch (2014: 32–3) suggests that states' limited presence, especially in rural areas, brings extractive companies into closer contact with surrounding communities. As companies assume state responsibilities, however, they might also be held accountable for shortcomings of the state more broadly (also Owen and Kemp 2017: 187). This means that:

> The mining industry can no longer assume that the state is its only negotiating partner or that rural or indigenous communities lack the resources or capacity to challenge their operations, whether by forging international alliances with NGOs and other partners, through legal action in the countries in which the corporations are based, or by force—albeit at a terrible cost. (Kirsch 2014: 232)

My ethnographic account of Queensland's CSG developments supports this assessment, with the important caveat that the 'reversibility of relations of authority' (Rose 2005: 159) is contingent upon community members' various capitals and capacities that enable or constrain them from formulating challenges to the state or private companies. These capitals may be considerably higher for many citizens in the Western Downs, compared to other communities discussed in this volume. Across these diverse contexts, however, it is important to pay attention to social inequalities, power imbalances and unequal risk positions that can be (re)produced in local responses to a limited state presence.

In outlining the dialectic of presence and absence within CSG risk debates, I discussed the encounters between a subset of host community members and the Queensland state. The ethnographic account of their experiences points towards the roles citizens might assume within resource development contexts that are seemingly driven by neo- or advanced liberal policies. Considering the ecological and techno-scientific complexities of CSG developments, I echo Beck's (2009: 45–6 and 195) concern over whether individual citizens can close the gap left by, at least partially, absent states. This is, of course, not to suggest that all citizens perceive such a gap or are equally affected by it, nor to follow Beck's (ibid.) gloomy outlook that they are 'ultimately left to their own devices'. Nonetheless, it is important to critically consider what happens when, for some, states appear to fail in their task to safeguard against risks and harm, especially against a backdrop of an uneven distribution of different forms of uncertainty, risk and sociocultural as well as economic capitals. What, then, are the responsibilities and accountabilities of state departments, community members, private companies, NGOs, universities or research institutes in regard to the risks of extractive projects such as CSG developments? Drawing out some of the challenges local community members experienced around the ongoing reconfiguration of the public and private within the dialectic of the Queensland state's presence and absence can contribute towards addressing this pressing question.

References

AFG (Australian Federal Government), 2018. 'Australian Energy Resources Assessment: Gas.' Canberra: Geoscience Australia.

Auyero, J. and D. Swistun, 2008. 'The Social Production of Toxic Uncertainty.' *American Sociological Review* 73: 357–379. doi.org/10.1177/0003122408 07300301

Bainton, N. and J.R. Owen, 2018. 'Zones of Entanglement: Researching Mining Arenas in Melanesia and Beyond.' *Extractive Industries and Society* 6: 767–774. doi.org/10.1016/j.exis.2018.08.012

Ballard, C. and G. Banks, 2003. 'Resources Wars: The Anthropology of Mining.' *Annual Review of Anthropology* 32: 297–313. doi.org/10.1146/annurev.anthro.32.061002.093116

Barclay, M.A., J. Everingham, L. Cheshire, D. Brereton, C. Pattenden and G. Lawrence, 2012. 'Local Government, Mining Companies and Resource Development in Regional Australia: Meeting the Governance Challenge.' St Lucia: University of Queensland, Centre for Social Responsibility in Mining.

Beck, U., 2009. *World at Risk*. Cambridge and Malden: Polity Press.

Beck, U. and P. Wehling, 2012. 'The Politics of Non-Knowing: An Emerging Area of Social and Political Conflict in Reflexive Modernity.' In F.D. Rubio and P. Baert (eds), *The Politics of Knowledge*. London: Routledge. doi.org/10.4324/9780203877746

Bille, M., F. Hastrup and T.F. Sørensen (eds), 2010. *An Anthropology of Absence: Materializations of Transcendence and Loss*. New York: Springer. doi.org/10.1007/978-1-4419-5529-6

Bourdieu, P., 1986. 'The Forms of Capital.' In J. Richardson (ed.), *Handbook of Theory and Research for the Sociology of Education*. Westport (CT): Greenwood Press.

Brereton, D., 2003. 'Self-Regulation of Environmental and Social Performance in the Australian Mining Industry.' *Environmental and Planning Law Journal* 20: 261–274.

Checker, M., 2007. '"But I Know It's True": Environmental Risk Assessment, Justice, and Anthropology.' *Human Organization* 66: 112–124. doi.org/10.17730/humo.66.2.1582262175731728

Cleary, P., 2011. *Too Much Luck: The Mining Boom and Australia's Future*. Collingwood (Vic.): Black Inc.

Cronshaw, I. and R. Quentin Grafton, 2016. 'A Tale of Two States: Development and Regulation of Coal Bed Methane Extraction in Queensland and New South Wales, Australia.' *Resources Policy* 50: 253–263. doi.org/10.1016/j.resourpol.2016.10.007

Das, V. and D. Poole (eds), 2004. *Anthropology in the Margins of the State*. Oxford and New York: Oxford University Press.

Dashwood, H., 2012. *The Rise of Global Corporate Social Responsibility: Mining and the Spread of Global Norms.* Cambridge: Cambridge University Press. doi.org/10.1017/CBO9781139058933

De Rijke, K., 2013a. 'The Agri-Gas Fields of Australia: Black Soil, Food, and Unconventional Gas.' *Culture, Agriculture, Food and Environment*, 35(1): 41–53. doi.org/10.1111/cuag.12004

———, 2013b. 'Hydraulically Fractured: Unconventional Gas and Anthropology.' *Anthropology Today* 29(2): 15–17. doi.org/10.1111/1467-8322.12017

———, 2018. 'Drilling Down Comparatively: Resource Histories, Subterranean Unconventional Gas and Diverging Social Responses in Two Australian Regions.' In R.J. Pijpers and T.H. Eriksen (eds), *Mining Encounters: Extractive industries in an overheated world.* London: Pluto Press. doi.org/10.2307/j.ctv893jxv.10

Douglas, M. and A. Wildavsky, 1982. *Risk and Culture.* Berkeley: University of California Press.

Dwyer, P.D. and M. Minnegal, 2006. 'The Good, the Bad and the Ugly: Risk, Uncertainty and Decision-Making by Victorian Fishers.' *Journal of Political Ecology* 13: 1–23. doi.org/10.2458/v13i1.21675

Edwards, G., 2006. 'Is There a Drop to Drink? An Issue Paper on the Management of Water Co-Produced with Coal Seam Gas.' Brisbane: Queensland Department of Mines and Energy.

Eriksen, T.H., 2016. *Overheating: An Anthropology of Accelerated Change.* London: Pluto Press.

———, 2018. *Boomtown: Runaway Globalisation on the Queensland Coast.* London: Pluto Press. doi.org/10.2307/j.ctv3mt8xn

Espig, M., 2018. Getting the Science Right: Queensland's Coal Seam Gas Development and the Engagement with Knowledge, Uncertainty and Environmental Risks. St Lucia: University of Queensland (PhD thesis). doi.org/10.14264/uql.2018.345

Espig, M. and K. de Rijke, 2016. 'Unconventional Gas Developments and the Politics of Risk and Knowledge in Australia.' *Energy Research and Social Science* 20: 82–90. doi.org/10.1016/j.erss.2016.06.001

———, 2018. 'Energy, Anthropology and Ethnography: On the Challenges of Studying Unconventional Gas Developments in Australia.' *Energy Research and Social Science* 45: 214–223. doi.org/10.1016/j.erss.2018.05.004

Everingham, J.A., N. Collins, W. Rifkin, D. Rodrigues, T. Baumgartl, J. Cavaye and S. Vink, 2014. 'How Farmers, Graziers, Miners, and Gas-Industry Personnel See Their Potential for Coexistence in Rural Queensland.' *SPE Economics and Management* 6(3): 122–130. doi.org/10.2118/167016-PA

Everingham, J.A., V. Devenin and N. Collins, 2015. '"The Beast Doesn't Stop": The Resource Boom and Changes in the Social Space of the Darling Downs.' *Rural Society* 24: 42–64. doi.org/10.1080/10371656.2014.1001480

GCQ (GasFields Commission Queensland), 2018. 'Queensland's Petroleum and Gas Industry Snapshot.' Brisbane: GCQ.

Giddens, A., 1991. *The Consequences of Modernity.* Cambridge: Polity Press.

Gill, F., 2011. 'Responsible Agents: Responsibility and the Changing Relationship Between Farmers and the State.' *Rural Society* 20: 128–141. doi.org/10.5172/rsj.20.2.128

Gillespie, N., C. Bond, V. Downs and J. Staggs, 2016. 'Stakeholder Trust in the Queensland CSG Industry.' *Australia Petroleum Production and Exploitation Journal* 56: 239–246. doi.org/10.1071/AJ15018

Gluckman, M. (ed.), 1972. *The Allocation of Responsibility.* Manchester: Manchester University Press.

Hepburn, S., 2015. *Mining and Energy Law.* Melbourne: Cambridge University Press. doi.org/10.1017/CBO9781107480025

Jasanoff, S., 2002. 'Citizens at Risk: Cultures of Modernity in the US and EU.' *Science as Culture* 11: 363–380. doi.org/10.1080/0950543022000005087

———, 2015. 'Future Imperfect: Science, Technology, and the Imaginations of Modernity.' In S. Jasanoff and S.H. Kim (eds), *Dreamscapes of Modernity: Sociotechnical Imaginaries and the Fabrication of Power.* Chicago: University of Chicago Press. doi.org/10.7208/chicago/9780226276663.001.0001

Kirsch, S., 2014. *Mining Capitalism: The Relationship between Corporations and their Critics.* Oakland: University of California Press.

———, 2018. *Engaged Anthropology: Politics beyond the Text.* Oakland: University of California Press.

Lloyd, D.J., H. Luke and W.E. Boyd, 2013. 'Community Perspectives of Natural Resource Extraction: Coal-Seam Gas Mining and Social Identity in Eastern Australia.' *Coolabah* 10: 144–164.

Makki, M., 2015. Coal Seam Gas Development and Community Conflict: A Comparative Study of Community Responses to Coal Seam Gas Development in Chinchilla and Tara, Queensland. St Lucia: University of Queensland (PhD thesis). doi.org/10.14264/uql.2015.1105

Makki, M. and K. van Vuuren, 2016. 'Place, Identity and Stigma: Blocks and the "Blockies" of Tara, Queensland, Australia.' *GeoJournal* 82(6): 1–15. doi.org/10.1007/s10708-016-9730-2

Malin, S., 2014. 'There's No Real Choice But to Sign: Neoliberalization and Normalization of Hydraulic Fracturing on Pennsylvania Farmland.' *Journal of Environmental Studies and Sciences* 4: 17–27. doi.org/10.1007/s13412-013-0115-2

Measham, T.G., D.A. Fleming and H. Schandl, 2016. 'A Conceptual Model of the Socioeconomic Impacts of Unconventional Fossil Fuel Extraction.' *Global Environmental Change* 36: 101–110. doi.org/10.1016/j.gloenvcha.2015.12.002

Mercer, A., K. de Rijke and W. Dressler, 2014. 'Silences in the Boom: Coal Seam Gas, Neoliberalizing Discourse, and the Future of Regional Australia.' *Journal of Political Ecology* 21: 279–302. doi.org/10.2458/v21i1.21137

Ortner, S.B., 1995. 'Resistance and the Problem of Ethnographic Refusal.' *Comparative Studies in Society and History* 37: 173–193. doi.org/10.1017/S0010417500019587

Owen, J.R. and D. Kemp, 2017. *Extractive Relations: Countervailing Power and the Global Mining Industry.* London: Routledge.

Povinelli, E.A., 1993. '"Might Be Something": The Language of Indeterminacy in Australian Aboriginal Land Use.' *Man* 28: 679–704.

Pyysiäinen, J., D. Halpin and A. Guilfoyle, 2017. 'Neoliberal Governance and "Responsibilization" of Agents: Reassessing the Mechanisms of Responsibility-Shift in Neoliberal Discursive Environments.' *Distinktion* 18: 215–235. doi.org/10.1080/1600910X.2017.1331858

QSG (Queensland State Government), 2017. 'Mining and Extractive Resources.' Brisbane: Department of Infrastructure, Local Government and Planning.

———, 2020. 'Managing Coal Seam Gas Activities.' Brisbane: State Audit Office (Report 12: 2019–20).

Reid, S. and G. Cann, 2016. 'The Good, the Bad and the Ugly: The Changing Face of Australia's LNG Production.' Deloitte Australia.

Rose, N., 2005. 'Governing "Advanced" Liberal Democracies.' In A. Sharma and A. Gupta (eds), *The Anthropology of the State: A Reader*. Oxford: Blackwell.

Sharma, A. and A. Gupta, 2005. 'Introduction: Rethinking Theories of the State in an Age of Globalization.' In A. Sharma and A. Gupta (eds), *The Anthropology of the State: A Reader*. Oxford: Blackwell.

Stilgoe, J., 2005. 'Controlling Mobile Phone Health Risks in the UK: A Fragile Discourse of Compliance.' *Science and Public Policy* 32: 55–64. doi.org/10.3152/147154305781779704

———, 2007. 'The (Co-)Production of Public Uncertainty: UK Scientific Advice on Mobile Phone Health Risks.' *Public Understanding of Science* 16: 45–61. doi.org/10.1177/0963662506059262

Swayne, N., 2012. 'Regulating Coal Seam Gas in Queensland: Lessons in An Adaptive Environment Management Approach.' *Environmental Planning and Law Journal* 29: 163–185. eprints.qut.edu.au/49293/

Taylor, M. and T. Hunter, 2019. *Agricultural Land Use and Natural Gas Extraction Conflicts: A Global Socio-Legal Perspective*. Abingdon: Routledge. doi.org/10.4324/9780203702178

Trigger, D., 1997. 'Mining, Landscape and the Culture of Development Ideology in Australia.' *Cultural Geographies* 4: 161–180. doi.org/10.1177/147447409700400203

———, 2000. 'Aboriginal Responses to Mining in Australia: Economic Aspirations, Cultural Revival, and the Politics of Indigenous Protest.' In P. Schweitzer, M. Biesele and R. Hitchcock (eds), *Hunters and Gatherers in the Modern World: Conflict, Resistance, and Self-Determination*. New York: Berghahn.

Trouillot, M.-R., 2003. *Global Transformations: Anthropology and the Modern World*. New York: Palgrave Macmillan. doi.org/10.1007/978-1-137-04144-9

Wehling, P., 2006. *Im Schatten des Wissens? Perspektiven der Soziologie des Nichtwissens [In the Shadow of Knowledge? Perspectives of the Sociology of Nonknowledge]*. Konstanz: UVK Verlagsgesellschaft mbH. doi.org/10.1007/s11614-008-0007-y

Wynne, B., 1996. 'Misunderstood Misunderstandings: Social Identities and Public Uptake of Science.' In A. Irwin and B. Wynne (eds), *Misunderstanding Science? The Public Reconstruction of Science and Technology*. Cambridge: Cambridge University Press. doi.org/10.1017/CBO9780511563737

7

The State's Selective Absence: Extractive Capitalism, Mining Juniors and Indigenous Interests in the Northern Territory

Sarah Holcombe

Introduction

Junior mining companies are an elusive group. They are by far the most numerically significant mining companies to operate globally, but most of them have a limited public profile and very few are publicly listed. Unlike the majority of major mining companies and some mid-tier companies, junior miners are unlikely to subscribe to international standards for corporate social responsibility (CSR), nor are they likely to have the internal resources and capability to demonstrate alignment with these standards.[1] We know comparatively little about how junior miners engage with local and prior interests, especially those of Indigenous people.

---

1   International standards for the extractive industry include the International Finance Corporation (IFC) Performance Standards, and those that the International Council for Mining and Metals (ICMM) members endorse, which mirror the IFC standards. To gain membership of the ICMM, companies have to commit to establishing particular policies and procedures to ensure that they align with the standards. However, compliance measures to ensure implementation of these policies is a significant gap.

For instance, do they simply follow the minimum regulatory standard for the jurisdiction and, if so, are these adequate for the task of ensuring that Indigenous peoples can access their rights and interests?

While there is no set definition for a junior mining company, 'juniors' may be defined by their size or by their role within the industry (e.g. as companies that discover and develop new resources to the point at which they can be taken over by larger operators). The latter appears to be a function of the decision by many of the 'senior' companies to cut back on exploration spending around the turn of the millennium. However, what matters for the purpose of this chapter is the way that juniors are defined by their entitlement to specific state subsidies; yet they typically underperform in terms of social and environmental responsibility compared to larger mining companies (Dougherty 2013: 3; Lyons et al. 2016). Noting, of course, the obvious caveat that major companies do not always meet their own CSR commitments.

This chapter will explore the case of one such company seeking to operate in the Northern Territory (NT) of Australia, which, for ethical reasons I will refer to as 'Top End Mining', as they are still seeking approval to develop a mine. I argue that a focus on junior miners provides an important (and largely overlooked) window into the strategic presence and absence of the state in resource extraction contexts, and helps to reveal the structural factors that shape the relationship between miners and Indigenous communities. In this chapter, I set out to achieve three things. In the first part of the chapter, I provide an overview of this particular mining project, and the key features of junior miners, addressing a gap in anthropology of mining literature (see Bainton 2020). I then consider the regulatory limitations within the environmental and social impact assessment processes in the NT, in comparison with other Australian jurisdictions, and the way that junior mining companies have been able to take advantage of this 'absence' to the detriment of Aboriginal communities in mine-affected areas. This provides the basis to consider the disconnection between this impact assessment process and the agreement-making process for which provision is made under the *Aboriginal Land Rights (Northern Territory) Act 1976* (Cth) and subsequent native title legislation. This disconnection is by no means unique to the NT; it is also found in other parts of Australia, and in Papua New Guinea.

The political context of the NT includes the regulatory frameworks at state and federal levels; local-level agreements between Indigenous groups, the mining company, Indigenous representative bodies; and, importantly, the structural inequality that frames this 'Top End' remote Indigenous region. The broader sociopolitical context is framed by the paradigm of extractive capitalism, exemplified by the Australian Government's *White Paper on Developing Northern Australia* (AFG 2015). This narrative draws on the language of developing the 'untapped promise [of], abundant resources', 'enabling infrastructure [for] agriculture, aquaculture and previously stranded energy and minerals resources' and 'making it easier to use natural assets' (2015: 4–5). While there is a nod towards consulting with Indigenous interests, Indigenous people are regarded as another set of stakeholders rather than a distinct majority interest (in remote areas) in this development approach, which has been robustly critiqued (Morrison 2017; Chambers et al. 2018; Archer et al. 2019).[2] Likewise, neither does this approach to Indigenous interests specifically recognise or focus on the distinct and unique cultural and social conditions of this Indigenous domain, as the mainstream or standard approach to permitting and approvals processes for extractive projects is transplanted to this context.

In this chapter I am especially concerned with the limitations of the current 'environmental impact assessment' (EIA) regulations and processes, which also include the requirements for undertaking an 'economic and social impact assessment' (ESIA) for new mining projects (NTG 2013). Conventional approaches to ESIA (and social impact assessment or SIA generally) are based on a 'no harm' approach, where the intention is to ensure that projects do not inflict additional harm upon local environments and communities. However, this raises critical questions about how conventional forms of ESIA can address existing socioeconomic disadvantage. This also draws attention to the relationship between the ESIA process and the agreements that are negotiated between the proponents and the Indigenous groups for mining projects. These regulations and processes concomitantly highlight the multiple ways in which the inherent deficiencies of junior mining companies are exacerbated in remote Indigenous contexts. The sociocultural elements of project impact, which are arguably the most complex, are demonstrably the least regulated. Yet, it is in this domain where Indigenous rights and interests are most at risk, and the possibilities for local forms of development are most pressing. In this context, the limitations of the usual 'no harm' approach to EIA for extractive capitalism become apparent.

---

2   In response to the criticism from Indigenous leaders based in the north, including Yu and Morrison, an Indigenous reference group was established in late 2017, which included these two leaders among the eight (Canavan and Scullion 2017).

Tracing the key aspects of the regulatory context within which the Top End Mining company is seeking to operate sheds light more broadly on the political economy within which the extractive industries in general operate across this jurisdiction—for which, of course, there may be parallels in other jurisdictions. Likewise, I argue that the approach of this particular company to the task of seeking project approval is in some ways representative of other junior companies that are also operating, or have done so, in the NT. And, as will be elaborated, although the absence of the state is apparent in the channelling of particular models of extractive development on its remote Indigenous citizens, the state is present for these same citizens in other governance contexts (see also Lewis, Chapter 8, this volume). This sectoral dimension of the state's selective engagement is felt through the structural inequalities that mark out the daily lives of Indigenous Territorians in a diverse range of punitive governance and conditional social welfare measures, where the state is very present.

Top End Mining: Reviving a 'Stranded Asset'

Top End Mining is a subsidiary of a larger international company that has multiple business interests, including trading and vehicle distribution. This mining company was established through the purchase of existing mining leases, their associated 'stranded assets' and the existing exploration and mining agreements that had been negotiated between the previous company and the Northern Land Council (NLC), as the statutory representative body for Indigenous interests under the Aboriginal Land Rights Act (ALRA). The previous company, also a junior miner, had gone into voluntary administration. The stranded assets include a trial pit (where bulk sampling was undertaken by the previous company) in an open-cut operation, a laboratory and an accommodation village, both in the form of demountable buildings, for up to 60 staff. This trial pit remains in 'care and maintenance': there is a rotating skeleton staff of two on/two off, as it has not been producing since 2014 and operated for less than two years. The other mining leases and exploration licences do not have any infrastructure, while the total 'mine area disturbance footprint' is potentially around 2,400 hectares, according to the 'notice of intent' lodged for the environmental impact statement (EIS).

As a result of the resuscitation of this mining project, the ESIA process did not start from scratch, so the company had effectively purchased a range of approvals. The existing local agreements had been moved across (or 'novated') to this new company at a meeting convened by the NLC for the company to discuss their proposal to reopen the mine and develop the other leases with the Traditional Owners (TOs), as customary land owners are termed under the ALRA. At that meeting, however, the additional aspects of the project, which subsequently required the ESIA processes, were not raised. Yet, as it unfolded, approval for the new aspects of the project were essential for its economic viability and, as will be discussed, the proponent appeared keen to ensure that the only approvals required for this additional stage were those of the state, rather than the TOs as well.

The agreements entail a range of company commitments, which include: cultural heritage protection, payment of mining royalties, and preferential employment for TOs as agreement beneficiaries. Likewise, if the project proceeds (and the new aspects gain the appropriate approvals) then, as part of the agreement, Top End Mining has to prepare and complete an ESIA and associated management plan *with* the input of TOs and the NLC. This second ESIA is in addition to that required by the regulatory regime, and the focus would be on Indigenous interests. This ESIA, including a baseline assessment, is not only essential for assessing the social impact of the operation over time, but, if done well, could also be a very useful tool for the region in negotiating their service delivery needs with government (see for instance Taylor et al. 2000; Taylor and Scambary 2005; Taylor 2018).[3] Indeed, the possibility of being engaged in establishing such a baseline for the region was a catalyst for my involvement in this project through the Centre for Social Responsibility in Mining at the University of Queensland. As I will discuss below, I was engaged to prepare the section of the ESIA on 'culture and heritage' and, as an extension of this, the draft 'cultural heritage management plan' for the project. Hence, my growing recognition that the predominant focus

---

3   However, such an ESIA will require significant resourcing and good faith from the company, while the NLC will need to ensure that the research occurs. There is no evidence that the previous company—under the same agreement—undertook an ESIA. Furthermore, in remote areas especially, the methodology of Australian Bureau of Statistics (ABS) census collection has been strongly critiqued as not reflective of Indigenous households and patterns of mobility. Likewise, in some regions there is an acknowledged undercount by as much as 40 per cent. Such an undercount has direct implications for service delivery (Martin et al. 2002; Taylor 2018).

of the regulatory system that I had to work through was to assess and manage for *environmental* impacts. Minimal state guidance was provided for the social impacts, which are embedded within the ESIA process.

As this mining company was specifically established to develop several mineral deposits, it is not surprising that they do not have existing 'social performance management systems' and staff in place. This term refers to the resources and capabilities that companies need in order to manage a broad range of socioeconomic and political issues, and risks, in the context of their operations (see Johnston and Kemp 2019). In this case, the entire company consists of two directors, one of whom lives internationally, a secretary and the four staff members responsible for the care and maintenance of the site—and as we shall see, this is more or less representative of many junior miners.

The Precarity and Culture of Junior Miners

There is no universal definition of a junior mining company and considerable debate surrounds the criteria that classify a company as a junior miner. Much of the critical literature (Dougherty 2011, 2013; Lyons et al. 2016) and the online financial guidance (Anon. 2019) focuses on junior gold miners. The boundaries between the different classifications of senior, mid-tier and junior firms are ambiguous, though they are usually categorised by their numbers of sites, levels of capitalisation and sources of revenue. According to Cranstoun, senior firms derive their revenue from production and sale, whereas juniors are essentially venture capital companies, while mid-tier firms derive capital from both equity financing and production and sale (quoted in Dougherty 2013: 343). Another definition also indicates that junior miners are neither producing companies nor the recipient of significant income from production or from some other business venture (Anon. n.d.a). These definitions do not fit the current case study, nor the neighbouring operation, which will also be briefly discussed, as they are both producing companies. This reinforces the fact that very little research seems to have been undertaken on junior companies, and that they are an extremely diverse bunch.

The Australia-based 'Junior Miners' website (www.juniorminers.com/about.html), which began in 2007, lists over 2,000 junior miners globally. To gain perspective, this compares with the Minerals Council of

Australia (MCA), which has 45 members,[4] and the International Council on Mining and Metals (ICMM), which has 27. The largest transnational companies, or the 'majors', and some aspirational mid-tier companies seek membership of these peak professional bodies. While these majors also extract the bulk of the earth's minerals, there has been a proliferation of junior companies globally (Dougherty 2011). One reason cited for this proliferation has been the rent-seeking behaviour of juniors, as incentivised by the state (Dougherty 2013), as well as the recent uptick in the mining industry.

Similar to Canada (see Dougherty 2013: 341), the Australian state incentivises the formation and proliferation of juniors, as it does more generally the mineral and fossil fuel industries. A 2017 federal government scheme involved 'tax incentives for junior exploration companies to encourage investment and risk taking' (Turnbull 2017). This entailed a government commitment of AUD100 million to secure additional private investment, referred to as the Junior Mineral Exploration Tax Credit.[5] Likewise, on a significantly larger scale, the industry more generally receives significant support. According to Peel et al. (2014), over a six-year period, Australian state and territory governments spent AUD17.6 billion supporting the mineral and fossil fuel industries. Of this total, between 2008 and 2015, the NT Government spent almost AUD407 million on minerals and fossil fuel expenditures and concessions. Ironically, given the deficit position of the NT budget, in 2013–14, of the AUD113 million in mining royalties received, AUD88 million was returned in assistance to the minerals and fossil fuel industries (Peel et al. 2014: 4–12). Powerful industry lobby groups, including the MCA and the Association of Mining and Exploration Companies (AMEC) are active in pursuing these cost savings on behalf of industry.[6]

---

4   It also has approximately 15 'associate members' who are not mining companies but consultancy firms.
5   The Junior Mineral Exploration Tax Credit is said to be an improvement on previous pilot programs assisting junior mineral explorers and has been developed based on industry feedback. Offering incentives for investment is of course not limited to junior mining companies. Glencore, which owns the MacArthur River mine in the NT, is known to pay zero royalties in most years. The only known royalty payment, of AUD13 million, came in 2008 after a historic peak in the zinc price. At best the mine would contribute just one third of 1 per cent of NT Government revenue (Campbell et al. 2017).
6   According to AMEC's 2019 annual report, they have 'saved the industry millions of dollars by being willing to challenge public policy issues and advocate crucial initiatives to reduce the cost of doing business for the industry ... AMEC has ready access to ministers, leading politicians and government decision makers in territory, state and federal levels' (AMEC 2019: 5).

The absence of the state therefore manifests as a fiscal absence, reflected in the failure to claim taxes from the industry at both territory and federal levels. The NT is the only jurisdiction in Australia with a profit-based royalty system and because many extractive industry operations are capital and infrastructure-intensive, this often means that no royalties are paid at all by industry. This approach also assumes that any infrastructure and development the industry builds will have flow-on and broader benefits to communities and the region, in a 'trickle-down effect', both during and beyond life-of-mine.

Although some junior companies may subscribe to particular industry standards or may have developed their own policies, online 'networks' seem to be a major forum for information and discussion.[7] These online services, some of which are fee-based (for subscriptions to magazines such as *Junior Mining Monthly*), provide insight into the attitudes and approach of juniors.[8] Delving into several of these online sites is a window into the capital logic and cultural drivers that underpin the junior mining entrepreneur. An American investment site called the 'Outsider Club' (that also publishes the *Junior Mining Monthly* and the *Junior Mining Trader*), which runs the byline 'because you'll never be on the inside', speaks to an idealised frontier spirit that resists regulation and state oversight. As they state on their website: '[W]e'll … help you shield your civil liberties and freedoms. We pledge allegiance to no political party. The *Outsider Club* will inform you of suspicious laws and policy—and show you how to stand up to and fight against them' (Outsider Club n.d.).

Perhaps unsurprisingly, the editors of this site are six white American men, suggesting that the 'outsiders' as a frame of reference is a limited reference point. The business logic of junior miners is framed around risks and uncertainty. As the Australian junior mining online page states: 'those uncertainties and risks can range from everything from government politics to commodity prices and everything in between' (Anon. n.d.b). Though such risks are not unique to junior companies, they are typically very reactive to the volatility of the global minerals market and can disappear just as fast as the mineral stocks plummet. This is one reason

---

7   See for instance: www.juniorminingnetwork.com. Note that mid-tier and major companies also engage in networks and forums.
8   Note, however, that there are blurry and shifting boundaries between these 'categories'. For instance the Australian company Newcrest is listed as a junior on the 'Junior mining' site, but according to its website it is one of the largest gold mining companies in the world and it is also a member of the MCA and the newest member of the ICMM.

why it is commonplace for mines operated by junior companies to become abandoned or placed in 'care and maintenance' until another company picks up the operation. This was the case of the 'stranded asset' purchased by the Top End Mining company. Likewise, during the same period another neighbouring mine, and its associated stranded assets, were also purchased by another junior company.[9]

The parent company that established Top End Mining is not readily searchable online and no information could be located about their corporate business standards. Of note however, is that mining does not seem to be a significant element of their diversified business. Since incorporating, however, and during the process of applying for the EIS approvals (still ongoing as I write), Top End Mining has developed a one-page 'Sustainable Development Policy' (available online as an appendix in their Mining Management Plan). Their priority is clear in the first line: 'A fundamental objective of [the company] is to create long-term value for its owners through the discovery, development and delivery to its customers of mineral and related products'.[10] The model of sustainability at play here is thus of the corporate financial kind. I am not suggesting that there is anything exceptional about the fact that these companies do not have social performance standards in place or have not employed the rhetoric of corporate social responsibility (CSR). However, the significant chasm between the approach to CSR of such companies and those of major companies and many mid-tier companies is of particular note.

Major companies, including those who are members of the ICMM, and some who are members of the MCA, generally have 'social performance management systems'. This term refers to the suite of integrated policies, procedures and guidelines that assist the company to manage the social dimensions and impacts of their operations.[11] I suggest that we need to recognise that a major feature of junior mining companies is a consistent lack of CSR standards and associated governance systems. As Margaret Lyons and her co-authors note: 'little is known about the meaning and practice of CSR in junior and mid-tier companies' (Lyons et al. 2016:

---

9   According to one often cited estimation, there are approximately 50,000 abandoned mines in Australia—from major open-cut mines to small mine shafts. No major open-cut mine has been fully rehabilitated and relinquished in Australia (Roche and Judd 2016).

10   Note, that in the interests of not identifying this company I will not provide a reference for this.

11   These typically include policies, guidelines and standards that cover: 'communities', 'stakeholder relationship management', 'human rights', 'Indigenous peoples', 'local procurement and employment', 'social baseline and impact assessment', and 'local and community grievance' mechanisms.

204). Though there has been a significant global push to develop such standards across most corporate sectors, juniors seem to have missed this trend and largely continue to fly under the radar. As observed by a staff member of a major mining company:

> a lot of the juniors … think community engagement is just a kind of luxury or add on [because] it costs money … and they just want to get on with digging ore out of the ground … the digging the hole bit is easy, its what's outside the mine fence and engaging successfully there that's the key to business going forward. (in Doyle and Cariño 2013: 48)

In other words, for junior miners, the 'social' is an externality, far removed from core business,[12] whereas more mature companies tend to consider the risk profile of new purchases and conduct their due diligence to understand what legacies they might be buying into. As a result, the ability of a junior company to sell the project to a larger development partner, once the project has gained the relevant approvals and established viability, can prove hazardous. The risk is that a development can be worthless if the project is encumbered by negative social and environmental legacies; a point also made by one particular mining major when it considers purchases (Doyle and Cariño 2013).

Nevertheless, even though major companies may have systems and standards in place, these are inconsistently developed and applied. As a result, many of the criticisms directed at juniors also hold for some mid-tiers and major companies. And, likewise, I readily acknowledge that there is a performative element to both the implementation of voluntary and mandatory industry standards, and public reporting (Fonseca 2010; Boiral 2013). While recognising these caveats and limitations, communities stand a stronger chance of gaining better social, economic and environmental outcomes with a company that has some social performance systems in place (and associated resourcing), and a reputation to protect, than a company that does not. Against this background to the operating logic of junior mining companies, I will trace the context within which this company, and the industry, operates in the NT.

---

12 See also Kemp and Owen (2013) who discuss this concept of 'core and non-core' for social performance in mining, noting that 'community relations is core to business, but not "core business"'. I suggest that this issue is significantly more pronounced for junior miners.

The Political Economy of the NT and Indigenous Social Development

Unlike the Pilbara or Goldfields regions in Western Australia (WA) that are known as 'mining regions', and that effectively underpin that state's economy, the NT does not have a comparable regional or big infrastructure economy.[13] Though mining is one of the three dominant industries in the NT, contributing 12.9 per cent of gross state product in 2019, the largest 'industry' is 'government and community services' at 23 per cent of gross state product, followed by service industries (which includes tourism) at 19.4 per cent (NTG 2020). This sector comprises public administration and safety, education and training, healthcare and social assistance. This most significant sector is also predominantly funded by the Commonwealth Government, to the effect that approximately 80 per cent of the NT's income is funded federally (see Zillman 2018). One reason the NT receives such a significant proportion of commonwealth funding is primarily because of its large Indigenous population.

Approximately 30 per cent of the total population of the NT is Indigenous and it is recognised that on the major indicators—education, health, housing and employment—they experience high levels of disadvantage, relative to the mainstream population. More than 70 per cent of this total Indigenous population live in remote or very remote areas with significant historical deficits in civil society organisations (such as not-for-profits, educational and advocacy bodies) and public infrastructure. Although a significant proportion of this Commonwealth grant is funding for Indigenous Territorians, because the money is calculated by accounting for Indigenous disadvantage and remote servicing, only about 50 per cent of the allocation is actually spent on them.[14]

The mendicant position of the NT Government is raised here as it has implications for their engagement with proponents of large-scale infrastructure projects, including mining. And it is in tension with the fact that the NT is one of the few jurisdictions globally with legislation recognising collective Indigenous rights to land that includes rights to

---

13   Mining makes up 36 per cent of gross state product in WA (WASG 2020).
14   This is because there are 'no strings attached' to the money when it leaves Canberra and flows into the NT Government's accounts. This issue has been a major and ongoing source of contention and debate in the NT, and between the NT and the Commonwealth governments, notably for Indigenous parliamentarians and Indigenous representative bodies (Davidson 2018; Walsh 2018).

free prior and informed consent (FPIC) over development on this land and inalienable freehold title.[15] More than 50 per cent of the NT is now recognised as Aboriginal Freehold Title, under the ALRA, while much of the remaining land is subject to native title rights and interests, including 85 per cent of the coastline. The *Native Title Act 1993* (Cth), which applies to the whole of Australia, is a poor cousin to the ALRA. Native title is the recognition that Aboriginal and Torres Strait Islander people have rights and interests to land and waters according to their traditional law and customs, as set out in Australian law. However, it comes in two forms: non-exclusive possession and exclusive possession.[16] The vast majority of native title recognised Australia-wide is non-exclusive and only offers the right to negotiate, not the right of veto or to protect places of significance (McGrath 2016). However, in the NT there have been significant cases where exclusive possession has been recognised, including Blue Mud Bay No. 2 (AIATSIS 2016).

While this land tenure in the NT would appear to provide Indigenous landowners with a powerful and strategic political position, the regulatory and governance environment diminishes these rights at both territory and federal levels. A significant mechanism leading to marginalisation of Indigenous interests in land occurs principally through the state (Territory) level EIA process (which encompasses ESIA). This is triggered when a proponent submits a 'notice of intent' to develop a project—including mining operations. The NT Government Environmental Protection Authority then provides terms of reference for the preparation of an EIS for the specific project.

In the case discussed here, and similar to other projects, this requires the proponent to identify all of the processes and activities associated with the project and the ways in which they will mitigate and monitor

---

15   The ALRA is Commonwealth (federal government) legislation. The right to FPIC is a qualified one, however, as a 'yes' to exploration is also a 'yes' to mining. While a 'no' to exploration is only for 5 years, when the company can return to ask again. There have been some strong campaigns by TOs that 'no means no'; that the company is not welcome to return. Of note, other states globally that have legislated Indigenous peoples' land rights, including the right to FPIC, are the Philippines (with the Indigenous Peoples Rights Act of 1997) and some Latin American states that are signatories to ILO Convention 169 (which recognises Indigenous and Tribal Rights) and that have adopted domestic laws to implement the Convention, including Colombia, Chile, Brazil and Peru.

16   In the case of non-exclusive possession, native title is found to coexist with non-Indigenous property rights such as pastoral stations or areas where there is a shared interest with another party (being the most common form). Exclusive possession native title includes the right to possess and occupy an area to the exclusion of all others. Exclusive possession native title can only be granted across certain areas such as unallocated crown land or areas that were previously held or owned by Aboriginal people, which was found to be the case for the Blue Mud Bay native title claim in Arnhem Land.

environmental and social impacts, as well as outlining the proposed potential economic and social benefits. Because the reporting requirements are so specific, the proponent typically engages a consultancy firm, who then engages experts to prepare sections for the EIS addressing particular areas that require impact assessment, such as hydrology and the biological and physical environment. In the NT context, archaeological experts are routinely engaged to address the tangible cultural and historical aspects of the project, but it is less standard for other social scientists, such as social anthropologists, to be engaged in the cultural and social aspects of the impact assessment.

The Top End Mining company has to deal with a challenging socio-political context of poor mining legacies in the region from previous juniors, including abandoned mines and associated agreements, complex land tenure arrangements and customary governance structures. These land tenures include Aboriginal Land (both granted and not yet finalised), native title, pastoral lease, national park, marine reserve and a special purpose lease. All of these formal land tenures are underlain by customary land tenures, which, in the case of Aboriginal Land, intersect. The rest of this section begins to outline the social and political context within which this company is seeking to operate and the existing vulnerabilities for the Aboriginal population.

As a very remote area in the NT, the region where the company is seeking to operate is an Aboriginal domain. The population, the languages and the social structures are Aboriginal, while the formal land tenure is Aboriginal land, recognised as native title land or under native title claim. More specifically, in this local government area, as one of five in the regional shire, 80 per cent of the population identify as Indigenous.[17] Only 30 per cent speak English at home, with nearly 50 per cent of the population speaking Kriol and 15 per cent local and other languages. In administrative terms, this region is part of a regional shire with a total population of 7,500 people over a geographic area that covers 186,000 square kilometres.[18] This regional shire includes nine Aboriginal communities, many more outstations (small family-focused homelands) and four towns, which are mostly populated by non-Aboriginal people.

---

17 To avoid identifying the proponent, I do not name this region.
18 Key industries in the region are pastoralism, some agriculture, extractive industries, and tourism and government services. The Council is one of the largest employers in the region. Like all the other remote areas of the NT, this shire is dependent on public sector grants and funding, having only a low rates base from which to derive income for municipal services (via privately owned houses/private property).

According to the 2016 census, the average annual income in this region is AUD23,500.[19] This average income is low, reflecting the fact that almost half of the population is unemployed. The basic Newstart (government unemployment) allowance is just under AUD15,000 per year, which is less than half of the national minimum wage. A majority of people living in households whose reference person receives a social security benefit fall below the poverty line, according to the Australian Council of Social Services (Davidson et al. 2018).[20] Other barriers to mainstream forms of development are also high rates of substance abuse and the use of violence as a form of social regulation.[21]

This demographic profile is indicative of a vulnerable and marginal population. The regional education profile also suggests limited employment, as only 7.6 per cent of the population have completed year 12, while the most substantial cohort, at 21.1 per cent, have completed year 9 or below.[22] The region is also very young demographically, with a median age of 24 years, while 31.5 per cent of the total population is under 15 years of age. Aboriginal people over the age of 65 make up less than 5 per cent of the population.

This set of structural factors clearly requires an innovative approach, not only to forms of development, but also to the ESIAs that are required as part of the EIA process before developments proceed—and, if developments do proceed, to the ongoing role of the company in supporting said forms of development. Yet, with few other mainstream economic opportunities in remote communities, there has been growing pressure and a normative assumption by the state (in this case the NT Government) that mining is a preferred development path. The recent government approvals for non-conventional gas fracking to proceed is also indicative of this approach (Chlanda 2017; NTG 2018a; see also Espig, Chapter 6, this volume).[23]

---

19   See Martin et al. (2002) for a critical engagement with the appropriateness of the ABS census method for remote and very remote populations.

20   Housing tenure also has a major impact on poverty. The majority of people below the poverty line are also renting, another characteristic of this regional Aboriginal population (Davidson et al. 2018), where there are very few home owners. Note that there are no incentives to own your own home (hence take on a mortgage), as the market economy barely exists in remote communities (Altman et al. 2005).

21   In 2005–06 the rate of alcohol-attributable deaths in the NT among non-Aboriginal people was double the national rate and, among Aboriginal Territorians, 9 to 10 times the national rate (D'Abbs 2017).

22   These unemployment rates and the education profile is comparable to other remote shires in the Northern Territory.

23   Fracking has been banned in Victoria and there are moratoriums in place in NSW and WA. In NSW there have also been buybacks and cancellations of exploration licences (Cox 2018).

This extractive capitalist approach to development has been reinforced by the diminishment of the welfare or social state, so that market-driven development has begun to present the dominant option for Aboriginal people. Likewise, access to social security is increasingly conditional via various 'work for the dole schemes'. The abolition of the Community Development Employment Program in 2015 was part of a broader policy approach enforcing the mutual responsibility paradigm as an element in citizenship entailments. The Community Development Employment Program was purpose-built for remote Indigenous contexts and provided flexible adaptive forms of employment. Its replacement with the Community Development program, in which people effectively do the same work but are classed as unemployed and now under the surveillance of the social security system, was perhaps one of the most obvious shifts towards a punitive order of federal state governance (see Kral 2017; Sanders 2017).

Though extractive capitalism is often acknowledged as intensifying neoliberal pressures, for instance in mining regions such as the Pilbara (Peck 2013; Taylor 2018), it has been also promoted by some as the panacea for the socioeconomic issues of remote Indigenous communities (Langton 2013, 2017). Marcia Langton has argued that post the Native Title Act, and the subsequent wave of negotiated agreements and improvements in the culture of mining companies, the extractive industry generates local benefits that exceed local costs (Langton 2013). Within Australia and Canada, as settler colonies, there has also been consideration that the retreat of the state has created space for Indigenous people to exercise their agency (for instance, Slowey 2008; Feit 2010), notably in relation to negotiating agreements for resource development (Langton et al. 2003; O'Faircheallaigh 2011). The state is rarely a party to these agreements and, as will be discussed, they usually contain a complex mix of training, employment and community development initiatives.

Like Canada, Australia does not have '[a] public policy framework that guides [agreements] negotiation, terms of reference or implementation' (Cameron and Leviton 2014: 26). Although mining land use agreements can potentially be positive for the beneficiaries, any benefits hinge on their successful implementation. This includes the capacity of the company, the Aboriginal communities and the Indigenous representative body to uphold the conditions of the agreements. Given the significant level of disadvantage that requires addressing and, thus, what these agreements are expected to achieve, this lack of public policy is both surprising and a major gap.

A Selectively Present State

As I have begun to outline, in the NT, the absence or retreat of the state has been selective. On the one hand, the state has never been present in relation to the many negotiated agreements between its Aboriginal citizens and mining interests,[24] while we have seen a retreat of social welfare, such as the Community Development Employment Program and a mainstreaming of Indigenous services such as housing. On the other hand, we have seen an increase in the presence of the state in forms of punitive governance and surveillance, and a coupling of responsibilities and duties with citizenship entitlements (Holcombe 2018; Spiers Williams 2019). While Shared Responsibility Agreements, which were introduced by the conservative Howard government and tied welfare entitlements with discretionary development funding, have fallen out of the policy toolbox, the 'policy settings' of shared responsibility have continued (Sullivan 2011: 33), as can be seen in the 2015 NT *Homelands Policy: A Shared Responsibility.* Memoranda of understanding (MOU) have also become commonplace, such as the community safety MOU between local communities, their regional shire and the local police service in relation to community safety issues. This attachment of responsibilities and duties to (Aboriginal) citizenship entitlements is now an intrinsic element of public services and policies.

The NT Emergency Response of 2007 was a clear marker of this restructuring of Aboriginal citizen/state relations.[25] Since 2007, various legal reforms have specifically targeted Aboriginal citizens in a selective increase in state interventions. These include mandatory sentencing and mandatory reporting laws in relation to domestic and family violence issues, leading directly to an escalation of male and female Aboriginal incarceration rates and the removal of 'at risk' children. The result is that Aboriginal people now make up approximately 85 per cent of the NT prison population (NTG 2018b), while there was also a threefold rise in child removals since 2007 (Gibson 2015). Other punitive laws

---

24 As far as I am aware, one of the very few negotiated mining land use agreements in Australia where the state is also a party is the Century Agreement in Queensland (see Everingham et al., Chapter 9, this volume).

25 The Northern Territory Emergency Response was triggered by widespread allegations of child sexual abuse within Aboriginal communities. A raft of extreme measures were implemented that, taken together, constituted a governmental intervention unmatched by any other policy declaration in Indigenous affairs in the last 40 years (see Altman and Hinkson 2007).

criminalise an increasing range of Aboriginal behaviours with a focus on 'public' offences, such as alcohol consumption and driving infringements. As most small communities now have police stations, post 2007, the risk of over-policing is routine (Pilkington 2009; Pyne 2012). This punitive governance is an element in the state's attempt at coercing a particular type of disciplinary civility (Hindess 2002; Holcombe 2018).

This active state presence in very particular areas of Aboriginal policy, programming and legislation, conversely brings into relief the state's absence in other areas of legislation for these same peoples. This absence is in relation to the *Environmental Assessment Act 1982* (NT), which establishes administrative procedures to address development interests across the NT that also affect the Aboriginal estate and Aboriginal rights and interests. This Act stipulates the range of procedures a developer will need to address in an EIS for a project.

How the State Minimises Indigenous Interests through the EIA Processes

As discussed, the NT is predominantly comprised of Aboriginal-owned land under the ALRA. Significant portions of the remaining areas are either recognised as native title, or under native title claim. Yet, the interests of these Aboriginal landowners and managers are barely visible in the EIA processes for development on these lands—especially in the extractive industries.

The predominant focus of the EIA process is to assess and manage for environmental impacts. Minimal guidance is provided for the social impacts that are embedded within the EIA process. The NT 'Guidelines for Preparation of an Economic and Social Impact Assessment' are six pages long (NTG 2013).[26] This compares with the Queensland SIA Guidelines at 24 pages (QSG 2018) and the New South Wales (NSW) SIA Guidelines at 55 pages (NSWSG 2017). The NSW SIA Guidelines are also specific to the extractive industries, and the NSW Government has developed a comprehensive 'Guide to Community and Stakeholder Engagement', as part of their Draft Environmental Impact Assessment

---

26   They guide the proponent in their collation of the range of data that needs to be provided to government and the public about the potential economic and social 'contributions' and the impacts and benefits of the project.

Guidance Series, as well as a 'Scoping Tool'. By way of comparison, in the EIS Terms of Reference for this particular project, while there was reference for the need to establish a 'stakeholder communication strategy', there was no guidance offered as to how to develop this strategy.

For TOs, the interdependence between the environment, customary law, governance and identity makes separating out the social impacts from the environmental impacts neither meaningful nor possible. Though as previously indicated, Aboriginal people own more than half of the NT and there are statutory requirements that underpin the ALRA, as well as other applicable acts, including the Commonwealth Native Title Act of 1993 and the NT Aboriginal Sacred Sites Act of 1989, the NT ESIA guidelines barely mention Indigenous peoples' interests. Rather, they state that 'consultation with Indigenous peoples has particular needs [sic]' and point practitioners to an obscure reference from the United States (NTG 2013: 4).

This obscure reference to a three-page document as a generic piece on 'Indigenous peoples', rather than the specific context of the NT, exemplifies the disconnection from the reality that Aboriginal people have significant interests under law for much of this jurisdiction, while most of the major mines are also on these lands. Aboriginal people are *the* major stakeholder. Yet, this lack of recognition and regulatory guidance diminishes these rights and interests. Whether this is the result of ignorance of what good guidelines look like, or yet another form of 'administrative violence' (Lea et al. 2018) to marginalise Aboriginal interests, or perhaps a combination of both, results in the same effect. As this legislation does little to accommodate the particularities of Aboriginal interests, so it becomes a form of absence.

In some ways, the minimal guidance (and thus expectation) from the NT Government can be addressed for each specific project, as the proponent's 'notice of intent' to develop the project then triggers a more detailed Terms of Reference (ToR) for the preparation of the specific EIS. However, in this particular project, only 3.5 pages of the 36 pages of the ToR for this EIS were allocated to how the 'cultural and historic' issues and the 'socioeconomic environment' should be addressed. A list of 'suggestions that may assist with highlighting the social and economic value of the project' was provided. Thus, when it comes to working with a junior mining company that is compliance-driven, pushing for higher standards was, I realised, a challenge. This became clear when I was discussing the draft SIA scope of works with the company. As the company preferred to maintain minimum compliance, rather than going beyond compliance, we discontinued this element of the project, and it was subsequently taken up by a consultancy firm. As will be discussed, while major mining companies may well have similar compliance-driven approaches to permitting, it did seem to me at the time that I was operating in not only a regulatory void, but also a void of corporate social standards. There were no other internal policy levers that we could lean on to direct the company towards more responsible practices.

I had initially only been engaged to prepare the section of the EIS which became the public documents for comment on 'Culture and Heritage' and the draft 'Cultural Heritage Management Plan'. The consultancy firm managing the entire EIS process had requested that I submit a scope of works for the SIA (excluding the economic component) and had also actively encouraged me to draw on the stronger practice standards of NSW, for instance. Among other issues, I had proposed to develop a 'Community and Stakeholder Engagement Plan' to augment the 'stakeholder communications strategy' that was required for the ESIA as part of the EIA process. This engagement plan was proposed as a resource that, though driven initially by the EIA, also looked post-EIA and would be reviewed as part of the project life cycle, rather than an exercise focused on ensuring the company would get 'over the line' in community approvals. It seemed especially important that this company develop an engagement plan, given they did not have any community engagement processes in place. That Indigenous interests are externalities to the business of ensuring that the project proceeds seemed to be facilitated by the NT Government during the EIA processes.

Social Impact Assessment and Shifting the Baseline

In industrialised countries, such as Australia, the standard approach to social impact assessment is to document the existing socioeconomic conditions within which a proposed development is to occur, assess its likely impacts and identify strategies to minimise and mitigate these negative effects (O'Faircheallaigh 2011). This baseline is then used to evaluate ongoing impacts from the project. However, there is an emerging approach to SIA that is critical of this method—especially when it is applied in regions with high levels of social vulnerability such as the NT. While the lack of attention to pre-existing negative impacts may be acceptable in some mainstream SIA contexts (such as regional or rural areas in Australia with a predominantly Anglo population), maintaining the status quo in contexts of marginality and vulnerability is not good enough. Applying a 'no harm' approach in this context renders invisible the particularities of local Aboriginal needs and interests. As Ciaran O'Faircheallaigh indicates:

> existing [Aboriginal] social conditions are not simply a baseline against which subsequent changes caused by development are measured and deemed positive or negative. Rather they constitute a fundamental problem, and both social impact assessment (SIA) itself and the proposed development that is being assessed are judged in terms of their potential to change existing social realities. (2011: 138)

Attempting to carve out an approach that moves beyond mitigation and trade-offs recognises that the political ecology of extractive capitalism in the NT is currently limited to the two-sided conflict of economy versus environment (Escobar 2008). The state has been absent in recognising and facilitating alternative modes of production that celebrate or at least promote Indigenous interests, strengths and existing capabilities (see Altman and Kerins 2012; Chambers et al. 2018). There has been considerable critical analysis of the limitations of large-scale natural resource development, including mining, to fundamentally shift the well-being of the Indigenous peoples on whose lands mining operates (e.g. Trigger 1997; Langton and Mazel 2008; Altman and Martin 2009). However, these academic critiques (many of which focus on the NT) do not seem to have influenced NT resources policy, including EIA processes. There have, however been attempts over the last decade, co-funded by the federal government, to consider what a cultural and conservation economy could look like for northern Australia (Hill et al. 2008). Nevertheless, the only element of this approach to establishing sustainable forms of livelihood that has been actively supported by government are Indigenous ranger programs, which have been very successful and of which there are several in this region (AFG 2015; Mackie and Meacheam 2016).

An important feature of O'Faircheallaigh's 2011 case study mentioned above (of an SIA undertaken for James Price Point in the Kimberley region of WA) was that while the proponent (Woodside) funded it, it was produced as part of an independent process. The goals of the SIA were driven by Indigenous interests and managed by the regional representative body, the Kimberley Land Council. As he noted, in this context 'SIA is inextricably bound up with achieving basic change in underlying social conditions; that is with social development [and] … sustainability' (O'Faircheallaigh 2011: 139). Such an approach actively manages for the inadequacies of conventional SIAs and ensures that the process is useful for those most impacted. This is markedly distinct from the majority of SIAs that are undertaken as a state-initiated element of a development project's permitting and compliance, which often lead to them being perceived as compromised—as they are funded by the proponent, specifically because they seek development approval. Likewise, Anthony Hodge, focusing on the extractives, uses a framework he terms 'contribution analysis' (Hodge 2018). He argues that a conceptual shift from the 'current deeply entrenched practice of focusing on the identification and mitigation of negative impacts (or effects) to a "higher test" based on the achievement (or not) of a net positive contribution to human and ecosystem wellbeing over the long term' needs to occur (Hodge 2018: 370; see also Bebbington and Humphreys Bebbington 2018).

Other relevant approaches to SIA, for this remote Indigenous context, include a human rights–based approach that draws on the core values of human rights as the framework (MacNaughton and Hunt 2011). Deanna Kemp and Frank Vanclay (2013) argue that in some contexts it is more effective to integrate human rights impact assessment with the social impact assessment process, rather than have them as stand-alone assessments. The linking of business and human rights initially emerged in response to a series of egregious cases of human rights abuses by business entities driven by market-based notions of access and entitlement to resources (for some notable cases see Kemp and Vanclay 2013). The mounting international pressure subsequently led the UN to establish the 'Protect, Respect and Remedy Framework' (Ruggie 2008) to promote business-related engagement with human rights, followed by the UN Guiding Principles on Business and Human Rights to 'operationalise' the framework (Ruggie 2011). Though I am not aware of any human rights impact assessment being undertaken in Australia, one was recently undertaken for the first time in Papua New Guinea for an SIA for a proposed liquefied natural gas project (Götzman and Bainton 2019).

Indigenous-led SIA, contribution analysis and human rights impact assessment highlight the limitations of applying the usual 'no harm' approach in the context of a low socioeconomic baseline and challenging social conditions. This is further compounded by a political context that actively marginalises Indigenous interests. Clearly, if the aim is only to mitigate in order to maintain this baseline, then the opportunity for mining to contribute or intervene in positive ways into these socio-economic contexts is severely compromised. In this context, both the capacity and the attitude of the mining company is crucial, because if the company does not have capacity and positive intent then the SIA will be further compromised.

Choices in a Context of Limitation

Though the statistics of high unemployment and limited formal education do not define individuals, they do provide an indicator of the ways in which the parameters for agency and choice are established. This concept of 'agency' has gained popular currency in the development literature where it is sometimes defined as 'empowerment' and articulated through the related concept of 'voice'.[27] As a 'process freedom', the ability to exercise agency is now accepted as part of mainstream development thinking (World Bank 2014: 6). In this context, limited mainstream education and work experience in a market economy act as constraining elements to engaging on equal terms with the proponent, even as mediated by the NLC. The options and limits for TOs, as social actors, are also mediated by the interrelation between an individual's subjectivity as a person and the subsequent agency they are able to manifest in any given context. As cultural responsibilities and obligations assert a determining role in what sort of 'agent' an individual can be, these cultural parameters need to be taken into account, as well as the structural effects of material poverty.

Against this backdrop of apparent material need, structural disadvantage and political marginalisation, there are two major enticements being offered as the 'big ticket' items, if the mine proceeds. The first is that the proponent proposes to upgrade a significant section of the currently unsealed road to dual lane bitumen. This all-weather road upgrade, to enable the transport of the ore on road trains to the export market, will make the road safer in the long term; the trade-off being sharing it with large numbers of road trains (trucks hauling two to three carriages) and

---

27   In the anthropological literature, the concepts of agency and personhood are understood as deeply intertwined. See, for instance, Wardlow (2006) and Holcombe (2018).

also making the region more accessible to tourists and other interests.[28] Another enticement being discussed, and as part of the original local agreement, is the possibility of jobs during construction and operation. As a potentially 20-year mine, this possibility of employment for local TOs and other Aboriginal people from the region is very significant. This 'opportunity', in particular, is a major element in TOs considering reopening the mine, even though the haulage route has changed—and in a way that is seen to compromise the integrity of both the customary economy and the spiritual values of the land, river and sea countries.

In 2017, the local ABC radio station ran a story about the project and interviewed several senior women from the largest and closest 'affected community'.[29] According to this report, the women actively opposed reopening the mine for environmental and spiritual reasons. Yet, having since spoken with these same women, I can say that their opposition is a qualified one. If certain elements, including the dust from the transport of the ore can be mitigated and it can be can be guaranteed that certain other environmental changes do not occur, then *perhaps* jobs for their sons and grandchildren will sway their opposition.

Without having undertaken extensive qualitative research in relation to TOs' and affected community members' hopes, aspirations and concerns about the potential of the mine, some insight can still be drawn from comparable contexts, in both Australia and Canada.[30] In British Colombia, when First Nations were asked what was needed in order for them to support an industrial-scale liquified natural gas plan, they stated that education, training and employment was the most important issue (Stokes et al. 2019). Likewise, in this NT context, the major challenge will be *enabling* employment opportunities through the company investing in pre-employment and work-ready programs and ensuring that the workplace facilitates a culturally amenable environment. Workplace retention for Indigenous employees is a major issue in all remote mines (Sarker and Bobongie 2007; Brereton and Parmenter 2008). Although these issues are an element of the SIA for the project, it will be down to the

---

28 There are other issues as well that are currently unknown, including the length of time it will take to upgrade the road and if and how long the road trains will be using the road before it is sealed.
29 In the interests of maintaining the anonymity of the proponent this ABC report will remain un-referenced.
30 The extent of fieldwork for my role in the EIS was a five-day trip to communities in the region.

effective implementation of the agreement as to whether the employment outcomes and other potential social benefits are met (see Everingham et al., Chapter 9, this volume for comparable discussion).

However, the *disconnection* between the legislation that promotes Aboriginal rights and interests and guide agreement making, and NT Government legislation relating to EIA, does little to facilitate the opportunities that are contained in these agreements. Thus, we find there is very little relationship between the SIA for the development project and the subsequent negotiated agreement.

Likewise, there is no trigger in the SIA process as to the appropriate time to engage with and include the Aboriginal representative body in dialogue about development projects. The role of the regional land council as the mediator and conduit for identifying and protecting Aboriginal interests is not defined other than in the ALRA legislation. There is no consideration of this role in the NT Environment Assessment Act. Rather, the only mention made in the 'Guidelines for the Preparation of Economic and Social Impact Assessment' is to bundle the ALRA and the Native Title Act with the other Commonwealth legislation. Guidance as to what stage to engage Indigenous interests (and the representative body and 'community stakeholder' groups), or the most appropriate mechanisms for engagement, is not provided. To what extent it is the role of the Land Council to inform the TOs of the project prior to the proponent undertaking the SIA is unclear. While the general understanding is that it is the Land Councils' responsibility to inform TOs and affected community groups about the proposed project, at what exact stage this occurs and in what form (i.e. a large community meeting and/or smaller meetings with particular TOs), is not specified in the ToR that proponents receive from the Environmental Protection Authority.

Likewise, the fact that the NT Government often fails to collect royalties from extractive industries, instead encouraging extractive industries to develop infrastructure, appears to be based on the assumption that any infrastructure and development the industry builds will create broader flow-on effects and benefits for communities and the region both during and beyond life-of-mine (see also Everingham et al., Chapter 9, this volume). This ad-hoc opportunistic approach is, again, directed by mainstream external interests with little systematic consideration of long-term Aboriginal interests.

Unfinished Business: The ALRA and TO Rights and Interests

When I was initially engaged on the ESIA for this mining project, I was of the understanding that TOs would have veto rights over a key element of it. This key element was in relation to proposed infrastructure on a major river as a transport corridor. My understanding was based on my being engaged, some years previously, as the lead author of the anthropologists' report for the 'Beds and Banks' claim over this river, under the ALRA. The Aboriginal Land Commissioner had recommended the grant of Aboriginal freehold title and, like other people familiar with the claim, especially Aboriginal claimants, I had assumed that it had progressed to the 'finding' (the technical term) of the grant of the river-beds and banks.[31] However, as I now know, and as I have since informed some of the claimants that I worked with all those years ago (during a field visit for this project), this was not the case. Due to the high degree of 'detriment' (the technical term for opposing interests), the final sign-off by the Commonwealth Minister for Aboriginal Affairs that is required under the ALRA did not occur. This was principally in relation to the opposition of the amateur fishing industry, as was also the case for another 15, mostly beds and banks, claims in the 'Top End' of the NT.

However, coincidently during 2017 and 2018, Minister Scullion, then Federal Minister for Indigenous Affairs, requested the Aboriginal Land Commissioner undertake a review of outstanding Aboriginal land claims (ALC 2019). The ToR for the review indicated that 'a range of political, legal and procedural factors are likely to have contributed to the delayed resolution of detriment issues associated with the … land claims and therefore warrant consideration in this review.'[32] For over 20 years, these claims had been in limbo and, as such, TOs were not able to exercise their rights as landowners over these customary areas, which also includes the right of veto over development interests (including mining), under the ALRA.

---

31 The Aboriginal Land Commissioner is a statutory officer of the Commonwealth appointed to perform functions outlined in section 50 of the ALRA. The Commissioner's functions include undertaking inquiries into traditional land claims in the NT.

32 According to media reports, the Federal Minister for Aboriginal Affairs Scullion granted almost $500,000 to three industry groups—the NT Seafood Council, the NT Cattlemen's Association and the NT Amateur Fishermen's Association—to help resolve their detriment cases in outstanding land rights claims (Allam 2018; Gibson 2018). See: www.abc.net.au/news/2018-10-31/indigenous-advancement-strategy-funds-given-to-lobby-groups-nt/10451664.

This was not surprising, and it reflects the disconnection between Commonwealth legislation and NT-based legislation. As the ALRA is Commonwealth legislation, the NT Government had an adversarial approach, consistently opposing land claims during the hearings in pursuing detriment and engaging anthropologists to find inconsistencies in the Aboriginal evidence. This approach to opposing Aboriginal interests was initially driven by the conservative Country Liberal Party who held power in the NT from self-government in 1978 until 2001, when the first Labor Government was elected. It was notable that during the 2002 hearing for the river claim at issue here, the then new Labor Government took an almost facilitative approach to the claim. However, this approach of accommodation has since changed. The 2018 *Report on the Review of Detriment* outlines the current NT Labor Government's approach. Of particular relevance is the submission made by the Department of Primary Industries and Resources in relation to minerals and energy. As summarised by the *Detriment Review Report*:

> DPIR [Department of Primary Industries and Resources] claimed detriment on behalf of [4] energy interests in respect of negotiating agreements with traditional Aboriginal owners and the uncertainty that a future agreement would be made [i.e. that veto rights might be exercised]. They also claimed that failure to reach agreement poses a strong risk that proposed patterns of land use associated with [extractive] exploration and production could be adversely affected if access is withdrawn or restricted [i.e. the land is granted]. (ALC 2019: 181)

The Commissioner was dismissive of the Department of Primary Industries and Resources claim, noting that 'whatever the validity of speculative claims, the ALRA was not intended to accommodate them; much less to permit them to stand in the way of a grant of land to traditional owners' (ALC 2019: 182). Nevertheless, the federal minister still has the final sign-off on the grant of land of all of these claims. As Minister Scullion stated: 'It is critical that this many years into the process we do all we can to help settle outstanding claims to provide certainty and opportunity to all Territorians' (Scullion 2019).[33] That these other interests have had

---

33   As the Minister stated in the press release about the report: 'All parties with an interest in the land claims reported on by the Commissioner will be given the opportunity to comment on the report before I make any decision about whether to proceed with the recommendations for grants of land under the Aboriginal Land Rights (Northern Territory) Act 1976 ... I also maintain my commitment to work with all stakeholders to ensure that detriment issues are resolved' (Scullion 2019). It is noteworthy that they also had ample opportunity to comment prior to the report.

ample opportunity (including government funding), to make their cases of detriment heard, suggests that Aboriginal interests are not compatible with those of 'all [other] Territorians'; and that his focus is ensuring that these 'opportunities' are not curtailed. This response makes clear that the state is more present for some Territorians than others, when it comes to facilitating particular forms of development. It is also of note that Minister Nigel Scullion was, at that time, also an NT Senator. Here the 'state' includes both NT and federal levels.

Unfortunately, the Top End Mining company also submitted a case of detriment to the enquiry, indicating that their business model was threatened by any extension of Aboriginal rights and interests. This was the case, even though it was made clear to them (by myself and the other consultants engaged) that the SIA would also need to include this opposition to the granting of the land as a negative social impact, which could materially compromise their project. Their oppositional approach compounded the need for me to withdraw from further discussions in relation to undertaking additional elements of the SIA (recall that I had attempted to encourage higher standards beyond minimum compliance).[34]

Customary Rights and Interests: Contesting Ecologies of Knowledge

What are these Aboriginal interests? The detriment review report quoted the original land commissioner's finding for this river claim: 'there are few areas in Australia where the traditional attachment of the Indigenous people to their land exceeds that of the present claim' (ALC 2019: 174). These interests understand the river as sentient, embodying not only customary harvest potential, but also the ancestral travels of the *Gilyirringgilyirring* (mermaids), and many other totemic species. Sections of the river are associated with the regional *Kunapipi* ceremonial tradition (Berndt 1951). This necessarily means that those who hold the role *Mingirringi* (attachment through father's father) and *Junggayi* (attachment through mother's father) for these places will have to consider a much wider group

---

34 Likewise, the irony for myself in proceeding in such a role with a company that was opposing the land claim that I had played a major role in preparing was perverse, not to mention professionally and ethically compromising.

in their decision making. Ultimately, it will be these particular TOs that will carry the decision-making responsibility for river-based aspects of this project and any other development interests.

That TOs do not perceive the human/environment dichotomy that modern Western science and Western society in general practices represents a core challenge to the state, as exemplified by the ESIA process under the EIA. This other set of intangible spiritual interests confounds the state's commitment to extractive capitalism and land as alienable resource empiricism. Such spiritual attachments to land are a direct risk to this enterprise, which may explain why the state practices a local form of 'epistemicide' in this neo-colonial 'global south' (Santos 2014). The marginalising and attempted diminishment of Aboriginal knowledges, as alternative ways of knowing that exist outside of the dominant epistemological norms, presents as an ongoing contest between the valued and undervalued, the recognised and under-recognised, the visible and invisible—even while these Indigenous ethno-epistemologies are increasingly named, and recognised as valid and valued (Maffie 2009).[35]

Yet, as one would also expect, *within* the regional Aboriginal population there is also contestation over decision-making rights and a diverse range of interests and intersecting responsibilities (see also Trigger 1997). These include those TOs who have specific rights and responsibilities under Aboriginal customary law to maintain and protect particular significant sites. There are also Aboriginal rangers, as the region just to the north of the major community at hand was recently declared an Indigenous Protection Area and the local land and sea rangers work out of this community. These rangers work with modern Western science and their ethno-ecological knowledges, and they bring their TO rights and responsibilities to their work. The third loose grouping are those who are interested in the potential work that will come from the mining construction and operation, and the local Aboriginal Development Corporation is very interested in encouraging these. All of these interests—which are not mutually exclusive: individuals can endorse several positions—have to be weighed in the final decision by the TOs. Noting that, as I write, the extent of their legal decision-making rights in relation to the river is still unfolding.

---

35 As seen, for instance, in the growing body of research on ethnobotany (Janke 2009; Robinson 2010; Ens 2012; Ens et al. 2012) and the expansion of the Indigenous Protected Area program, which jointly recognises the cultural and ecological values of the landscape (Hill et al. 2012).

The State and Negotiated Mining Agreements

The local agreements that Top End Mining took over incorporate a range of benefits that will be in the TOs' minds when they meet to discuss the project, post the EIA finalisation. These agreements, which were effectively abandoned along with the previous mine, were picked up by Top End Mining. Of note, at a regional full council meeting of the NLC, then CEO Joe Morrison stated that global market forces had played a part in the mine's demise and had made the community 'realise that we've just got to be a bit more aware of commodity prices and the impact that it has on local economies'. This issue is exacerbated with junior companies who are more likely to seek to develop smaller, and thus riskier, deposits.

As O'Faircheallaigh indicates, land use agreements can offer Aboriginal groups a degree of autonomy from the state in their provision of access to mining income and other benefits (O'Faircheallaigh 2010). The state does not play any role in these agreements: they are between the Indigenous representative body, on behalf of TOs, and the company. However, this raises questions as to whether there should be some sort of public policy framework to guide their negotiation and implementation. These questions have been explored in the Canadian First Nations context. In the Northwest Territories of Canada it was found that 'the roles of mining companies, Aboriginal organisations and government are not properly defined and some inappropriate off-loading of responsibilities by government is occurring' (Kennett, quoted in Cameron and Leviton 2014: 29).

In the NT, this notable absence of the state from any role in agreement making is a reflection of their broader absence in not establishing public policy and institutional reform to address the raft of land rights that Aboriginal people have been successful in achieving. The ALRA and the Native Title Act seem to operate in a parallel, and still contested, sphere alongside NT Government legislation and programs as a form of segmentary state logic. It is entirely up to the capability of the Indigenous representative body to ensure that the agreement encompasses all that is sought by their constituents, including forms of community development. As it is standard for agreements to be commercial-in-confidence, the possible relationships between state-based social programs and the social programs offered within the agreements are not drawn. This includes features such as education and training opportunities offered by the mining company through the agreement, and those that may already exist as public programs within the community. The capacity of a mining junior to deliver on these commitments, or the Land Council to ensure they are implemented, is not tested by the state.

Likewise, there is also the risk that the NT Government will cut its expenditure on communities in receipt of royalties and other benefits under agreements, negating some of their positive impact (O'Faircheallaigh 2004, 2010). This 'substitution funding', where citizenship entitlements—such as health, housing and education—are effectively privatised, has been found to be the case in mining regions such as the Pilbara (Holcombe 2009; Taylor 2018). Likewise, in the NT, even before there were comprehensive agreements that included education and training and other negotiated community benefits, Jon Altman records early cases from the 1980s and 1990s where substitution funding was standard for outstation services in the region serviced by the Gagadju Association that was established from the Ranger uranium mine (Altman 1998 in O'Faircheallaigh 2004). In light of the precedence of substitution funding elsewhere in the NT and remote WA, the presence of the company is likely to further facilitate the absence of the state in terms of infrastructure and other public services in this region.

Conclusions

This discussion of a junior mining company in northern Australia has been nested within the broader political economy of the ongoing marginalisation of Aboriginal rights and interests. While the Aboriginal citizens of the NT have consistently won court battles in relation to their land, those victories have not translated into public policy and institutional reform to ensure that development is inclusive of the majority permanent population of these lands (Morrison 2017: 4). Likewise, though the land use agreements that are routinely negotiated between Aboriginal interests and developers encompass a range of mechanisms that address disadvantage, public policy has not been developed to guide their negotiation, terms of reference or implementation. This absence of the state is also mirrored in the mainstream EIA processes that have been transplanted to these remote Aboriginal domains. Even though Aboriginal interests are the major landholding group, they are treated as another set of stakeholders.

For the junior miner discussed in this chapter, the EIA processes function as a default corporate social performance system, and the inadequacy of the EIA process highlights this significant gap in the operations of junior companies more generally (although any mining company in this context is also likely to pursue a compliance-based approach). The SIA elements of the EIA regulatory framework, though a lesser standard than other states, reflect a conventional SIA approach that seeks to 'do no harm', focusing on mitigating negative impacts. Such an approach is not appropriate in contexts where the majority interests are Aboriginal, as it not only misses opportunities to contribute to genuine collaboration with these interests, but they also emerge from the presumption that these interests are aligned with the dominant mainstream population and nested among the other stakeholder groups.

The possibility of an extractive-led pathway to Aboriginal development, levering off the local agreement, is a hopeful one. The targeted support for big project development, such as mining, however still functions on the theory of trickle-down economics. As Peter Yu stated, this won't work because it is not accompanied by a range of structural reforms to support Indigenous economic inclusion (Yu quoted in Morrison 2017: 3). That the operations of this junior miner fit into this neocolonial paradigm was notably apparent in their adversarial approach to the granting of the beds and banks land claim, as potentially impacting their operation by granting TOs free prior informed consent rights. Such an approach has Indigenous people as obstacles to development, rather than development partners.

This example from the NT has revealed a selectively absent state. Although my focus has been on absence in the context of extractive capitalism, in the broader socioeconomic context within which Aboriginal citizens live, the state looms ever present in forms of punitive governance and surveillance in a coupling of responsibilities and duties with citizenship entitlements. On the other hand, there has been a notable retreat of social welfare, such as the Community Development Employment Program and a mainstreaming of Indigenous services, such as housing. The state is notably absent in relation to the intersection of Indigenous interests and development interests, as market forces are enabled to be directive in shaping social outcomes in the interests of growth in state revenue. In the NT, the state has never been present in relation to the many negotiated mining land use agreements between its Aboriginal citizens and mining companies. In this context of development, while it may appear that these Aboriginal interests are appropriately served by Land Councils and specific legislation such as the ALRA, without intersecting and supportive state legislation the contest between Indigenous and non-Indigenous interests continues.

Ironically, this absent state is also reflected as a form of fiscal absence: in the absence of claiming taxes from the industry, at both territory and federal levels. This approach assumes that any infrastructure and development the industry builds will have flow-on and broader benefits to communities and the region both during and beyond life-of-mine. Yet, market-based solutions to social problems, as reflective of less state, is a failure of the state to meet its obligations and commitments to its most vulnerable citizens.

References

AFG (Australian Federal Government), 2015. *Our North, Our Future: White Paper on Developing Northern Australia.* Canberra: Commonwealth of Australia.

AIATSIS (Australian Institute of Aboriginal and Torres Strait Islander Studies), 2016. 'Native Title Information Handbook: Northern Territory.' Canberra: AIATSIS.

ALC (Aboriginal Land Commissioner), 2019. *Report on Review of Detriment: Aboriginal Land Claims Recommended for Grant But Not Yet Finalised.* Canberra: Office of the ALC.

Allam, L., 2018. 'Indigenous Groups Call for Investigation into Scullion Fund Stoush.' *The Guardian*, 3 November.

Altman, J. and M. Hinkson (eds), 2007. *Coercive Reconciliation: Stabilise, Normalise, Exit Aboriginal Australia.* Melbourne: Arena Publications.

Altman, J. and S. Kerins (eds), 2012. *People on Country: Vital Landscapes, Indigenous Futures.* Melbourne: Federation Press.

Altman, J.C., C. Linkhorn and J. Clarke, 2005. 'Land Rights and Development Reform in Remote Australia.' Canberra: The Australian National University, Centre for Aboriginal Economic Policy Research (Discussion Paper 276).

Altman, J. and D. Martin (eds), 2009. *Power, Culture, Economy: Indigenous Australians and Mining.* Canberra: ANU E Press (Centre for Aboriginal Economic Policy Research Monograph 30). doi.org/10.22459/CAEPR30. 08.2009

AMEC (Association of Mining and Exploration Companies), 2019. *Annual Report*. Viewed 20 October 2020 at: secureservercdn.net/198.71.233.51/0h5.0cf.myftpupload.com/wp-content/uploads/2020/03/AMEC-2019-Annual-Report.pdf

Anon., 2019. 'What Criteria Classify a Company as a Junior Gold Miner?' Investopedia blog, 25 June. Viewed 28 September 2020 at: www.investopedia.com/ask/answers/040815/what-criteria-classify-company-junior-gold-miner.asp

Anon., n.d.a. 'Mineral Resource Stocks—Junior vs. Senior Mining Companies.' UndervaluedEquity blog. Viewed 28 September 2020 at: undervaluedequity.com/mineral-resource-stocks-junior-vs-senior-mining-companies/

Anon, n.d.b. 'Why Mexico is the Go-To Place.' Junior Miners blog. Viewed 28 September 2020 at: www.juniorminers.com/news/mining-exploration-mexico.html

Archer, R., J. Russell-Smith, S. Kerins, B. Costanza, A. Edwards and K. Sangha, 2019. 'Change and Continuity: The North Australian Cultural Landscape.' In J. Russell-Smith, G. James, H. Pedersen and K.K. Sangha (eds), *Sustainable Land Sector Development in Northern Australia: Indigenous Rights, Aspirations, and Cultural Responsibilities*. Boca Raton (FL): CRC Press. doi.org/10.1201/9780429471056

Bainton, N., 2020. 'Mining and Indigenous Peoples.' In *Oxford Research Encyclopedia of Anthropology*. Oxford: Oxford University Press. doi.org/10.1093/acrefore/9780190854584.013.121

Bebbington, A. and D. Humphreys Bebbington, 2018. 'Mining, Movements and Sustainable Development: Concepts for a Framework.' *Sustainable Development* 26: 441–449. doi.org/10.1002/sd.1888

Berndt, R., 1951. *Kunapipi: A Study of an Australian Aboriginal Religious Cult*. Melbourne: Cheshire.

Boiral, O., 2013. 'Sustainability Reports as Simulacra? A Counter-account of A and A+ GRI Reports.' *Accounting, Auditing and Accountability Journal* 26: 1036–1071. doi.org/10.1108/aaaj-04-2012-00998

Brereton, D. and J. Parmenter, 2008. 'Indigenous Employment in the Australian Mining Industry.' *Journal of Energy & Natural Resources Law* 26: 66–78. doi.org/10.1080/02646811.2008.11435178

Cameron, E. and T. Levitan, 2014. 'Impact and Benefit Agreements and the Neoliberalization of Resource Governance and Indigenous–State Relations in Northern Canada.' *Studies in Political Economy* 93: 25–52.

Campbell, R., J. Linqvist, B. Browne, T. Swann and M. Grudnoff, 2017. 'Dark Side of the Boom: What We Do and Don't Know about Mines, Closures and Rehabilitation.' Canberra: The Australia Institute.

Canavan, M. and N. Scullion, 2017. 'First Nations Peoples Critical to Developing Northern Australia.' Media release, 11 December. Viewed 28 September 2020 at: www.minister.industry.gov.au/ministers/canavan/media-releases/first-nations-peoples-critical-developing-northern-australia

Chambers, I., J. Russell-Smith, R. Costanza, J. Cribb, S. Kerins, M. George and G. James, 2018. 'Australia's North, Australia's Future: A Vision and Strategies for Sustainable Economic, Ecological and Social Prosperity in Northern Australia.' *Asia Pacific Policy Studies* 5: 615-640. doi.org/10.1002/app5.259

Chlanda, E., 2017. 'Fracking Interim Report: Two Bob Each Way?' *Alice Springs News*, 17 July.

Cox, L., 2018. '"Not Safe, Not Wanted": Is the end of NT Fracking ban a taste of things to come?' *The Guardian*, 18 June.

D'Abbs, P., 2017. 'Alcohol Policy in the Northern Territory: Toward a Critique and Refocusing.' Submission to the Northern Territory Alcohol Policies and Legislation Review.

Davidson, H., 2018. '$500m Meant for Indigenous Services was Spent Elsewhere by NT Government.' *The Guardian*, 11 January.

Davidson, P., P. Saunders, B. Bradbury and M. Wong, 2018. 'Poverty in Australia, 2018.' Sydney: Australian Council of Social Services and University of New South Wales, Poverty and Inequality Partnership (Report 2).

Dougherty, M., 2011. 'The Global Mining Industry, Junior Firms and Civil Society Resistance in Guatemala. *Bulletin of Latin American Research*. Society for Latin American Studies.

Dougherty, M.L., 2013. 'The Global Gold Mining Industry: Materiality, Rent-Seeking, Junior Firms and Canadian Corporate Citizenship.' *Competition and Change* 17: 339–354. doi.org/10.1179/1024529413z.00000000042

Doyle, C. and J. Cariño, 2013. *Making Free, Prior & Informed Consent a Reality: Indigenous Peoples and the Extractives Sector.* London: Philippine Indigenous Peoples Links.

Ens, E., 2012. 'Conducting Two-Way Ecological Research.' In J. Altman and S. Kerins (eds), *People on Country: Vital Landscapes, Indigenous Futures.* Melbourne: Federation Press.

Ens, E., M. Finlayson, K. Preuss, S. Jackson and S. Holcombe, 2012. 'Australian Approaches for Managing "Country" Using Indigenous and Non-Indigenous Knowledge.' *Journal of Ecological Management and Restoration* 13: 100–107. doi.org/10.1111/j.1442-8903.2011.00634.x

Escobar, A. 2008. *Territories of Difference: Place, Movement, Life, Redes.* Durham (NC): Duke University Press.

Feit, H., 2010. 'Neoliberal Governance and James Bay Cree Governance: Negotiated Agreements, Oppositional Struggles, and Co-Governance.' In M. Blaser, R. de Costa, D. McGregor and W.D. Coleman (eds), *Indigenous Peoples and Autonomy.* Vancouver: University of British Columbia Press.

Fonseca, A., 2010. 'How Credible are Mining Corporations' Sustainability Reports? A Critical Analysis of External Assurance under the Requirements of the International Council on Mining and Metals.' *Corporate Social Responsibility and Environmental Management* 17: 355–370. doi.org/10.1002/csr.230

Gibson, J., 2018. 'Indigenous Advancement Funds Given to Lobby Groups Impacted by Aboriginal Land Claims.' Australian Broadcasting Corporation, 31 October. Viewed 28 September 2020 at: www.abc.net.au/news/2018-10-31/indigenous-advancement-strategy-funds-given-to-lobby-groups-nt/10451664

Gibson, P., 2015. 'Removed for Being Aboriginal: Is the NT Creating Another Stolen Generation?' *The Guardian*, 4 March.

Götzman, N. and N. Bainton, 2019. 'Papua LNG Human Rights Impact Assessment: Focus on Gender, Security and Conflict.' Copenhagen: Danish Institute for Human Rights.

Hill, R., C. Grant, M. George, C. Robinson, S. Jackson and N. Abel, 2012. 'A Typology of Indigenous Engagement in Australian Environmental Management: Implications for Knowledge Integration and Social Ecological System Sustainability.' *Ecology and Society* 17: 23. doi.org/10.5751/es-04587-170123

Hill, R., E.K. Harding, D. Edwards, J. O'Dempsey, D. Hill, A. Martin and S. McIntryre-Tamwoy, 2008. *A Cultural and Conservation Economy for Northern Australia: A Proof-of-Concept Study.* Canberra: Land and Water Australia.

Hindess, B., 2002. 'Neoliberal Citizenship.' *Citizenship Studies* 6: 127–143.

Hodge, A.R., 2018. 'Towards Contribution Analysis.' In T. Addison and A.R. Roe (eds), *Extractive Industries: The Management of Resources as a Driver of Sustainable Development.* Oxford: Oxford University Press.

Holcombe, S., 2009. 'Indigenous Entrepreneurialism and Mining Land Use Agreements.' In J. Altman and D. Martin (eds), *Power, Culture and Economy: Indigenous Australian and Mining.* Canberra: ANU E Press (Centre for Aboriginal Economic Policy Research Monograph 30). doi.org/10.22459/caepr30.08.2009.07

———, 2018. *Remote Freedoms: Politics, Personhood and Human Rights in Aboriginal Central Australia.* Stanford (CA): Stanford University Press. doi.org/10.1515/9781503606487

Janke, T., 2009. 'Indigenous Ecological Knowledge and Natural Resources in the Northern Territory: Report on the Current Status of Indigenous Intellectual Property.' Darwin: Northern Territory Natural Resource Management Board.

Johnston, S. and D. Kemp, 2019. 'Organising for Social Performance in the Global Mining Industry: A Snapshot Study.' St Lucia: University of Queensland, Centre for Social Responsibility in Mining.

Kemp. D. and J.R. Owen, 2013. 'Community Relations and Mining: Core to Business, But Not "Core Business".' *Resources Policy* 38: 523–531. doi.org/10.1016/j.resourpol.2013.08.003

Kemp, D. and F. Vanclay, 2013. 'Human Rights and Impact Assessment: Clarifying the Connections in Practice.' *Impact Assessment and Project Appraisal* 31: 86–96. doi.org/10.1080/14615517.2013.782978

Kral, I., 2017. 'Submission to the Inquiry into the Appropriateness and Effectiveness of the Objectives, Design, Implementation and Evaluation of the Community Development Program (CDP).' Canberra: Parliament of Australia.

Langton, M., 2013. *The Quiet Revolution: Indigenous People and the Resources Boom.* Sydney: Harper Collins (The Boyer Lectures 2012).

———, 2017. 'Australian Mining Industry Annual Lecture.' Viewed 28 September 2020 at: files.ozblogistan.com.au/sites/4/2017/06/11132429/LANGTON_2017_MINING_INDUSTRY_LECTURE.pdf

Langton, M. and O. Mazel, 2008. 'Poverty in the Midst of Plenty: Aboriginal People, the "Resource Curse" and Australia's Mining Boom.' *Journal of Energy & Natural Resources Law* 26: 31–65. doi.org/10.1080/02646811.2008.11435177

Langton, M., O. Mazel, L. Palmer, K. Shain and M. Teehan (eds), 2003. *Settling with Indigenous Peoples: Modern Treaty and Agreement Making.* Melbourne: Federation Press.

Lea, T., K. Howey and J. O'Brian, 2018. 'Waging Paper Warfare: Subverting the Damage of Extractive Capitalism in Kakadu.' *Oceania* 88: 305–319. doi.org/10.1002/ocea.5203

Lyons, M., J. Bartlett and P. McDonald, 2016. 'Corporate Social Responsibility in Junior and Mid-Tier Resources Companies Operating in Developing Nations: Beyond the Public Relations Offensive. *Resources Policy* 50: 204–213. doi.org/10.1016/j.resourpol.2016.10.005

Mackie, K. and D. Meacheam, 2016. 'Working on Country: A Case Study of Unusual Environmental Program Success.' *Australasian Journal of Environmental Management* 23: 157–174. doi.org/10.1080/14486563.2015.1094752

MacNaughton, G. and P. Hunt, 2011. 'A Human Rights Based Approach to Social Impact Assessment.' In F. Vanclay and A.M. Esteves (eds), *New Directions in Social Impact Assessment: Conceptual and Methodological Advances.* Cheltenham: Edward Elgar. doi.org/10.4337/9781781001196.00034

Maffie, J., 2009. '"In the End, We have the Gatling Gun, and They Do Not"': Future Prospects of Indigenous knowledges.' *Futures* 41: 53–65. doi.org/10.1016/j.futures.2008.07.008

Martin, D.F., F. Morphy, W.G. Sanders and J. Taylor, 2002. *Making Sense of the Census: Observations of the 2001 Enumeration in Remote Aboriginal Australia.* Canberra: ANU E Press (Centre for Aboriginal Economic Policy Research Monograph 22). doi.org/10.22459/caepr22.03.2004

McGrath, P.F. (ed.), 2016. *The Right to Protect Sites: Indigenous Heritage Management in the Era of Native Title.* Canberra: AIATSIS Research Publications.

Morrison, J., 2017. 'Keynote Address.' Presentation to the Northern Australian Development Conference, Cairns, 19 June.

NSWSG (New South Wales State Government), 2017. 'Social Impact Assessment Guideline.' Sydney: Department of Planning and Environment.

NTG (Northern Territory Government), 2013. 'Guidelines for the Preparation of an Economic and Social Impact Assessment.' Darwin: Environmental Protection Authority.

———, 2018a. *Final Report of the Scientific Enquiry into Hydraulic Fracking in the NT.* Darwin: NTG.

———, 2018b. 'Northern Territory Correctional Services Annual Statistics: 2016–2017.' Darwin: Criminal Justice Research and Statistics Unit.

————, 2020. 'Northern Territory Economy.' Darwin: Department of Treasury and Finance.

O'Faircheallaigh, C., 2004. 'Denying Citizens their Rights? Indigenous People, Mining Payments and Service Provision.' *Australian Journal of Public Administration* 63(2): 42–50. doi.org/10.1111/j.1467-8500.2004.00377.x

————, 2010. 'Aboriginal–Mining Company Contractual Agreements in Australia and Canada: Implications for Political Autonomy and Community Development.' *Canadian Journal of Development Studies* 30: 69–86. doi.org/ 10.1080/02255189.2010.9669282

————, 2011. 'Social Impact Assessment and Indigenous Social Development.' In F. Vanclay and A.M. Esteves (eds), *New Directions in Social Impact Assessment: Conceptual and Methodological Advances.* Cheltenham: Edward Elgar. doi.org/ 10.4337/9781781001196.00020

Outsider Club, n.d. 'About Us.' Viewed 16 October 2020 at: www.outsiderclub. com/about

Peck, J., 2013. 'Polanyi in the Pilbara.' *Australian Geographer* 44: 243–264. doi.org/ 10.1080/00049182.2013.817037

Peel, M., R. Campbell and R. Dennis, 2014. 'Mining the Age of Entitlement: State Government Assistance to the Minerals and Fossil Fuel Sector.' Canberra: Australia Institute (Technical Brief 31).

Pilkington, J., 2009. *Aboriginal Communities and the Police's Taskforce Themis: Case Studies in Remote Aboriginal Community Policing in the Northern Territory.* Darwin: North Australian Aboriginal Justice Agency.

Pyne, A., 2012. 'Ten Proposals to Reduce the Over-Imprisonment of Aboriginal Men in NT Prisons.' *Australian Indigenous Law Review* 16(2): 2–17.

QSG (Queensland State Government), 2018. 'Social Impact Assessment Guideline.' Brisbane: Department of State Development, Manufacturing, Infrastructure and Planning.

Robinson, D.F., 2010. 'Traditional Knowledge and Biological Product Derivative Patents: Benefit Sharing and Patent Issues relating to Camu Camu, Kakadu Plum and Açaí Plant Extracts.' Tokyo: United Nations University, Institute of Advanced Studies (Traditional Knowledge Bulletin).

Roche, C. and S. Judd, 2016. 'Ground Truths: Taking Responsibility for Australia's Mining Legacies.' Sydney: Mineral Policy Institute.

Ruggie J., 2008. 'Report of the Special Representative of the Secretary-General on the Issue of Human Rights and Transnational Corporations and Other Business Enterprises.' New York: United Nations General Assembly, Human Rights Council. doi.org/10.1163/2210-7975_hrd-4208-0204

———. 2011. *Guiding Principles on Business and Human Rights: Implementing the United Nations Respect, Protect and Remedy Framework.* United Nations Human Rights Office, Human Rights Commission.

Sanders, W., 2017. 'Submission to the Inquiry into the Appropriateness and Effectiveness of the Objectives, Design, Implementation and Evaluation of the Community Development Program (CDP).' Canberra: Parliament of Australia.

Santos, B.S., 2014. *Epistemologies of the South: Justice Against Epistemicide.* Boulder (CO): Paradigm Press.

Sarker, T. and G. Bobongie, 2007. 'Survey of Aboriginal Former Employees and Trainees of Argyle Diamond Mine.' St Lucia: University of Queensland, Centre for Social Responsibility in Mining.

Scullion, N., 2019. 'Statement on the Aboriginal Land Commissioner's Report on Review of Detriment.' Media release, 21 February. Viewed 28 September 2020 at: www.indigenous.gov.au/news-and-media/announcements/minister-scullion-statement-aboriginal-land-commissioner%E2%80%99s-report

Slowey, G., 2008. *Navigating Neoliberalism: Self Determination and the Mikisew Cree First Nation.* Vancouver: University of British Columbia Press.

Spiers Williams, M., 2019. 'Innervating Colonialism: Exploring the Retraction of Indigenous Rights through Two Sentencing Provisions.' *Australian Feminist Law Journal* 44: 203–220. doi.org/10.1080/13200968.2018.1547102

Stokes, D., B.G. Marshall and M.M. Veiga, 2019. 'Indigenous Participation in Resource Developments: Is it a Choice?' *Extractive Industries and Society* 6: 50–57. doi.org/10.1016/j.exis.2018.10.015

Sullivan, P., 2011. *Belonging Together: The Politics of Disenchantment in Australian Indigenous Policy.* Canberra: Aboriginal Studies Press.

Taylor, J., 2018. 'Change in Aboriginal Social Indicators in the Pilbara: 2001–2016.' Report to the Pilbara Regional Implementation Committee.

Taylor, J., J. Bern and K. Senior, 2000. *Ngukurr at the Millennium: A Baseline Profile for Social Impact Planning in South-East Arnhem Land.* Canberra: The Australian National University, Centre for Aboriginal Economic Policy Research (Monograph 18).

Taylor. J. and B. Scambary, 2005. *Indigenous People and the Pilbara Mining Boom: A Baseline for Regional Participation.* Canberra: ANU E Press (Centre for Aboriginal Economic Policy Research Monograph 25). doi.org/10.22459/IPPMB.01.2006

Trigger, D., 1997. 'Century Mine: Preliminary Thoughts on the Politics of Indigenous Responses.' In D. Smith and J. Finlayson (eds), *Fighting Over Country: Anthropological Perspectives.* Canberra: The Australian National University, Centre for Aboriginal Economic Policy Research (Monograph 12).

Turnbull, M., 2017. 'Investing in the Future Strength of the Australian Resources Sector.' Press release, 2 September. Viewed 22 September 2020 at: www.minister.industry.gov.au/ministers/joyce/media-releases/investing-future-strength-australian-resources-sector

Walsh, C., 2018. 'Northern Territory Government in Financial Crisis, Will Seek Bailout from Canberra.' Australian Broadcasting Corporation, 14 December. Viewed 28 September 2020 at: www.abc.net.au/news/2018-12-14/nt-government-financial-report-commonwealth-bailout/10621136

Wardlow, H., 2006. *Wayward Women: Sexuality and Agency in a New Guinea Society.* Berkeley: University of California Press. doi.org/10.1111/j.1548-1352.2010.01116.x

WASG (Western Australia State Government), 2020. 'Western Australia Economic Profile, August 2020.' Perth: Department of Jobs, Science, Tourism and Innovation.

World Bank, 2014. *Voice and Agency: Empowering Women and Girls for Shared Prosperity.* Washington (DC): World Bank.

Zillman, S., 2018. 'Why is the Northern Territory in So Much Debt?' Australian Broadcasting Corporation, 19 December. Viewed 28 September 2020 at: www.abc.net.au/news/2018-12-19/why-is-the-northern-territory-in-so-much-debt/10632654

8

Broken Promise Men: The Malevolent Absence of the State at the McArthur River Mine, Northern Territory

Gareth Lewis

> The open-cut is right in the place where the Rainbow Serpent rests. They cut open not just the snake, but us ceremony people too. They pushed us aside just like they did first time when they invaded our Countries with their guns and poison to take our land.
>
> Jacky Green (Green et al. 2017)

Introduction

The McArthur River in the southeast of Australia's Northern Territory flows from its source in the dry Barkly tablelands 520 km northwards to the tropical southwestern Gulf of Carpentaria where it meets the ocean near the Sir Edward Pellew group of islands. The township of Borroloola is situated on its western bank, 45 km upstream from the coast. The McArthur is a river which has economically and spiritually sustained countless generations of Aboriginal peoples including, for its middle and lower reaches, the Gurdanji and Yanyuwa along with their Mara and Garrwa neighbours. Replete with food and utilitarian resources, a means of travel, trade and communication, the river's importance is reflected in its central place in an associative cultural landscape, being the foci for many narratives, ceremonies, songs and sites of spiritual and sacred significance for Aboriginal people across the Gulf region.

For recent generations, the river has been transformed. Following European arrival into the region in the second half of the nineteenth century, Aboriginal people have experienced massive upheavals from violence, disease, dislocation from and usurpation of their traditional lands, and the devastating host of health, social and cultural impacts wrought by towns, pastoral stations and mining development. At the time of European 'settlement' the McArthur River flowed with the blood of the Gulf country's Aboriginal peoples, massacred at the hands of European police and 'settlers'. Currently it flows with contamination from the McArthur River Mine located some 50 km upstream from Borroloola. Throughout this period the state in its various guises has played a critical role, usually from afar, in shaping the policies and practices of non-Indigenous people towards and regarding Aboriginal people in the Gulf region.

This chapter examines the role of the state as both sponsor and regulator of resource extraction in this region. I explore recent Aboriginal perspectives of the McArthur River Mine, its operator Glencore, and by default the state, who together have dominated and disrupted the local and regional Aboriginal world and literally reshaped Aboriginal people's connections to their traditional lands and waters. Reflecting on the Indigenous historical experience and memory of state-sanctioned violence and dispossession in the region, I conclude that the state emerges as a consistently malevolent absence in Aboriginal peoples' lives, bridging past violence with current environmental, social and cultural impacts of the state-sanctioned—yet questionably assessed and monitored—resource extraction activities at the McArthur River Mine.

I review the history of the development of the McArthur River Mine against a backdrop of shifting Australian Indigenous policy and mining company practices, then apply Sawyer and Gomez's (2008) work on transnational governmentality and resource extraction, and Owen and Kemp's (2017) work on social licence and countervailing power in the global mining industry, to explore the interplay of the state and mining company throughout. From recent Aboriginal voices expressed in word, song and imagery I conclude that the Aboriginal peoples of the Gulf have been brutalised as colonial subjects and are now disempowered as 'extractive subjects' (Fredericksen and Himley 2019: 59) through the actions of a malevolently absent state.

The Malevolent Absence of the Colonial State

From the 1870s the first wave of sustained European economic and physical presence in Gulf Country of the Northern Territory commenced with the rapid expansion of the pastoral industry from north Queensland into the Gulf region and beyond. Aboriginal inhabitants of the Gulf country felt the full force of this pastoral expansion from the late 1800s through to the early decades of the twentieth century. Referred to by Gulf Aboriginal peoples as the 'Wild Times', this was an era of rampant frontier violence with police, pastoralists and civilians taking to their guns against them with impunity. The contemporary Gulf Aboriginal perspective is simply put:

> the government cleared us off our lands by shooting us and putting chains around our necks and dragging us off. (Kerins and Green 2016: 124)

Conditions were described in the 1887 Northern Territory Government Resident's Annual Report:

> The whites look well to their Winchesters and revolvers, and usually proceed on the principle of being on the safe side. It is an affectation of ignorance to pretend not to know that this is the condition of things throughout the 'backblocks' and the 'new country' of Australia. (South Australia 1887: 17)

The politicians, their policy, and decision makers in government were well removed from the action in the corridors of power far away in Adelaide, which still administered the Northern Territory, or in the northern administrative outpost of Darwin. Extensive historical research into this frontier violence has been undertaken by Tony Roberts. Former public servant Brian Stacey notes Roberts' description of the enormity of the dispossession and violence that overwhelmed Aboriginal people in the short space of three decades between 1881 and 1910. He writes:

> On the back of a pastoral boom engineered by the South Australian colony, more than 600 Aboriginal men, women, children and babies, or about one sixth of the population, were killed in the Gulf Country.

> In just four years … the Aboriginal population of at least 4000, composed of 15 tribes or language groups, was dispossessed of every inch of land. (Stacey 2018: 13)

Stacey goes on to describe this practice as:

> terra nullius applied in a savage, illegal and secret way in Australia's last frontier by South Australia's colonial politicians. The evidence is there were dozens of massacres in the Gulf Country which South Australia's Attorneys-General and Premiers were alert to but refused to prevent, or to prosecute those responsible. (Stacey 2018: 13)

Robert's 2009 essay *The Brutal Truth* details the gruesome original records of state-endorsed and sponsored 'nigger hunts', dispersals, and 'picnics with the natives'. He notes the equipping of police and civilians in the Gulf Country with the latest powerful Martini-Henry rifles and unlimited ammunition:

> Successive governments issued the police with unlimited ammunition and the authority to use it; they also issued arms to civilian punitive parties.

> They kept sending the ammunition in massive quantities to the Territory, knowing it was being used for a single function: to shoot the blacks. They asked no questions, did no stocktakes, took no count of the dead … There is evidence of at least 50 massacres in the Gulf Country of the Territory up to 1910. It is impossible to estimate how many more went unrecorded. (Roberts 2009)

On the McArthur River some 60 km upstream from the McArthur River Mine is Dunggunmini: a permanent waterhole which is a Gurdanji sacred place (see Figure 8.1). John Avery recorded a narrative of a massacre at this place from a land claim informant during 1977:

> There was a mob of Aboriginals camping here in the old times, poor buggers, when Top Station was up. Frank Meagan came out with pack horses, plant, trailing horses and rifles. They been hunting around for people to shoot. They left their horses north of the spring and swung around the spring. They into them and shot them all. They shot the whole mob. Some fellas got out, some got up the steep cliffs. The waterhole was all blood—girls and boys, old women and men were shot. They did the same all around right down to Kilgour and Amelia Spring. Frank Meagan also poisoned people at Warunguri. (Avery and McLaughlin 1977: 4)

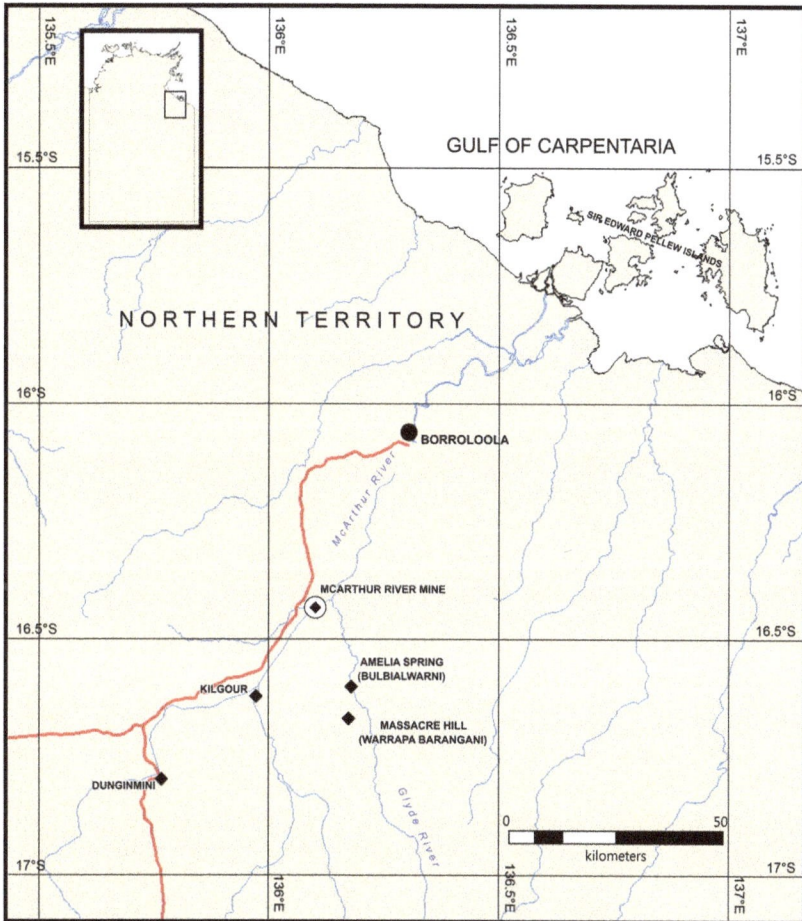

Figure 8.1 McArthur River Mine and surrounding region, Northern Territory

Source: Map by the author, background data supplied by the Department of Infrastructure, Planning and Logistics © Northern Territory Government

Kilgour is downstream again on the McArthur River closer to the mine. Amelia Springs is a site on the Glyde River where a rainbow snake travels northwards to meet its northern counterpart at the mine site. Nearby, upstream on Lamont Creek, a tributary of the Glyde, marked on the current topographic maps is Massacre Hill, unashamedly marking this rapid violent transformation of the local traditional geography and knowledge systems developed over countless generations by Aboriginal people.

The historical records attest to the malevolent role the colonial state played in this violence. Administered from Adelaide by the South Australian Government, nearly 3,000 km away, up until 1911 and then by the Commonwealth Government even further away in Canberra, the physical distance of the state could not have been greater. Such physical distance combined with the lack of any administrative presence left the implementation of state policy on the frontiers in the hands of police, pastoralists and other civilians who were heavily armed by the state. It created the illusion of an absent state on the frontier when in actuality the state was embodied by those whom it armed. Andrew Lattas surmised similar processes in colonial New South Wales:

> democracy makes every man an expression of reason … and thus makes him part of the state. It was precisely in these terms that pastoralists on the frontier justified their use of violence, they saw themselves as an extension of the rational ordering effects of the state. They appropriated often the state's right to a monopoly of violence because they claimed to embody the principles of the state. (Lattas 1987: 50)

Such a 'democratisation of violence' during these Wild Times allowed the state to occupy with ambiguity the role of policy maker while, whenever convenient, feigning absence and ignorance of the violent implementation of their policies. Similar to Lattas' observations on the 'privatisation of the penal system', here we witness the 'privatisation of genocide' in the Gulf Country through a malevolent and feigned absence of the state.[1]

McArthur River Mine and the Challenge of Land Rights

The 'Here's Your Chance' deposit, one of the largest known sedimentary stratiform lead, zinc and silver deposits in the word, was identified by Mount Isa Mines geologists in 1955 on the McArthur River some 50 km upstream from Borroloola, on land traditionally owned by the Gurdanji (MRM 2018: 1.1). Surface mineralogy had been noted as early as 1887 at McArthur River station and lead outcrops in the area had been

---

1   Borrowing from Lattas' (1987: 56) descriptions of the privatisation of the penal system in colonial NSW, where convicts were being transferred to the control of pastoralists at the same time as police and pastoralists were carrying out a 'policy of genocide' against Aborigines.

sporadically mined since around 1910, but the nature of the orebody identified in 1955 made the project unviable for any development until mining technology had sufficiently advanced in the 1990s. This time, between the identification of the orebody and its development in the early 1990s, saw the political landscape of Australia and the Northern Territory undergo massive changes especially in relation to the Indigenous policy arena.

During the 1960s the fight for equal wages for Aboriginal station workers culminated in Gurindji station workers going on strike during 1966 in what become known as the Wave Hill Walk-Off, and in 1963 the Yolngu people of northeast Arnhem Land delivered the Yirrkala Bark Petitions to the Commonwealth parliament in Canberra protesting the grant of mineral leases on land excised from the Arnhem Land Aboriginal reserve (Walker 2013: 33). In 1967 a national referendum determined that Indigenous people be included in the Australian census and that the Commonwealth be empowered to create laws for Indigenous people. By 1968 the sentiments expressed in the Yirrkala Bark Petitions escalated into legal action. The Gove Land Rights case heard in 1971,[2] although unsuccessful, led to the Woodward Royal Commission inquiry into Aboriginal land rights in 1973–74 and the enactment by the Commonwealth of the *Aboriginal Land Rights (Northern Territory) Act 1976* (ALRA), using its post-referendum powers.[3]

This growth of rights-based protests and legal action was reflected in a major shift in Australian Indigenous policy in the early 1970s, with Commonwealth-led self-determination policies replacing decades of assimilationism, which since the 1950s had sought to economically, culturally and physically centralise and integrate Indigenous peoples into the majority Australian mainstream society (Altman 2009: 21). A national Indigenous representative organisation, the Aboriginal and Torres Strait Islander Commission or ATSIC, was created and funded to administer Indigenous-specific programs, while arguably the high-water mark of self-determination policy, the ALRA, delivered the ability for Aboriginal

---

2   *Milirrpum and Others v Nabalco and the Commonwealth of Australia.*
3   There is abundant literature regarding the ALRA. For an early yet still relevant analysis of the principles of land rights as enacted in the Northern Territory see the introduction and relevant chapters in Peterson and Langton (1983).

people in the Northern Territory to claim and win inalienable freehold title to their traditional lands and to control and negotiate benefits from any mining or other developments on such lands.

Mining companies seeking to explore and/or mine on granted Aboriginal freehold land were for the first time required to negotiate agreements with and obtain the consent of traditional Aboriginal owners before they could proceed. The ALRA established Land Councils as largely independent Commonwealth statutory bodies governed by elected Aboriginal community members (see Altman 2009: 22).[4] These Land Councils administer the legislation, with the largest two, the Northern Land Council and the Central Land Council, employing considerable bureaucracies with professional legal, anthropological and technical staff. The ALRA also enabled the Northern Territory Government to set in place complementary legislation to protect Aboriginal scared sites in the Territory, which it did by enacting the Northern Territory Aboriginal Sacred Sites Act (NT Sacred Sites Act) soon after self-government was granted to the Northern Territory in 1978.[5] Although it was intended by the Whitlam Commonwealth Government as a model to be applied on a national scale, the idea of land rights never had any political chance of survival in the mineral-rich state of Western Australia (Altman 2009: 22).

In 1992 Mount Isa Mines announced its intention to commence underground mining at McArthur River and an environmental assessment process was undertaken. An initial 1993 Northern Territory Government approval for the mine signified the start of a series of approvals made by both the Commonwealth and Territory governments for the project and its subsequent expansions, aimed at securing the development and ensuring its longevity. These approvals were remarkable in that they sought to curtail potential and perceived threats of opposition to the mine posed by Aboriginal land rights, and the newly recognised native title rights.

To put this into further context: the early 1990s saw an electorally entrenched conservative Country Liberal Northern Territory Government vociferously opposed to land rights and desperate for an investment and economic development 'win' after the controversial Coronation Hill project in southern Kakadu National Park, the prevention of mining at Guratba by the then Bob Hawke–led Commonwealth Government, and

---

4   For details of the structure of the Northern Land Council, the most relevant land council in relation to this chapter, see their website: www.nlc.org.au/about-us/our-structure/our-people.

5   See the Aboriginal Areas Protection Authority website: www.aapant.org.au/about-us/history.

the uncertainty following the High Court recognition of native title in the Mabo case. The Coronation Hill decision was formed around a range of processes undertaken via the NT Sacred Sites Act, the ALRA, the Commonwealth *Aboriginal and Torres Strait Islander Heritage Protection Act 1984* and a Resource Assessment Commission inquiry.[6] The High Court Mabo case saw the broadening of the recognition of pre-existing forms of Indigenous land ownership under the ALRA to a national level, transforming Australian property law and creating a new set of rights for those Indigenous peoples able to demonstrate a continuity of their society with that of their forebears at the time of European arrival. For governments and the mining industry it imposed new requirements, enshrined in the Commonwealth *Native Title Act 1993*, to negotiate with and compensate Indigenous peoples recognised as holding or potentially holding native title rights over land sought for development (O'Faircheallaigh 2016).

Soon after the Coronation Hill decision, the Keating-led Commonwealth Government, also wary of a post-Mabo negative investment environment, fast-tracked the Mount Isa Mines proposal to commence the McArthur River Mine development. So, by 1993, unique enabling Northern Territory legislation[7] had been passed to validate the McArthur River mineral leases, and the Native Title Act had been amended to ensure the validity of the McArthur River mineral leases and to apply a non-extinguishment principle and future right to compensation (Young 2009: 8). In this manner, the state-enabled McArthur River Mine commenced operations in 1995.

As early as 1976 the Gurdanji, Yanyuwa, Mara and Garrwa peoples had sought support from the Aboriginal Land Fund Commission to acquire the McArthur River pastoral lease on their behalf.[8] Instead, and despite protests by the Department of Aboriginal Affairs in Canberra, the Foreign Investment Review Board approved the sale of McArthur River to Mount Isa Mines in 1977 (Avery and McLaughlin 1977: 3). Avery and McLaughlin wrote:

---

6   See discussion in Lewis and Scambary (2016: 232).

7   *McArthur River Project Agreement Ratification Amendment Act 1993* (NT).

8   The Aboriginal Land Fund was established in 1974 by the Whitlam Commonwealth Government during the development of land rights legislation to assist Aboriginal groups around the nation to purchase alienated or private lands.

> The attempted purchase of McArthur Station represented for many Gurdanji the last chance to develop their own community in the area. The Station covers core religious sites owned by a wide range of clans. Ownership of the land presented the real possibility for these people conducting an economic cattle enterprise, coupled with traditional foraging practices. (Avery and McLaughlin 1977: 3)

Given that the pastoral lease contained all of the mineral leases and port facilities required to operate the McArthur River Mine, and that had the pastoral lease passed into Aboriginal ownership an almost definitely successful ALRA land rights claim would have ensued, the 1977 sale to the mine operator can only be seen as a strategic move by the Commonwealth and Territory governments to prevent such a land rights claim over the station.

Instead, and nearly 20 years later, in 1994 the Commonwealth entered into the McArthur River Agreement as a compensation deal with the broader Aboriginal community impacted by the project, which included the purchase of the Bauhinia Downs pastoral lease which adjoins McArthur River station some 35 km east of the mine. This purchase allowed for a subsequent land rights claim to be run over that station, benefiting certain Gurdanji families associated with the McArthur River mineral lease areas, but not all of those whose traditional estates were associated with the Yulanji dreaming tracks along the McArthur River itself and at the mine site well to the east of Bauhinia Downs.

Former senior Commonwealth public servant Brian Stacey notes:

> My impression, as a public servant in ATSIC [Aboriginal and Torres Strait Islander Commission] at the time, involved in the administration of the Land Rights Act, was that Aboriginal people were bullied into the McArthur River Mine. I also thought it was a grave injustice to the Aboriginal people of the Gulf who were denied the Right to Negotiate under the Native Title Act 1993. (Stacey 2018: 13–14)

With no land claim, there was no requirement for the company to establish a legal agreement with the traditional owners, and the representative organisation, the Northern Land Council (NLC), was hamstrung in its ability to negotiate with the mining company. In this sense the Commonwealth and Territory governments colluded to create a vacuum from which they absented themselves, leaving the mining company free to operate and determine its own community engagement processes with no accountability to any agreement provisions or statutory processes afforded under land rights or native title laws. The traditional owners were left as impoverished bystanders effectively exposed to the whims of the company (Stacey 2018: 14).

The company took this opportunity to strategically apply their substantial resources to control and divide the community, isolating individual senior traditional owners, manipulating local Aboriginal organisations and undermining opposition to the mine. As noted by Kerins:

> Reports from the 1990s, when the mine was first being developed, tell how MRM [McArthur River Mine] employees actively worked to separate Aboriginal people from their representative organisations and to create mistrust between them. (Kerins 2017: 12)

Mine Expansions

In 2002 mine expansion plans for the McArthur River Mine were announced that included a massive open-cut pit on the river-bed of the McArthur River, which would require a 5.5 km diversion of the river and a bund wall or levee some 100 m thick and 60 m high to block the river's wet season flows (Young 2009: 9). These controversial plans soon gave rise to a new political turn. In 2003 global mining giant Glencore purchased the McArthur River Mine and a six-year highly politicised environmental assessment and approvals war raged in the courts, once again pitting the Northern Territory and Commonwealth governments against local Aboriginal people and their cultural, social and environmental concerns represented by the NLC, as well as considerable environmental opposition from environmental groups and broader society. In reaction to various legal defeats over the approvals process the Northern Territory once again enacted remarkable legislation in 2007 to validate the approvals process—the *McArthur River Project Amendment (Ratification of Mining Authorities) Act 2007* (NT). When the Federal Court declared the Commonwealth Minister's approvals invalid, the Minister simply reissued his approval in a valid manner, allowing the open-cut development to proceed in February 2009.[9]

---

9   For a more thorough summary see Young (2009: 9–15).

Mirroring the 1970s and 1990s, no agreements were in place and Aboriginal rights to and interests in their lands were again ignored, thwarted and overridden by a state that was negatively present in relation to Aboriginal people but very much positively present and active on behalf of the miners (see also Holcombe, Chapter 7, this volume). In 2013 the Northern Territory Government approved the Phase 3 expansion at the McArthur River Mine which allowed for expanded production, enlargement of the open cut and an increase in the size of the tailings storage facility and the footprint and height of the northern overburden emplacement facility (the 'waste rock dump') (Young 2015). Later that year, the waste rock dump began to combust and plumes of sulphur dioxide smoke poured off the mine site for over a year, drawing increased public attention to the project and concern about downstream contamination of the McArthur River itself (Bardon 2016). Glencore's assessment of their waste rock material had proven to be wrong, forcing them into a new environmental assessment process which continues at the time of writing.

In August 2018 the Northern Territory Environment Protection Authority[10] approved the expansion, purportedly favouring ongoing mining over the risk that the company would abandon the project. The mine still requires authorisation from the Territory and Commonwealth governments, including from the Aboriginal Areas Protection Authority,[11] which will need to seek Aboriginal site custodians' views regarding the proposed raising of the height of the waste rock dump from 80 m to 140 m to match the height of the nearby hill which is a barramundi dreaming sacred site called Damangani (Bardon 2018).

Broken Promises

I first engaged with the traditional owners of the McArthur River Mine when working as an anthropologist for both the NLC in the mid-2000s, and then in more recent years with the Aboriginal Areas Protection Authority. This latter work involved conducting sacred site clearances in response to applications for Authority Certificates from Glencore, which were precursors to the approvals process for the most current mine

---

10 The NT independent statutory authority established by and to administer the *Northern Territory Environment Protection Authority Act 2012*.
11 The NT independent statutory authority established by and to administer the *Northern Territory Aboriginal Sacred Sites Act 1989*.

expansion proposals (as described above). Based on this work, which involved various mine site visits, aerial and ground-based sacred site surveys and work on surrounding matters, I came to know the key traditional Aboriginal owners and site custodians for the mine and its surrounds and gained insight into company practices and their community relations.

During this time a senior man, Mr Coolibah, was active as a senior Gurdanji *nimirringi* (or 'owner') for the Mambaliya Wurrini estate on the McArthur River at the mine site. A regular informant for my work, he held strong knowledge of his estate and worked closely with a cohort of other *nimirringi* and *junggayi* ('managers') for his own as well as neighbouring estates and others with traditional connections to the mine area.[12] He had emerged as a senior figure as the 2000s progressed and with the passing of his predecessors and senior Aboriginal figures of the region such as Harry Lansen and Bruce Joy. Throughout my work with him, Coolibah was a quiet but firm man seeking to improve his and his family's lot while clearly being caught between the manipulations and promises made by the McArthur River Mine 'company men' and his broader cultural obligations to sacred sites, country and his fellow 'countrymen'. Trapped by the historic chicanery of the state and the company, Coolibah and the rest of the Gurdanji have been marginalised in relation to the mine while other Northern Territory Aboriginal groups with mining projects on their lands have been able to negotiate benefits under the ALRA. The Gurdanji have experienced this powerlessness in the face of incomprehensible expansions to the mine, destruction of their lands, disturbance and contamination of their waters, and constant threats to their sacred sites.

In 2016 Coolibah was diagnosed with terminal cancer and after treatment in Royal Darwin Hospital he returned home and died in February 2017 surrounded by family. During my many field trips, numerous phone conversations and other engagements with him he regularly spoke at length about the mine and its impacts on him, his people and his society.

---

12   Aboriginal society in the Gulf region (and many surrounding regions) recognises complementary affiliation to land which involves the transmission of rights in country along lineages of patrilineal descendants, who are considered to be 'owners' of an estate of land and its sites and dreamings. Complementary rights are transmitted via matrilineal connections to the same estate with matrifiliates considered to be 'managers' or 'police', whose role it is to ensure that the owners are caring for their country, sites and ceremonies appropriately. In the Gulf region the respective label for 'owner' is *nimirringi* and for 'manager' is *junggayi*. Also see further discussion in the chapter.

The following is a selection of Coolibah's reflections which I collected across my various times with him which I have paraphrased:

> They gave me my block, dongas, motor car, bus, money, tank and fencing—that made me a little bit happy. But we didn't understand—we knew about the old mine, but didn't know it [the open cut and diversion] was going to be such a big mess.

> Me, Bruce and Ronnie tried to stop it [the mine expansion] with a lot of other people. Didn't know it was going really open like that, change that river—you can't do that! We could see they gone in like a mob of worms under the ground, like a mob of termites or pigs digging everywhere [on seeing the underground operation being opened up].

> They ask me I must've been stupid—me and the others. Old Harry was right he tried to block it. We tried to push them to get royalty. They bin breaking my promise man Bruce.

> Mining company want to keep going higher for that dump. We said you asking too much—you need to give us good deal not rough deal. No higher than the barramundi they promised. Now they still asking. Everything changed and biggest mess since company came. Feel no good, sick, I can't stand it.

> Broken promises make me angry. That dreaming tree if that goes they gotta get something back.

> Muminini they already broke him with the bucket. Yakuwala they already broke that one.[13] Any more damage they gotta get a good deal. I'm sad, sick and worried. Gotta go this mine.

> That open cut—you have to blame me for that—I was stupid. My junggayi both said yes. I had to follow that line. Our way I have to follow them—old days you get belted if you ignore them.

> That river diversion the snake line been damaged. That bund wall the Garbula tree gonna get poisoned. Downstream—fish, water— can't eat fish anymore we frightened.

> That open cut—my head wasn't working that day. I had no power.

---

13 During earlier mine works a bulldozer ('the bucket') cut into Muminini, a sacred hill near the mine, and rocks were disturbed at Yakuwala another sacred site located close to current mine infrastructure.

Coolibah's Aboriginal English was good, but the difficulty of communicating the impacts of damage and loss in a brief set of statements is immense (a common challenge with other sacred site matters I have worked on). Not only are these intangible personal and spiritual matters, but they occur within a cultural framework and normative system that never considered industrial-scale damage to land or sites (see Lewis and Scambary 2016: 244–5). In order to explore what Coolibah was seeking to communicate in these paraphrased statements, they need context, expansion and some interpretation.

In the late 2000s the mining company gifted Coolibah a small block of land on McArthur River station near Garanbarini conservation reserve north of the mine. The company also provided some transportable buildings to establish a small outstation. My earliest interactions with Coolibah in the early 2000s were often about his frustrations with this promised outstation. His comments above indicate that he considered this outstation to be personal compensation for the initial 'old' mining operations and the 1994 Commonwealth compensation deal which gave Bauhinia Downs to other Gurdanji people, principally Harry Lansen's family. As Coolibah's traditional Wurrini estate does not extend westwards to Bauhinia Downs, he and his immediate family did not directly benefit from the return of Bauhinia Downs. In this sense Coolibah is saying that his outstation and the few ex-mine buildings that he received from the mine were part of what he had always expected and demanded from the miners (and by implication the state) as compensation from much earlier events in the mid-1990s.

Coolibah's words also express his and other traditional owners' lack of any comprehension of, let alone any free prior or informed consent for, the expansion of the McArthur River Mine to an open-cut project. He laments not following Harry Lansen's resolute opposition to the mine expansion, realising that he and others got caught out trying to get a royalty deal from the company. He notes the shock of the transformation of the landscape with the open cut and the river diversion and he notes that the company 'broke' his *junggayi* Bruce Joy, his now deceased 'promise man'. This was Coolibah's way of saying that it was Bruce who gave in to company pressure and said yes to the expansion work and that he himself had to follow his *junggayi*. Precisely what Coolibah meant by 'broke' remains conjectural: it could mean that the company broke Bruce's resolve, contributed to his death or both.

The *nimirringi–junggayi* relationship remains prominent in Gulf Aboriginal society.[14] Enshrined in ceremonial practices and in the codification of relationships to land this relationship extends into mundane life where the *nimirringi* will seek the approval of his or her *junggayi* for many decisions even in relation to apparently daily or trivial matters. It is always critical for the *nimirringi* to include their *junggayi* in land business, sacred site or cultural matters; failure to do so will usually result in acrimony and fines. In the context of McArthur River Mine approvals, Coolibah was implying a degree of self-redemption in that he was seeking to spread his own culpability for his decisions amongst his contemporaries.

Coolibah spoke of impacts of the latest expansion proposals on the Damangani sacred site, which is the hill associated with a barramundi dreaming immediately north of and downstream from the open cut. The waste rock dump is the subject of ongoing approvals processes with Glencore, who are seeking to increase the height of the current waste rock dump design. This was the subject of much earlier Aboriginal Areas Protection Authority Certificates where the original waste rock dump design was limited in height to 80 m in order to allay the concerns of Aboriginal custodians about perceived impacts on the integrity of the nearby Damangani sacred site. Attempts to dishonour this deal are read by Coolibah and others as a broken promise by the company.

The Garbula tree is another sacred site located within 50 m of the base of the open cut bund wall on the southern side of the McArthur River Mine pit. I visited this site on a few occasions with Coolibah and others and the decontextualisation of this and other sites from their natural state (and their proximity to the mine developments) have been a long-standing source of concern.[15] This tree and others upstream, along with nearby permanent waterholes on the McArthur River, are manifestations of two Yulanji rainbow serpent dreamings which converge at this point of the river. These and other sacred places associated with Gurdanji ceremony and ritual are encompassed within the mineral leases controlled by the mine, which has effectively restricted access to them by Gurdanji and other

---

14 See Footnote 12 for more detail.
15 See Bainton et al. (2012) for a comparable example of a decontextualised (or re-contextualised) sacred geography at the Lihir gold mine in Papua New Guinea.

Aboriginal people for a generation.[16] In my experience of site clearance work which required access to these sites with custodians, the irony is not lost on these custodians that the only time they can readily achieve access to these sites is when the company requires a site clearance for its own needs. Coolibah's words 'Broken promises make me angry' barely capture the frustration, anger and sense of loss and despair that was palpable during such visits. The threats posed to sacred sites by the evolving works at the McArthur River Mine echo previous experiences where other sacred sites, such as Muminini and Yakuwala, were damaged by earlier mining activity in the 1970–80s. Coolibah was expressing concern about any more damage to sites and what this meant to his personal health—he was 'sad sick and worried' by such threats. By 'a good deal' he is alluding to potential compensation for mining and site damage, which I read as an attempted warning and reassertion of his traditional authority as well as his need to be seen by others, especially his *junggayi*, to be exercising his traditional responsibilities by punishing transgressors. When Coolibah stated 'That open cut—you have to blame me for that—I was stupid', he was particularly tearful and it was the first time that I had heard him admit any fault or blame. While, as noted earlier, he qualified this confession at least in part by blaming Bruce Joy as his *junggayi*, it was clearly a powerful and painful realisation.

The diversion of the river has been regarded by many site custodians as damaging the dreaming track followed by the two Yulanji or rainbow serpent dreamings, one from the north and the other (Coolibah's) from the south which converge at the mine site. Coolibah notes the fears held by himself and his broader Aboriginal community about downstream contamination and impacts on their use of traditional resources.

The key theme which emerged consistently in my engagements with Coolibah, represented in the statements presented above, was the trope of the 'broken promise man' which I understood from him to be a multivalent figure that changes in identity as well as temporally to encompass a range of figures both real and conceptual who embody and enact moral uncertainty. The broken promise man is the mining company men who worked on Coolibah over many years manipulating and incentivising his tacit support when needed by making new promises, while delaying

---

16 Section 47 of the *Northern Territory Aboriginal Sacred Sites Act 1989* preserves the right of Aboriginal custodians to access sacred sites located on private land. This right has been heavily mediated at the McArthur River Mine as an active mine site with occupational health and safety risks.

on older promises for his outstation. The same mining company men damaged sacred sites at Muminini and Yakuwala in the past, and now threaten to destroy the Garbula tree and undo earlier promises (and agreements) made about limiting the height of the waste rock dump in relation to the barramundi dreaming at Damangani. The state (in the collective sense of both the Commonwealth and Northern Territory) is also the broken promise man who disempowered Coolibah and his fellow traditional owners by thwarting their land rights and native title claims, denying them the right to a full prior and informed consent process and therefore any negotiating power in relation to the McArthur River Mine.[17] Coolibah's *junggayi* Bruce Joy was also the broken promise man, in that according to Coolibah, he succumbed to the pressure and inducements made by the mining company and by approving the mine expansion, he forced Coolibah as his *nimirringi* to follow him. Finally and tragically, the broken promise man emerges as Coolibah himself, who through tears of regret faced his failures, realised that he too had been manipulated by the mining company, marginalised by the state, and ultimately had broken his own promise—his duty to care for his traditional lands and ancestral sites.

Coolibah's broken promise man ultimately embodies his personal powerlessness, and that of his regional Aboriginal community, in the face of a long history of events unfolding around him associated with the mine. The state's elusive role in facilitating the development meant that the face of Coolibah's protagonists were never fixed and were forever moving like dancers around him, underscoring the moral and institutional uncertainty of this figure. Throughout the years I knew him and despite his best efforts, he never managed to engage with the state as the power behind the mine, because unlike the company men, the state actors were never present or accessible to him (see also Skrzypek, Chapter 2, this volume). The power of the state, embodied and exercised in the activities of the company, have produced what Fredericksen and Himley (2019: 58–9) call extractive subjects, in the form of Coolibah and his countrymen, who are simultaneously shaped by the exercise of power relationships at this extractive frontier and integrated into regional, national and global extractive economies.

---

17   I think that while Coolibah probably understood that these layers of government were geographically and politically distinct holding different powers, like many Aboriginal people of his era and older, he did not differentiate between the two. Both were seen as 'Government mob', a unified form of externalised non-Indigenous power.

I was unable to attend Coolibah's funeral in Borroloola but a colleague recalled the event to me as one dominated by high-visibility work shirts, corporate logos and a funeral procession led by a Glencore-marked tray-back utility carrying Coolibah's coffin. The eulogy was read by the company man most associated with Coolibah, the Health, Safety and Environment Manager, who recalled his friendship and noted with apparent pride that Coolibah had given him an Aboriginal name. Upon hearing this I recalled Coolibah telling me during a site survey a few years earlier that he had given this man a name which was intentionally unflattering and reflective of his role in a litany of broken promises made by the company. So it was with deep irony that this company man, one of Coolibah's broken promise men, eulogised his 'friend'. In death Coolibah was fully appropriated by the company, at least symbolically via the funeral, transforming him from an excluded, vociferous and politically volatile character, into a remembered, revered and now silent unproblematic stakeholder. Glencore had, to use Fredericksen and Himley's terms (2019: 59), transformed Coolibah from an extractive subject into the perfect extractive citizen.

Open Cut I Cut Open

A few months after Coolibah's death, an exhibition of artwork and photographs protesting the McArthur River Mine was launched in Darwin's trendy market and gallery precinct of Parap. The modest exhibition of seven paintings by Jacky Green Warngkurli, accompanied by a series of black and white portrait photographs of Gulf Aboriginal people by Therese Ritchie, was orchestrated by Sean Kerins, a research fellow at the Centre for Aboriginal Economic Policy Research at The Australian National University (Figure 8.2). Jacky is a Garrwa Mambaliya man who I have worked with alongside other Gurdanji custodians, including Coolibah, on sacred site clearances at the McArthur River Mine. His long-standing public opposition to the project has meant that he is not popular with Glencore, who have resisted his access to the mine site and questioned his traditional connections to the area. In my experience Jacky's ceremonial connections to sites at the mine and his broader knowledge of regional matters has meant that he has been considered by all senior Gurdanji to be an essential part of the sacred sites work that I undertook.

The 2017 exhibition was entitled 'Open Cut'. One wall was occupied by a selection of Jacky's uncompromising paintings depicting the mine and environmental and cultural damage it has reaped upon the lands and people of the Gulf. Each painting was accompanied by a text from Jacky providing his insight into the mine and its irreconcilable conflict with his cultural values and those of his community. One example is the painting entitled *Red Country* (see Figure 8.3).

The exhibition catalogue for this piece reads:

> *Red Country*
>
> Right across the McArthur River region are The Dreaming tracks of the ancestral beings. The barramundi, the two snakes who travelled together and the one that come up from the south. The Rainbow Snake and the Stinking Turtle. They all there. So too are the places where they coiled or rested, or went down under the earth like at the place I have marked in the river. Big name places, important and sacred places, they are right across the region and they tie people to places and people together.
>
> Right in the middle of this sacred country is a torn-up place, right where the Sacred Tree is that forms part of the Rainbow Snake story. It's a big name place, right where the massive open cut pit now is. The black represents the hole that keeps getting bigger and bigger and the brown represents how the mining company is now talking about stuffing all the toxic waste rock back in the hole before they take-off with their money and leave us and generations to come with their toxic mess. (Green et al. 2017)

Such irreconcilable contestation over resource extraction in the Northern Territory has been a theme that I have noted elsewhere with Benedict Scambary:

> This pattern of conflicting interests demonstrates how such contests have commonly focussed on and been articulated around matters of Aboriginal sacred sites. They have involved the collision of interests where areas of geological and mineralogical significance in the form of mineral deposits and ore bodies coincide with places of cultural significance to Aboriginal people in the form of sacred sites. (Lewis and Scambary 2016: 222)

Figure 8.2 Jacky Green and Sean Kerins at Open Cut, Darwin, 2017
Source: Photograph by author

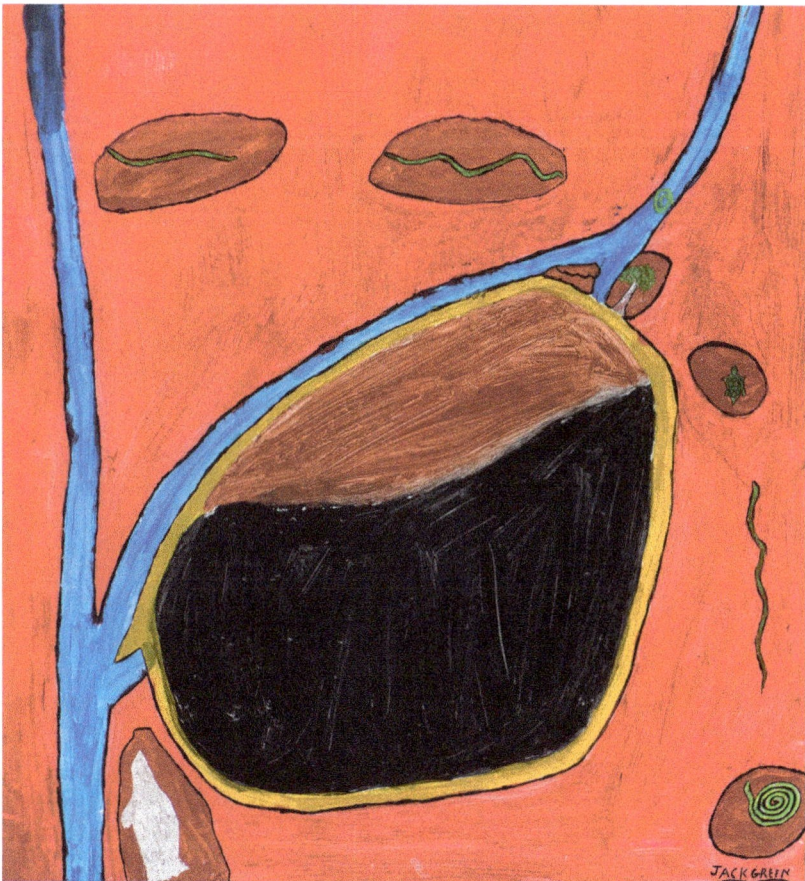

Figure 8.3 Jacky Green, *Red Country*, 2017, 96 x 88 cm, acrylic on canvas
Source: © Jacky Green

On the wall opposite Jacky's paintings were a set of black and white photographic portraits taken by Therese Ritchie of Gulf Aboriginal people with words and statements painted across their bodies in English and Garrwa. Jacky's portrait has 'Cut Open' across his chest, others include 'Intergenerational Despair', 'Hope', 'Lead', 'Sovereignty' and more. Kerins notes:

> Just as previous generations of Aboriginal people from the Gulf country have used artistic expression to make their voices heard and their feelings known, Aboriginal people in the Open Cut exhibition continue to do so, scarifying their bodies to speak to power. (Kerins 2017: 13)

Such use of the body to communicate with the broader community reflects the heart of this contest over resources. As Scambary and I previously wrote, and extending from the above quote:

> The phrase sacred bodies is used here with the intent of drawing focus onto the physical, and very much bodily, relationships which exist between place and people in Aboriginal Australia, as well the cultural, emotive and religious systems and traditions which underpin these relationships. We consider that such an emphasis is reflective of Aboriginal discourses about the impacts of damage to sacred sites … how the literal undermining of Aboriginal sites damages individuals, tears at the fabric of their societies and undermines the very ability of already vulnerable cultural groups to reproduce their traditions and thereby themselves. (Lewis and Scambary 2016: 222)

Prior to the 2017 exhibition, I was undertaking non-McArthur River Mine related fieldwork in Borroloola, Jacky took me aside and proudly showed me his 'little song' he had written about *Ngabaya* 'the devil'. The key lyric components as explained by Jacky were: *Ngabaya* [devil] *ngyirridji* [fiddles] *gunindjba* [digging] ('the devil that fiddles and digs in our country').

The *Ngabaya* is a generic dreaming figure prominent at sacred sites and dreaming tracks which cross the region. *Ngabaya*, often called 'devil', is more accurately an anthropomorphic spirit figure of ambivalent nature rather than embodying any evil in the Christian tradition.[18] I mistakenly

---

18   As John Bradley and Yanyuwa families note (2010: 195): 'The generic term *ngabaya* refers to any human-like spirit being, often translated as "ghost", "spirit" or sometimes by the Kriol term *debildebil* ("devil", which although probably originating in Christian teaching, was not generally thought of as a spirit of greater evil than other spirits)'.

interpreted Jacky's song to be demonising the mining company, but Jacky corrected me and described his more nuanced use of this character as referring to:

> anyone who messes about with country and damages things— it can be a mine operator, an Aboriginal person on a grader or a government bloke … anybody, they are dangerous because they don't understand, they interfere and cause trouble … (Jacky Green, personal communication, January 2017, Borroloola)

With a multivalence similar to Coolibah's 'broken promise man', this song about the *Ngabaya* was performed by Jacky along with other songs at the opening of the Open Cut exhibition in Darwin. On the end wall of the Open Cut exhibition was an annotated graphic timeline connecting the contact history of the colonial frontier 'wild times' with the policy environment of the present—which might be interpreted as a way of using memory to cope with uncertainty. Illustrated with photos and 'credits' for the notable actors in the history of the Gulf generally and the McArthur River Mine development in particular, this timeline presents like a wall of shame graphically demonstrating the connections between past and present for Aboriginal people of the Gulf region. Reflecting on the images displayed on the wall including their own photographic portraits, Gulf artists Stewart Hoosen[19] and Nancy McDinny[20] were quick to draw parallels between the violence of murderous frontier times to current experiences of marginalisation alongside the cultural and environmental damage inflicted by the McArthur River Mine. There is a consistency in this experience underpinned by the malevolent role of the state both past and present.

The Malevolent Absence of the Neoliberal State

For the Gurdanji, Yanyuwa and Garrwa in the Gulf, as for many Aboriginal people affected by other resource extraction projects across the Northern Territory, the state is often perceived as an absent actor. Engagement between miners and Aboriginal people has either been direct or preferably mediated by effective Indigenous representative organisations such as land

---

19 See Cross Art Projects (2017a).
20 See Cross Art Projects (2017b).

councils using legislative frameworks such as the ALRA or the Native Title Act. The Northern Territory experience of visiting politicians courting favour with and seeking project approvals from Aboriginal leaders during the early land rights era of the late 1970s and 1980s has faded. Safari suits have given way to high-vis shirts and hard hats as the neoliberal state has increasingly withdrawn from its remote servicing interests and responsibilities in favour of, and deference to, corporate community benefits packaged up as a social licence to mine, in what can be seen as an outsourcing of state responsibilities (see Scambary 2013: 9–10; see also Everingham et al., Chapter 9, this volume).

As noted earlier Australian Indigenous affairs has been the subject of major policy shifts over the last 40 years. The most recent of these shifts has been generated around the application of neoliberalism, which has witnessed the extension of market mechanisms into areas of the community previously organised and governed in other ways, especially in Indigenous communities, allowing for a broad reduction in the ambit and role of the state (Howlett et al. 2011: 312).[21]

From the self-determination era which ushered in land rights, Indigenous-specific policy making and a decolonising approach in the late 1970s, to the revival of assimilationism and mainstreaming policies of the Howard Government in the 2000s, current approaches to universalise, 'westernise' and 'incorporate' remote Aboriginal populations have, according to Jon Altman:

> been a mix of politics, ideology and cultural critique that has driven the current policy project to neoliberalise geographically remote Aboriginal people so as to free them from a relational ontology that emphasises commitment to family and kin, ties to place, and obligations in ceremony. (Altman 2014: 124)

For the mining industry operating across the physical and policy space of Indigenous Australia, the balance between unimpeded development, statutory requirements and social legitimacy or licence has both followed and influenced these spaces. From the dire warnings of industrial collapse in the face of land rights as companies were dragged into agreement-making processes with the Northern Territory land councils, to the seemingly enlightened community engagement approaches from the mid-1990s,

---

21   For thorough summaries of the Australian context see Altman (2014) and Howlett et al. (2011).

there have been many shifts in these balances.[22] Bound up in fluctuating terms such as 'community engagement', 'social licence', 'social licence to operate', 'social sustainability' or 'corporate social responsibility', the role and responsibility of mining companies in local as well as broader community affairs has become increasingly scrutinised in recent decades. Whether driven by demands from activists, civil critique, community pressure or by shareholder demand, the need to at least claim to be creating 'social value' or 'delivering benefits' has been driven more from within the industry rather than through any particular state intervention.

In my experience with the McArthur River Mine and elsewhere, social value via community engagement is a function often lumped in with the duties of other operational disciplines such as human resources, public relations, environment, or occupational health and safety. It was the Health, Safety and Environment Manager who played the key role in engaging with Aboriginal traditional owners such as Coolibah (including eulogising at his funeral, as noted earlier), and facilitating site visits at the McArthur River Mine. At the OM Manganese mine at Bootu Creek near Tennant Creek it was a staff member in a similarly titled role that organised liaison committee meetings, mine site visits, and fencing around an identified but ultimately ill-fated and desecrated sacred site (Lewis and Scambary 2016: 235–9).

The commentary I have received from the communities subjected to these engagement roles suggests an unsophisticated yet highly selective approach of capturing senior traditional owners. This commonly occurs via rudimentary support in the form of directing jobs to immediate family members, and/or providing valued resources such as spare parts or tyres for vehicles. Such forms of 'social value' or 'community benefit' appear to be driven purely by corporate self-interest seeking to appease and manipulate potentially problematic individuals—creating extractive subjects—rather than by any underlying ethic or desire to meaningfully support communities impacted by their operations or to mitigate these impacts.

John Owen and Deanna Kemp (2017: 33) identify the vagaries of terminology as one part of the problem. Being a poorly defined, metaphorical and intangible concept, social licence (or social licence to operate) in mining has been interpreted and applied inconsistently

---

22 For an excellent summary of these shifts in relationships between the mining industry and Indigenous peoples in Australia see Altman (2009: 2–27).

by industry, regulators, commentators and impacted communities. In their analysis of the global mining industry this manifests as a counterproductive clash between an industrial ethic (where companies are exclusively focused on profit) and countervailing powers which seek to resist corporate control, influence and impact. Building on Galbraith's modelling of such countervailing powers, they argue that an effective and mature 'social function' within mining companies is required to moderate their own self-interest (Kemp and Owen 2018: 492) and help to minimise some of the most perverse impacts upon local communities. The absence of this social function perpetuates self-interest and unleashes a range of consequences.

As detailed earlier, within this industrial and state policy framework, the McArthur River Mine has been allowed to occupy a unique space created by the state to ensure that the development of the project was unimpeded by Aboriginal land claims or native title claims. The state both actively removed what could have been a legitimate external countervailing power to the project, and allowed and legitimised the absence of any effective social function (or internal countervailing power) within the project. Kemp and Owen have observed a global decline in corporate commitment to the development of the social function and the capability to manage the social aspects and impacts of mining. They argue that 'the demise of the social function will only be reversed, or rebuilt, if and when the industry is subjected to … conflicts and scandals that cannot be concealed by the machinery of public relations' (Kemp and Owen 2018: 498). But in the case of the McArthur river mine, such scandals and conflicts have thus far had little impact upon the company.

In their work on transnational governmentality and resource extraction, Suzana Sawyer and Edmund Gomez state that 'Common wisdom holds that public–private collaborations among governments, IFIs [International Financial Institutions] and multinational corporations (MNCs) will enhance social well-being by eradicating poverty, promoting sustainable forms of economic development, protecting the environment and advancing the rights of indigenous peoples' (Sawyer and Gomez 2008: iii). Contrary to this common wisdom, it is recognised that there are 'monumental' and 'baffling' problems and growing concerns over the adverse impacts of resource extraction projects on Indigenous peoples across many nations. Sawyer and Gomez note that despite many government and corporate attempts to create inclusive consultative institutions aimed at affording Indigenous peoples property, veto and other empowering rights over resource extraction projects, the majority of Indigenous peoples find themselves increasingly subjected to discrimination, exploitation, dispossession and racism (2008: 1). The public–private partnerships they have reviewed, including case studies from Australia, demonstrate how a combination of institutional capture, clientelism and political positioning undermines the neutrality of the state and its capacity to protect Indigenous communities. They see this manifest most conspicuously through the funding of political parties by mining corporations and key appointments made to the boards of directors of international financial institutions. Mining companies have then used such donations and appointments to influence state policies and laws to serve their interests.

For the McArthur River Mine such capture is evident in the earliest stages of the project via the legislative support and approvals processes described earlier. In September 2017, Rod Campbell of the Australia Institute reviewed the economic modelling for the Phase 3 expansion of the McArthur River Mine which Glencore claimed will generate tax and royalty payments of over AUD1.5 billion. His findings deplore these claims as based on assumptions which he found to be 'almost comical'. He summarises that their modelling included payroll tax payments over a thousand years to the year 3017, estimates of AUD435 million in royalty payments—which are not actually paid at all in most years—and estimates of AUD1,035 million in company taxes when Glencore regularly pays no company tax at all in Australia.

Adding to this picture, the September 2017 disclosure that the size of Glencore's environmental bond for the McArthur River Mine, at AUD476 million, is estimated to only cover 25 years of mine life and rehabilitation when the actual costs are likely to extend for hundreds of years (Vanovac and Breen 2017).[23] Campbell concludes his analysis by pointing out that:

> The NT government estimates their cost of cleaning up abandoned mines in the NT would be over $1 billion. Given this history and the present reality of the McArthur River Mine's impacts on the environment and local community, it is likely that the environmental costs imposed by the mine will be far greater than the uncertain revenues that could be generated. From an

---

23   This disclosure was only possible after legal action taken by Jacky Green, represented by the Environment Defenders Office of the NT.

> economic perspective, it is likely that the best approach would
> be to close the mine and rehabilitate the site, and to ensure this is
> paid for by Glencore. (Campbell 2017: 17)

Barrister Anthony Young has also reflected on the long record of failings of the environmental assessment process at the McArthur River Mine by successive Northern Territory and Commonwealth governments. He cites deliberate and reckless political choices and the ignoring of scientific and expert advice, which amplify the potential for the mine to cause catastrophic environmental damage:

> The history of the McArthur River mine shows that governments
> are willing to sacrifice the health of the environment and of those
> directly reliant on it, such as the Aboriginal people of the region,
> to the mining industry and economic interest. (Young 2015)

> The Phase 3 process appears to follow this trend with Glencore
> forced into a new approvals process after its failings in its assessment
> of the composition of waste material being placed on the NOEF
> materials [the northern overburden emplacement facility, or waste
> dump]. (Bardon 2016)

This mirrors the international findings of Sawyer and Gomez (2008: 28) who note the failings of self-regulated corporate social responsibility where the state is too easily compromised by the overwhelming influence of transnational capital (see also Bainton and Macintyre, Chapter 4, this volume). The state cannot effectively regulate or monitor what it is authorising in the name of investment or development. Sawyer and Gomez identify transnational solutions:

> What is required is not public-private compacts but an effective
> arms-length and accountable relationship between governments
> and MNCs [multinational corporations] to deal with corruption,
> environmental degradation and violence. A viable institutional
> framework is thus required to honourably compensate local
> populations, including indigenous peoples, for the disruptive
> effects of resource extraction, not involving the state, given that it
> lacks neutrality, but an independent monitoring body, such as the
> UN (and possibly NGOs). (Sawyer and Gomez 2008: 28)

I have argued above that this accountability and a compensatory approach to the McArthur River Mine was available but avoided by the state from the start. Had a land purchase over McArthur River Station been allowed in the mid-1970s and had the ALRA been allowed to run its course, an agreement would have ensued between the traditional Aboriginal owners and the company mediated by the NLC, an independent Commonwealth statutory body. Such an agreement, while not necessarily addressing all environmental matters at the mine, nor addressing all matters for all local Indigenous peoples, would have still been an empowering move, the benefits of which are now impossible to quantify. Instead what emerged in relation to the McArthur River Mine is that the state, in the form of both the Territory and Commonwealth governments, has been captured by extractive capital interests and actively undermined readily available solutions of the very sort identified by Sawyer and Gomez. In removing itself as much as possible in order to acquiesce to extractive interests, the malevolence of the state is thoroughly discernible.

Adding to the analytical commentary above, the McArthur River Mine has also attracted regular negative media coverage and reporting on aspects of Aboriginal opposition, environmental failures and regulatory concerns. In the face of such criticism and concern, the state approvals for the project have flowed. It seems that despite such coverage, controversy and opposition, there has been little cause for alarm at Glencore headquarters and at the mine site: no discernible social function exists while their pro-mining commercials roll out regularly on Darwin television, which might be interpreted as a stronger corporate commitment to public relations than to actual local community relations.

The impoverished countervailing power of Indigenous interests expressed in the words of Coolibah and the protest art and imagery of people such as Jacky Green in Open Cut may have limited direct impact on corporate power, but it can at least draw this power to funerals and public protests, like the Open Cut exhibition. It demonstrates the unease which can be generated by the mobility of extractive subjects as unpredictable and uncontrolled characters. Just as Coolibah was his own broken promise man, Jacky Green in this sense embodies his *Ngabaya* figure, interfering with and disrupting the corporate power exerted on him and his fellow extractive subjects. Like other forms of Indigenous protest and opposition to mining, the mobilisation of art, words and story in social and other media in Open Cut appears to also be generative of reaffirmations of cultural beliefs and local discourses blending themes of cultural revival and resistance.[24]

24　See discussion in Lewis and Scambary (2016: 246–7).

Conclusions

The various operators of the McArthur River Mine have been the beneficiaries of a state that has been positively present for resource developers and negatively present for Aboriginal communities—evidencing a dialectical relationship between the absence and the presence of the state. The mine operators have benefited from Territory and Commonwealth government approvals and manoeuvring to thwart Aboriginal land claims and agreements and from unique legislative measures and amendments to negate all legal challenges to date. This situation also suggests a dialectical relationship between the institutional capture and the political positioning of the state. The captured state, whether located in Canberra or Darwin, has advocated at the highest possible levels for this development while absenting itself from the locale of development and divesting its functions into decentralised and distant forms of regulatory and corporate power.

For the Aboriginal people of the Gulf, the captured neoliberal state has maintained the malevolent absence of the colonial state. Bound up in the same development agenda, with a shift from cattle to base metals, and employing comparable if less violent techniques of divesting power over local affairs into the hands of remote self-interested parties, while absenting itself from the consequences. In the expressed perceptions, transmitted memories and experiences of the Aboriginal people of the Gulf, the state has exercised a malevolent absence across the region for 130 years. Absent decision makers have wielded power from afar directing punitive police raids, arming pastoralists and civilians, and approving mineral leases and mine expansions. The malevolent and selectively absent state looms as the agency behind Jacky Green's *Ngabaya* devil figure instantiated in art, song and protest, and emerges within Coolibah's dying attributions, confessions and absolutions of the broken promise men—it has subjectified Aboriginal people of the Gulf first as colonial subjects and now as extractive subjects.

Returning to the quote from Jacky Green prefacing this chapter, it was *they* that cut us open, *they* that pushed us aside, *they* who invaded. This collective and multivalent otherness of *they*—the police, the pastoralists, the miners, the Whites—has been consistently empowered by the state feigning absence yet looming malevolently present. As the authority behind massacres inflicted so recently that the victims include the grandparents of the remaining current senior generations of Gulf Aboriginal people—

and as the authority preventing current and future generations from protecting their cultural sites and from sharing in the benefits being extracted from their traditional lands—the state has ultimately failed to protect its Aboriginal subjects and citizens of the Gulf, breaking its most fundamental of promises.

References

Altman, J., 2009. 'Indigenous Communities, Miners and the State in Australia.' In J. Altman and D. Martin (eds), *Power, Culture, Economy: Indigenous Australians and Mining.* Canberra: ANU E Press (Centre for Aboriginal Economic Policy Research Monograph 30). doi.org/10.22459/caepr30.08.2009.02

———, 2014. 'Indigenous Policy: Canberra Consensus on a Neoliberal Project of Improvement.' In C. Miller and L. Orchard (eds), *Australian Public Policy: Progressive Ideas in the Neoliberal Ascendancy.* Bristol: University of Bristol, Policy Press. doi.org/10.2307/j.ctt1ggjk39.13

Avery, J. and D. McLaughlin, 1977. 'Submission by the Northern Land Council to Aboriginal Land Commissioner on Behalf of Traditional Aboriginal Owners in the Borroloola Region of the Northern Territory.' Darwin: Northern Land Council.

Bainton, N.A., C. Ballard and K. Gillespie, 2012. 'The End of the Beginning? Mining, Sacred Geographies, Memory and Performance in Lihir.' *Australian Journal of Anthropology* 23: 22–49. doi.org/10.1111/j.1757-6547.2012. 00169.x

Bardon, J., 2016. 'The Race to Avert Disaster at the NT's McArthur River Mine.' Australian Broadcasting Corporation, 12 February. Viewed 25 September 2020 at: www.abc.net.au/radionational/programs/backgroundbriefing/the-race-to-avert-disaster-at-the-nts-mcarthur-river-mine/7159504

———, 2018. 'EPA Recommends McArthur River Mine Expansion, Despite History of Environmental Incidents.' Australian Broadcasting Corporation, 11 August. Viewed 25 September 2020 at: www.abc.net.au/news/2018-08-11/mcarthur-mine-river-expansion-recommend-epa-glencore-environment/10 108052

Bradley, J and Yanyuwa families, 2010. *Singing Saltwater Country: Journey to the Songlines of Carpentaria.* Sydney: Allen and Unwin.

Campbell, R., 2017. 'Wishful Zinking: Economics of the McArthur River Mine.' Canberra: Australia Institute.

Cross Arts Projects, 2017a. 'Stewart Hoosan.' Video, 4 December. Viewed 13 October 2020 at: youtu.be/0BRqjOf3oI4

———, 2017b. 'Nancy McDinny.' Video, 4 December. Viewed 13 October 2020 at: youtu.be/etu9XJHpo4A

Frederiksen, T. and M. Himley, 2019. 'Tactics of Dispossession: Access, Power, and Subjectivity at the Extractive Frontier.' *Transactions of the Institute of British Geographers* 45: 50–64. doi.org/10.1111/tran.12329

Green, J., T. Ritchie and S. Kerins, 2017. 'Open Cut: An Exhibition by Jacky Green, Therese Ritchie and Sean Kerins.' Darwin: Northern Centre for Contemporary Art.

Howlett, C., M. Seini, D. McCallum and N. Osborne, 2011. 'Neoliberalism, Mineral Development and Indigenous People: A Framework for Analysis.' *Australian Geographer* 42: 309–323. doi.org/10.1080/00049182.2011.595890

Kemp, D. and J.R. Owen, 2018. 'The Industrial Ethic, Corporate Refusal and the Demise of the Social Function in Mining.' *Sustainable Development* 26: 491–500. doi.org/10.1002/sd.1894

Kerins, S., 2017. 'Open Cut: Speaking Truth to Power.' *Land Rights News— Northern Edition* 2017(4): 12–13.

Kerins, S. and J. Green, 2016. 'Indigenous Country in the Southwest Gulf of Carpentaria: Territories of Difference or Indifference?' In W. Sanders (ed.), *Engaging Indigenous Economy: Debating Diverse Approaches*. Canberra: ANU Press (Centre for Aboriginal Economic Policy Research Monograph 35). doi.org/10.22459/caepr35.04.2016.09

Lattas, A., 1987. 'Savagery and Civilisation: Towards a Genealogy of Racism.' *Social Analysis* 21: 39–58.

Lewis, G. and B. Scambary, 2016. 'Sacred Bodies and Ore Bodies: Conflicting Commodification of Landscape by Indigenous Peoples and Miners in Australia's Northern Territory.' In P. McGrath (ed.), *The Right to Protect Sites*. Canberra: Australian Institute for Aboriginal and Torres Strait Islander Studies.

MRM (McArthur River Mine), 2018. 'Overburden Management Project: Draft Environmental Impact Statement.' Submission to the Northern Territory Environment Protection Agency. Retrieved 13 October 2020 at: ntepa.nt.gov. au/environmental-assessments/register/mcarthur-river-mine/mcarthur-river-mine-overburden-management-project-2018/draft-environmental-impact-statement

O'Faircheallaigh, C., 2016. *Negotiations in the Indigenous World: Aboriginal Peoples and the Extractive Industry in Australia and Canada*. London: Routledge. doi.org/10.4324/9781315717951

Owen, J.R. and D. Kemp, 2017. *Extractive Relations: Countervailing Power and the Global Mining Industry*. London: Routledge.

Peterson, N. and M. Langton, (eds), 1983. *Aborigines, Land and Land Rights*. Canberra: Australian Institute of Aboriginal Studies.

Roberts, T., 2009. 'The Brutal Truth: What Happened in the Gulf Country.' *The Monthly*, November 2009.

Sawyer, S. and T. Gomez, 2008. 'Transnational Governmentality and Resource Extraction. Indigenous Peoples, Multinational Corporations, Multilateral Institutions and the State.' Geneva: United Nations Research Institute for Social Development, Identities, Conflict and Cohesion Programme (Paper 13).

Scambary, B., 2013. *My Country, Mine Country: Indigenous People, Mining and Development Contestation in Remote Australia*. Canberra: ANU E Press (Centre for Aboriginal Economic Policy Research Monograph 33). doi.org/10.22459/CAEPR33.05.2013

South Australia, 1887. 'Government Resident's Report on the Northern Territory.' Office of the Government Resident (Northern Territory). Adelaide: Government Printer.

Stacey, B., 2018. 'The Failure of Public Policy: An Anecdote about Borroloola.' Land Rights News—Northern Edition 2018(3): 13–15.

Vanovac, N. and J. Breen, 2017. 'McArthur River Mine: Toxic Waste Rock Ongoing Problem, Security Bond Inadequate, Report Finds.' Australian Broadcasting Corporation, 21 December. Viewed 25 September 2020 at: www.abc.net.au/news/2017-12-21/mcarthur-river-mine-toxic-waste-rock-ongoing-problem-new-report/9278922

Walker, E., 2013. 'Yirrkala Bark Petitions.' *Indigenous Law Bulletin* 8(7): 33.

Young, A., 2009. '"… Jealous for Our Country".' Darwin: William Forster Chambers.

———, 2015. 'McArthur River Mine: The Making of an Environmental Catastrophe.' Darwin: William Forster Chambers.

9

The State's Stakes at the Century Mine, 1992–2012

Jo-Anne Everingham, David Trigger
and Julia Keenan

Introduction

Century Mine, in the lower Gulf of Carpentaria region of far northwest
Queensland, was once the third largest zinc mine in the world. It provides
a distinctive example of the evolution of resource relations between
governments, miners and local communities. It also evidences tensions
in Australia between conceptions of the state's responsibilities to ensure
economic development on the one hand and the well-being of regional
populations, particularly Indigenous people, on the other hand. It is
therefore a suitable empirical case for exploring the complex sociopolitical
process that unfolds over time as multiple state actors engage with the private
sector and remote regional communities, including Indigenous Australians,
residents of regional towns and pastoralists. Reflecting upon his involvement
in this project, the lead negotiator for the company described Century Mine
as a classic example of the recent trend for 'social, environmental, cultural
and heritage issues [to play] … a significant part in the development of
major projects, particularly in mining' (Williams 1996: 1).

Over a century ago, colonial interactions in Northern Queensland were
characterised by considerable fighting and bloodshed between Aboriginal
people and others, and on occasions between different Aboriginal
groups in the regions. There were a number of massacres recorded
and settler diseases also took their toll on the Aboriginal population

(Evans et al. 1975).[1] It was a period of dispossession as land was occupied by pastoralists who often proceeded to force the Aboriginal people away from their land and waterholes. The police presence in the area focused on controlling attacks on settlers and their stock, but provided little protection or justice for Aborigines. Early legislation enabled controls over the lives of the Indigenous people through the establishment of the Protectorate system that gave administrators the power to remove people to government 'reserves' and church missions. The violence, dispossession and confinement of different language communities and family groups into common, small settlements, together with repressive missionary and colonial administrative practices all contributed to substantial Indigenous social crises and economic marginalisation (Rowley 1970).

The legacy of this historical experience, and the physical distance from key population centres, contributed to the situation whereby in the 1990s, despite an abundance of natural assets, development in the lower gulf region, especially for Aboriginal residents, lagged behind that of most other regions of Queensland. This was due to a combination of factors: sparse population; remoteness from urban centres; poor transport, communications and other infrastructure. As well, for Indigenous people there was entrenched welfare dependency, weak governance institutions and tenuous relationships with various levels and agencies of government. The fact that these conditions persisted after the Century Mine project was underway calls into question a central tenet of long-discredited 'trickle-down' concepts of development (Arndt 1983) and Australian national development, namely 'The proposition that large-scale natural resource development is unambiguously beneficial to all of the citizenry' (Trigger 1998: 155).

The discovery of the Century deposit in 1990 on the Lawn Hill Station pastoral lease spurred further exploration in the region and highlighted two factors shaping the government's role—the priority given to corporate interests as a foundation for regional development; and the historical record of officially denying and curtailing the rights of Indigenous people. In this chapter we argue that, because of distinctive sociopolitical conditions, these two motives spurred an unusually prominent state presence in the pre-approvals phase of the mine. However, after the initial operational years, the state became more absent and adopted a less direct role, with contracted agencies delivering many government-funded programs and Indigenous residents depending on the mine for many functions.

---

1   See also Lewis, Chapter 8, this volume, for comparable accounts in the Northern Territory.

This temporal pattern in the state's active presence had consequences for the widely held expectation that locals should benefit from development of a mine and that the state will ensure that mines are designed and operated in a way that builds local capacities, creates long-term economic and social opportunities, strengthens local communities and preserves or rehabilitates local heritage and the environment (Altman 2014).

This chapter examines the changing presence of the state experienced in the lower gulf region of Queensland during the 20 years from 1992 when the High Court made its historic Mabo decision[2] and negotiation of the Gulf Communities Agreement (GCA)[3] commenced in earnest. Martin notes that the GCA's 'formal signing followed protracted regional negotiations, initially outside the ambit of the Native Title Act 1993 (Cth) ("the Act") and involving the broader Gulf Aboriginal community' (1998: 4). The 20 years under discussion covers the period of these negotiations, mine construction and commissioning and the production phase until the final (i.e. 15-year) review of that agreement as zinc production began to wane. The GCA is central to this case as the legislated mechanism for dealings between the signatories—namely the Queensland Government, the mining company and the designated 'Native Title Groups'.[4] This period, encompassing the major landmarks in the planning, commissioning and operating of this large mine, provides clear illustration of the leverage of diverse regional actors, both Indigenous and non-Indigenous groups of residents and various sections of government, to procure the state's active presence and state action in the region. It also reveals that initial active engagement by the state faded to less direct involvement after production was underway and royalties were flowing.

Our research investigates how the state presence was experienced and understood among Aboriginal people with respect to Century Mine and how that changed over time. We present the argument in three main sections, beginning by sketching the background to the Century case and

---

2   The Mabo decision was the first time that an Australian court accepted that the pre-existing laws and customs of original inhabitants prior to European colonisation constituted a form of land and water tenure recognisable in Australian law.

3   The Gulf Communities Agreement (GCA) is a legal contract intended to cement an ongoing relationship between three native title parties (Waanyi, Mingginda and Gkuthaarn-Kukatj), referred to as native title groups in the GCA; Century Zinc Mine; and the Queensland Government. It came into effect in 1997. It is a comprehensive agreement about land use and benefit sharing in the lower Gulf of Carpentaria, Queensland, Australia covering a range of issues including compensation payments, land tenure arrangements, employment and training, and environmental and cultural concerns.

4   Groups of Aboriginal people claiming a historic and cultural attachment to parts of the land and sea affected by the proposed mining lease and mine development.

the economic and Indigenous context. The main body of the chapter follows and details a sequence of phases of the state's role in two pertinent policy areas: regional economic development and Indigenous affairs. It makes considerable reference to the GCA as a significant formal expression of the state's 'presence' in the mining project since this agreement departed from the usual Indigenous benefit agreement configuration by including the Queensland Government as an active party rather than being confined to community and company undertakings (Scambary 2009).

The evidence to underpin this argument is drawn largely from secondary sources and from a series of studies conducted by the authors in the region during the period under review (1992–2012). These studies employed diverse qualitative methods and allowed us to examine aspects of the state–community–company relationships that rely on the memories and perceptions of those directly involved in historic events.[5] The final section of the chapter discusses the sequence of phases in relation to governance theories and concludes by noting the disparate views revealed about how present the state should be in a development like Century Mine and the critical, yet fragile, role that formal agreements can play in engaging the state and compelling its presence in mining-induced development—especially in peripheral localities.

Background

Century Mine and its associated infrastructure (including a slurry pipeline, port and processing facilities) are located approximately 250 km north-northwest of Mt Isa, Queensland, in the lower Gulf of Carpentaria, on traditional lands of the Waanyi, Mingginda, Gkuthaarn and Kukatj people (see Figure 9.1). As noted above, the remote, monsoonal, hot and sparsely populated region experienced from early colonial times a turbulent history of relationships between the traditional inhabitants and settlers and a state intent on various kinds of development. The main communities within the lower gulf fall within two kinds of local government areas comprising the Aboriginal shire of Doomadgee (formed in 2004 under a Deed of Grant

---

5   For example, a roundtable in November 2014 hosted by the University of Queensland's Centre for Social Responsibility in Mining and MMG Century provided valuable historical details and insights. Eleven people who had played a prominent role in negotiating or implementing the GCA at various stages and who represented a mix of company, government and community perspectives attended the roundtable (proceedings and reflections of this roundtable are publicly reported by Brereton and Everingham 2016).

in Trust) as well as the surrounding shire of Burke and parts of adjacent shires of Mornington and Carpentaria. With the exception of the port town of Karumba, and a range of cattle stations located on pastoral leases, the constituents of the shires are predominantly Indigenous. The main communities are Doomadgee (more than 90 per cent Indigenous) and Mornington Island (more than 80 per cent Indigenous), each of which has a population of around 1,500; Normanton, a town of similar population where Indigenous people make up around 60 per cent of the population; and the smaller township of Burketown where about half of the total population of about 500 is Indigenous (QSG 2017). When referring to regional residents and regional communities, the chapter's focus is primarily the Indigenous population unless otherwise stated.

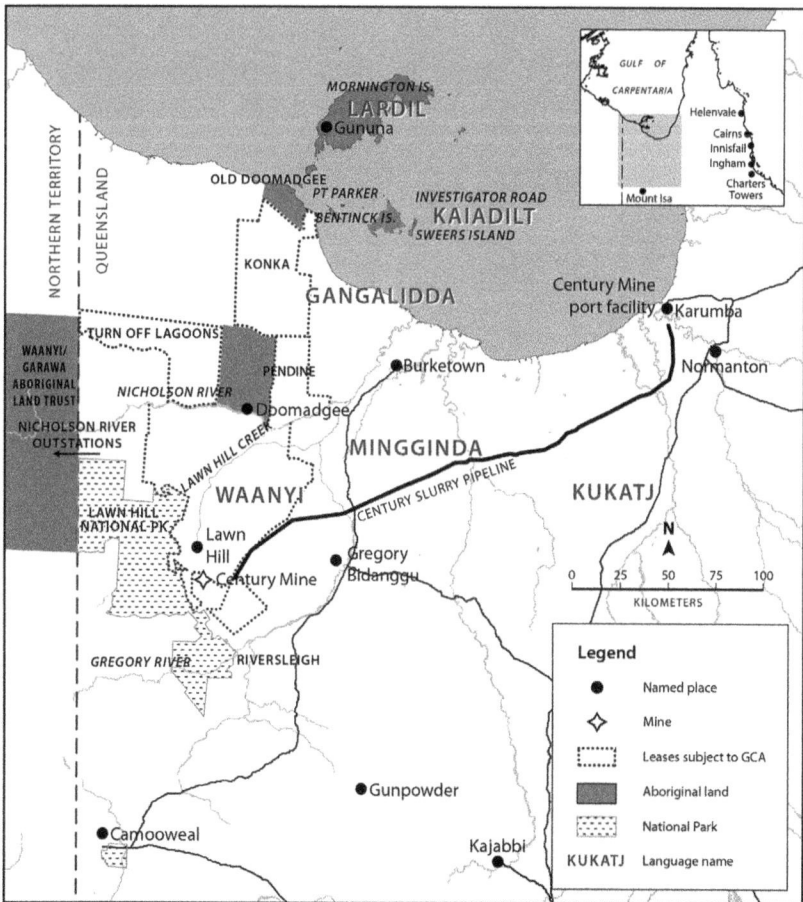

Figure 9.1 Map of lower Gulf of Carpentaria, northwest Queensland
Source: Scambary (2013: 189)

Prior to the construction of Century Mine in 1997, employment was mainly seasonal and non-Aboriginal pastoralists, the state government or mining companies controlled the majority of land. Road connectivity was poor, and communities could be cut off for months during the wet season. The entrenched socioeconomic disadvantage of the region at the outset subsequently proved a major obstacle to diversifying the local economy and meeting local aspirations for advancement. The company's lead negotiator at the time described the context as:

> [L]ess than 50 people within a 100 km radius of the mine and only 700 in Karumba, the project will impact, if only indirectly, the communities at Doomadgee, Burketown, Normanton and on the Wellesley Islands. The project spans an area about the size of Tasmania with a population of only 7,000 people of whom some 80 per cent are Aboriginal. The area supports an active pastoral industry and a fishing industry in the Gulf, but job opportunities for most people are non-existent. (Williams 1996: 2)

In the lead up to Century's development, the Indigenous land rights movement in Australia was drawing to an apex, challenging the principle of *terra nullius*,[6] which had underpinned crown control of land and sub-surface resources. The concept of 'native title' was gaining attention, and ultimately legal weight in the High Court in *Mabo v Queensland (No. 2) 1992* (the 'Mabo decision'), and the Commonwealth law in the Native Title Act of 1993.[7] These developments provided a legal pathway for recognition of Indigenous groups' rights to land and waters. 'Native title' claims are based on the original laws and customs exercised by the groups of Indigenous inhabitants of Australia from a time preceding colonisation. If native title rights are determined to exist,[8] that group of Indigenous people are recognised as the 'native title holders' over the specified area. Where the continuity of 'law and custom', albeit in adapted form, cannot be demonstrated, or where certain types of legal land tenure have been historically established, then native title rights are deemed to be 'extinguished'.

---

6   *Terra nullius* is a legal principle whereby land is legally deemed to be owned by nobody. British colonisation and settlement of the Australian continent was based on this principle, which became the focus of several important legal cases in the late twentieth and early twenty-first centuries.

7   The Native Title Act, its amendments and other court cases established and refined a process whereby groups of Aboriginal or Torres Strait Islander Australians can submit and register a 'native title claim' for determination by the Federal Court.

8   The rights that can be determined to exist vary between groups, depending on their traditions, adaptation since settlement, colonial history and other factors. Some native title rights also have the potential to coexist with some other types of land tenure, including pastoral and mining leases.

One right endowed by the Act is the 'right to negotiate'. This right requires that any party planning to undertake an activity that may impact native title rights (a 'future act') adheres to a three-month notification process, followed by a six-month (or longer) negotiation process with the native title holders (or registered claimants). Though this process is a regular feature of project development now (Trigger et al. 2014), at the time of Century's development these procedures were new, controversial and untested. Against this background, intricate tripartite relationships emerged around Century Mine between the state, the mining company and local communities—with none of these parties being unified and homogenous entities. These relationships are institutionalised by, and in part governed by, the agreements and legislation that essentially represent contracts between the parties—notably the GCA and the *Century Zinc Project Act 1997* (Queensland). These both privilege parties to the agreement (in sometimes asymmetrical ways) and exclude participation by others such as NGOs or broader society (Ballard and Banks 2003; St-Laurent and Le Billon 2015).

The shifts in the government involvement with the mine and the region are interesting to consider in the context of the state's role in the regulation and governance of an industry that is often hailed as presenting vital economic opportunities to Indigenous communities in regions with few other development options (Langton 2013). Rather than a dissociation between business and state/regulator interests, there were active links between state and company—including with respect to regional governance. This sort of relationship has prompted the claim that, in Australia 'state and capital have formed an enduring alliance that marginalises Indigenous people' (Altman 2009: 39).

In Australia's federal system, the subnational states and territories play a critical role in determining whether and how mining development takes place. This level of government owns the mineral wealth in their territory, and controls the institutional and legislative settings of resource development including most business, environmental and social regulation. These subnational entities are also key providers of public services, education and infrastructure. This is especially the case in remote regions, such as northwest Queensland, with sparse populations and a low revenue base for local-level government, which is responsible for local waste management, utilities, public facilities and liveability. In such locations, private companies such as mining corporations have been observed to assume some state responsibilities (Cheshire 2010; Cheshire et al. 2011; Bainton and Macintyre, Chapter 4, this volume).

While the construction and operation of the mine created significant employment opportunities for individuals and generated other benefits for communities in the region (such as improved infrastructure), after 15 years of mining, in 2012, health and education standards and infrastructure provision remained below Queensland and national levels. The communities—especially those with large Indigenous populations—faced ongoing social problems, including high rates of unemployment and welfare dependency (Everingham et al. 2014: 6).

Sequence of Events

While the shifts in relationships and strategies of the parties are complex and contested, the state's activities over this period can be understood in three phases, each involving different actors, key developments and differing degrees of state presence. To a large extent, these phases mirror the mine life cycle as it progressed through exploration and feasibility to approvals and construction, through operational years to end of production (ICMM 2019). Each of these phases is typically characterised by different activities, local impacts and relationships with local stakeholders. Other factors associated with the phases included the five-yearly reviews mandated in the GCA, changes of company ownership, political developments and fluctuations in zinc prices. A pre-development phase (from 1992 to the 1997 signing of the GCA and passing of the Century Zinc Project Act) was primarily a period of negotiations and marked by intervals of intense engagement between various parties from all levels of government, Indigenous and non-Indigenous lower gulf residents, those with traditional links to the land there, and the mining company.[9] During this period, there was a strong presence of national and subnational state actors. The second phase, from 1997 to the first five-year review of the GCA in 2002, covers what was initially an intense

---

9   This initial phase was conducted by the first of a series of five owning/operating companies (CRA-Rio Tinto). Pasminco assumed ownership after the GCA was signed, having reached an agreement to purchase in February 1997 that was conditional on the GCA being in place. Pasminco was not involved in the negotiations, except to assure the parties they would honour the agreement in full. After seven years Zinifex assumed control.

period of construction and early operation of the mine involving many of the same parties adopting strategies to implement their obligations, or achieve their aspirations under the GCA. This was also a period when direct state presence began to diminish and dissatisfaction with the lack of socioeconomic development emerged. The subsequent 10 years (2002–12) were the 'big zinc', mainly profitable years leading up to the 15-year review, a period of considerable turnover in actors involved but also of greater complacency from most parties and considerable distancing of various state agencies from active engagement. During this phase, the state's actions were often at a remove from the region, involving indirect advancement of its priorities through other actors. Key developments during each of these phases need to be understood in their social, economic and political context and, in particular, in terms of public policy and the state's position on two key issues: state–economy relations and Indigenous rights.

Phase 1 (1992–97): Pre-development, Negotiating Phase

Century was a high-profile project, because of its scale, and because of the long-standing pro-business stance adopted by Australian governments in the prevailing minerals development policy at the time (Trebeck 2007). As well, the 1990s was the decade that saw planning and registration of the initial suite of Indigenous native title claims over areas of land and sea, including in the southern Gulf of Carpentaria in the wake of the High Court's Mabo case (decided June 1992) and associated legal reforms recognising Indigenous rights. Century Mine came to be seen as an early test of the workability of the legislative response to the Mabo decision— the Commonwealth's *Native Title Act 1993*—and particularly of its 'right to negotiate' provisions (Howlett 2010). These factors meant that there was not only a high level of media interest in the project, it also attracted unparalleled attention from both the Queensland and Commonwealth (federal) governments, advocates and opponents of the new native title regime, as well as other miners, a range of politicians and activists.

After the discovery of the orebody was announced (late 1990), there were preliminary discussions in communities as well as anthropological and archaeological surveys of the proposed site commissioned by both the Indigenous parties and the company. Once Century Zinc applied to the Queensland Government for a mining lease, and the *Native Title Act 1993* was passed, negotiations between the company and Indigenous people were coordinated through the Carpentaria Land Council Aboriginal Corporation[10] and the United Gulf Regional Aboriginal Negotiating Team. The latter was the name adopted by a community group formed in the absence of a single representative body accepted by all Aboriginal interests. It involved members of the broad population of Aboriginal people living in the affected region, the Carpentaria Land Council, representatives of various local Aboriginal corporations, linguistic and territorial groups of traditional owners, and some of their advisors. At the time, the local communities, although inexperienced in dealing with large companies, had little trust in the mining industry. In fact there was considerable antagonism and concern that mining development was likely to be 'history repeating itself'[11] and that people would be further displaced from, and denied access to, their traditional lands. However, there was also considerable support for a mine given anticipation of jobs and training opportunities (Trigger 1997).

In 1994, the region's first native title claim was lodged on behalf of Waanyi people, over a small area known as the Camping and Water Reserve that was destined to be part of the mine development. Initially the claim was not accepted by the Registrar of the National Native Title Tribunal. The case was appealed as far as the High Court of Australia. A protest sit-in by Aboriginal people occurred in October that year at Lawn Hill National Park (subsequently named Boodjamala National Park), located adjacent to the small reserve and the proposed mine area. It was not until 1996 that the National Native Title Tribunal accepted the claim, as referred to them

---

10   In Australia, Land Councils are key representatives of the interests of Indigenous people and are closely engaged with the traditional owners of a particular region. They are community organisations, generally organised by region, that are formed to represent the Indigenous Australians who occupied that particular region before the arrival of European settlers. Their functions and responsibilities vary somewhat in different Australian jurisdictions but central to them are consultation with Aboriginal people and assistance with claims to and management of traditional lands. In the case of the region affected by Century Mine, the Carpentaria Land Council Aboriginal Corporation was formally determined to be a Native Title Representative Body in 1994. However, the right to negotiate provisions and Century negotiations applied more broadly than to those seeking recognition of Native Title Rights and there were some internal divisions among parties, so the broader group, the United Gulf Regional Aboriginal Negotiating Team, became involved.

11   Besides the history of colonial appropriation of land and displacement of people from their ancestral territory, there were more recent experiences and observations that coloured perspectives. These included reports of irresponsible mining company actions in the neighbouring Northern Territory, and contemporary actions during exploration whereby the company interested in developing Century Mine used its superior financial power to purchase pastoral properties in the region despite awareness of Indigenous interest in acquiring them.

by the High Court, significantly strengthening the Aboriginal bargaining position in negotiations with the company. Indigenous positions ranged across outright opposition to the mine, outright support and a lurching between more qualified expressions of each of those stances (Blowes and Trigger 1999). There were challenges in the diverse interests of local Indigenous people, and the lack of experience and resources of their groups and organisations, to negotiate on an equal footing with a mining company or governments (see Blowes and Trigger 1999 for a detailed account of key factors influencing the relative bargaining positions and powers of the various parties to the negotiations). The federal and Queensland governments and the company all supported establishing the United Gulf Regional Aboriginal Corporation, which would constitute a formalised (incorporated) representative structure to act on behalf of all Aboriginal organisations, communities and individuals across the lower gulf region in negotiations and which could be a party to any agreement and bind those organisations, communities and individuals. It involved some 15 constituent organisations and was avowedly focused on the region as a whole and on all Aboriginal people affected, not exclusively those with a native title connection. The governments and company therefore contributed tens of thousands of dollars to assist the transformation of the United Gulf Regional Aboriginal Negotiating Team into this incorporated body, which would then expand the range of Indigenous voices, inform Aboriginal residents, communities and interests about the project and its impact and facilitate equitable participation. The negotiations were described by one Indigenous participant as a sequence of community meetings, 'biting it off piece by piece and then taking it back to the mob. We never gave an answer on the spot' (Anonymous roundtable participant, personal communication, November 2014).

During the 1990s in Australia, prevailing rationalities of government echoed economic liberalism, with its lauding of a 'small state' and market forces (Palmer 2008), new regionalism (Everingham et al. 2006) and the contractual state (Plant 2012). This was evident in Queensland politics with strong support for Century Mine, initially from the Goss Labor Government as part of the vision for a Carpentaria–Mount Isa mineral province (Trigger 1998). Despite its strong desire to see Century Mine proceed, the nature and level of the Queensland Government's involvement in negotiations between 1992 and 1995 was unclear. Some reports claim that it remained in the background and stayed at arm's length from the intense interactions the company had with native title claimants and communities (Blowes and Trigger 1999), though others suggested it worked behind the scenes to encourage local and Indigenous support of the project (Howlett 2010). As progress stalled and confrontation mounted, the federal Minister for Aboriginal Affairs, Robert Tickner, intervened and appointed Hal Wootten (QC), of Aboriginal Deaths in Custody Commission[12] repute, to mediate between the company and Aboriginal communities—both native title claimants and other Indigenous residents. Another federal agency representing Aboriginal interests, ATSIC,[13] strongly supported the mine because it was convinced that if Century Mine proceeded it would prove that native title rights were not necessarily contrary to mining and economic development. Wootten's intervention achieved a resumption of negotiations.

In early 1995, the company requested that the Queensland Government initiate the 'right to negotiate' procedures as prescribed by the *Native Title Act 1993* (Williams 1996). However, the government declined to do so because of its view that native title had been extinguished by the pastoral lease history. It believed initiating the 'right to negotiate' procedures would create an unwelcome precedent for the future that could possibly prompt more claims (Howlett 2010). The mining company was more pragmatic in seeking a solution—eventually proposing, in July–August 1995, to the United Gulf Regional Aboriginal Corporation, an AUD60 million package of opportunities, cash payments and community investment in return for Indigenous support of the mine (Williams 1996: 12).

Once elected, in February 1996, Queensland Premier Borbidge enthusiastically maintained state support as he saw an operating Century Mine as a potential major achievement of his government, though he lacked specific plans for implementing his vision. Economic growth was seen as a greater imperative than economic redistribution and there was a winding back of welfare state commitments in favour of delegating delivery of state services through competitive, contractual arrangements (Palmer 2008; Plant 2012). To the Queensland Government, mining could provide a much-needed catalyst for socioeconomic development

---

12 This Royal Commission 1987–91 responded to concerns similar to those of the contemporary 'Black Lives Matter' movement about deficiencies in custodial care, both systemic and individual, and disproportionate rates of imprisonment for Indigenous Australians due to historical and social factors.
13 The Aboriginal and Torres Strait Islander Commission (ATSIC) (1990–2005) was established by the Hawke Labor Government as a national body of elected representatives through which Aboriginal and Torres Strait Islander Australians had a formal voice in the processes of government affecting their lives.

in this sparsely populated region distant from key manufacturing, service and agricultural centres and lagging behind the rest of the state on most indicators. Support for the mine at the federal level was similarly strengthened with a change of government in March 1996 to the Howard Liberal–National Coalition, which supported the market-oriented regime and championed an economic growth agenda.

After this series of political, legal and economic developments in late 1995 and early 1996, negotiations broke down and the Queensland Government presence peaked when recently elected Premier Borbidge offered to intervene. Discussion with a number of Aboriginal leaders in the region led to a 'summit' meeting to discuss the project and hear views firsthand in April 1996. Subsequent negotiations, facilitated by ATSIC, solidified the views of those Aboriginal representatives who regarded the mine favourably because of the foreshadowed benefits to be received. However, this became more muted when the Queensland Government publicly announced they would introduce special legislation to suspend Indigenous rights and allow the mine to proceed. Recognising the strong opposition to special legislation, and confident of a strong current of Indigenous support for the mine, the company asked the government not to proceed with that strategy and to accept the inevitability of the time-limited 'right to negotiate' process (Williams 1996). During this six-month period, the company sought to maintain a regional focus in the agreement (Blowes and Trigger 1999), though inevitably individual native title groups and their key spokespersons were focused on their particular interests and the possibility of negotiated benefits. Non-Indigenous residents of the region were not involved in the native title right to negotiate process.

The Queensland Government continued its close engagement with the appointment of ex-Governor General, Bill Hayden, to negotiate on its behalf while legal representation and anthropological advice, funded by ATSIC, was obtained for the claimant native title groups (namely Waanyi, Mingginda, Gkuthaarn and Kukatj peoples). A panel headed by the President of the National Native Title Tribunal, Justice French, was established and it appointed Rick Farley and Dr Mary Edmunds to mediate between the parties, though preparation of various positions took some months and negotiations really only commenced two months before the negotiation deadline. This negotiation, the first major one under the right to negotiate process, was being keenly followed by various interests in the Queensland and federal governments and the mining and pastoral industries, as well as Indigenous interests. Much internal discussion in the Queensland Government resulted in the government's chief negotiator outlining its position and making an offer which included significant Queensland Government commitments with an anticipated value of AUD30 million. This package was enshrined in the eventual agreement and included: an upgrade of the road from Nardoo to Gregory, servicing the mine and communities of Burketown and Doomadgee (AUD15 million); a comprehensive regional social impact study (AUD1.8 million) with funds to address issues identified (up to AUD5.7 million); AUD3 million for an outstation development at Gregory; AUD2 million for training and vocational education; and a number of other items relating to cultural and social issues (GCA 1997: Schedule 1—see summary in Blowes and Trigger 1999).

On 13 February 1997, the last day of the six-month negotiating period provided for by the Native Title Act, the native title groups pressured the company and the government to increase their respective offers, though few concessions were forthcoming and not all native title claimants agreed to sign (Brereton and Everingham 2016: 8). Given the failure to reach an agreement that required all signatures from the Indigenous parties (a disappointment to the company, government and many in the native title groups) the company announced it would initiate the arbitration provided for under the Native Title Act (Williams 1996).[14] This represented a third stage in the pre-approvals phase. A visit to the region by the Chairperson of ATSIC, a well-attended public meeting in Doomadgee and a petition and delegation of pro-mining native title claimants to Canberra to meet with the prime minister (John Howard), the leader of the opposition (Kim Beazley) and Cheryl Kernot (who was the leader of the Democrats and their spokesperson on Aboriginal and Torres Strait Islander affairs and reconciliation) all focused on overcoming the deadlock. In March, the deputy chairperson of ATSIC, Ray Robinson, engaged in a series of meetings with the claimants, including non-signatories, while the Commonwealth Government also involved senior public servant (and ATSIC Commissioner), Charles Perkins. After further revising of positions, Robinson eventually produced a fully signed copy of the agreement for the company and

---

14   Under the right to negotiate provisions of the Native Title Act, compulsory arbitration is instituted if negotiations reach an impasse after the designated notification of future acts and six-month negotiation period. The formality of the arbitration meetings added a new and disconcerting dimension and encouraged a greater willingness to sign.

Queensland Government to sign in May 1997 (Williams 1996: 22). Even before striking the agreement, the company was reported to be on site, constructing the pilot plant and employing hundreds of workers (Anonymous roundtable participant, personal communication, November 2014).

Shortly thereafter, the Queensland Government passed the *Century Zinc Project Act 1997* to enshrine its obligations. The wording of the Act reveals the Queensland Government's priorities: 'The project will improve the economy of Queensland and Australia' (Preamble, Clause 5); and 'The project will result in social and other benefits for all people of the mineral province' (Preamble, Clause 6). As well, Section 3 of the Act details the purpose as 'to facilitate certain aspects of the agreement made under the right to negotiate provisions of the *Native Title Act 1993* (Cwlth)'. It goes on to detail the government's commitments in terms of the mine, including compensated compulsory acquisition of land for mine infrastructure such as the pipeline and Karumba port and processing facilities and associated construction works. As well, it legislates for the Queensland Government to take over the development application by Bidunggu Aboriginal Land Trust proposing work at the Gregory outstation (which had gone to the relevant Shire Council/local government). These undertakings used state government powers to ensure land access and to smooth the way to granting leases under the *Mineral Resources Act 1989* (Queensland), such that they did not extinguish native title rights. Throughout Phase 1 (1992–97), there were active links between the company and a state government that was very sympathetic to a view of progress and regional development centred on mining expansion. This, in turn, fuelled active state government engagement with the region, primarily seeking to counter opposition to the mine and secure wide support for the project. Early efforts of both Century Zinc and the Queensland Government sought to confine the scope of the GCA to issues they regarded as flowing directly from the mine (Blowes and Trigger 1999: 109) and to maintain a distance between the mining company and the state government on the one hand and more nebulous issues of culture, welfare and social development on the other hand. This functioning of agreements to maintain power relationships and create distance between the state and other actors including 'affected Aboriginal peoples with territorial claims' has been remarked upon in Canada too (St-Laurent and Le Billon 2015).

Phase 2 (1997–2002): Early Operations

The Century Mine project was envisaged by all parties to the GCA as providing a means to reduce welfare dependency among Aboriginal people in the region. The state is responsible for welfare, with the Commonwealth providing social security payments and state-level governments providing much of the vital education, health and basic infrastructure that citizens need. Some of these responsibilities may take some years to fulfil, as do, for example, the subsequent consequences of educational opportunity.

With the signing of the GCA and the passing of the Century Zinc Project Act, there was an initial period that mining company participants describe as 'frenetic activity' on behalf of the mining company to begin production (Everingham and Keenan 2017). This involved construction of the mine, port, pipeline and infrastructure. For example, the company spent some AUD26 million to upgrade (though not seal) the continuation of a Queensland Government road from Gregory to the mine. In addition, the company funded a 220-KV transmission line from Mt Isa to the mine, with the understanding that it would be extended to Burketown, Gregory and Doomadgee (by the state-owned electricity authority of the time). The various construction and commissioning activities culminated in the first shipment of zinc concentrate in 1999. This was despite the company facing technical challenges, such as the propensity for the zinc concentrate to spontaneously combust in stockpiles and transport ships. Initially Century Zinc was diligent in ensuring the 'GCA was the bible' (Brereton and Everingham 2016: 15). In particular, they invested heavily in employment and training and signed a memorandum of understanding with the Queensland Department of Employment and Training. This demonstrated that the mine adopted a strategic approach to the GCA despite the socioeconomic constraints, with education and skill levels in the Indigenous communities proving significant hurdles to broadening access to employment opportunities. The company also moved promptly to implement some other parts of the GCA—notably returning partial title to pastoral holdings to Indigenous stewardship. For the major holding in this respect, the Lawn Hill and Riversleigh stations, a staged transfer was envisaged with initial joint management by a pastoral holding company set up under the GCA. These properties were seen as providing opportunities for Indigenous pastoral enterprises, cultural renewal on country, employment and training (Everingham et al. 2014).

Some state actors, likewise, engaged intensively at the outset of this period. Federally, ATSIC was active in establishing the institutional framework envisaged in the GCA. Considerable effort was invested in establishing the committees, Gulf Aboriginal Development Corporation, Aboriginal Development Benefits Trust and pastoral company specified in the GCA. For instance, Century Environment Committee, Century Education and Training Committee, Gulf Aboriginal Development Corporation, and Century Liaison and Advisory Committee were established with ATSIC facilitation after the GCA came into effect. As specified in the GCA, these bodies had varying membership of representatives from the mining company, government, native title groups and regional Aboriginal residents. After the withdrawal of ATSIC's direct support, however, many of these bodies were dormant by the turn of the century (Martin 2009). For instance, the Gulf Aboriginal Development Corporation did not hold an annual general meeting between 2001 and 2008 and the Century Liaison Advisory Committee, after only five meetings, did not reconvene until 2012. By 2002, the draft Five Year Review Report noted a need for renewed effort in institutional development—an early warning sign that the complicated governance structures of the GCA had not been sufficiently embedded. This was particularly true of the Gulf Aboriginal Development Corporation, which was noted as having a poorly defined role and lacking adequate financial resources after the start-up injection of funds to fulfil dual roles of administering GCA funds and furthering native title group rights and interests (Martin 2009). As a creature of the GCA, this corporation lacked capacity, influence and legitimacy with the disparate groups whose interests it was to further, and who were unused to working together for common goals (Everingham and Keenan 2017: 19). Just as the government and company struggled with the new native title regime and unclear legal and tenure issues, the Aboriginal groups had no familiarity with or experience of the new order and had not established decision-making processes aligned with the *Native Title Act 1993* (Martin et al. 2014).

The Howard Coalition Government that had come to power federally in 1996 did not enthusiastically embrace the GCA and held a less sympathetic view of Indigenous affairs than prevailed in the Hawke-Keating years. This was particularly evident in 1998 with the 'practical reconciliation' policy[15] (Gunstone 2008), and revision of the native title

---

15 Rather than even limited recognition of Indigenous rights and aims of structural change, this policy rejected 'symbolic' measures and defined government priorities as efficiently addressing the substantial socioeconomic disadvantage experienced by many Indigenous people in health, education, housing and employment, often through contracting out public services in these areas.

policy regime—particularly the *Native Title Amendment Act 1998* (Cth). The latter implemented much of prime minister Howard's 10-point plan that would curtail rights, including the right to negotiate about mining. As well, a Ministerial review of the Native Title Representative Bodies was proposed, and changes foreshadowed to encourage the negotiation of binding, voluntary agreements as an alternative to formal native title agreements (Altman 2009).

In the wave of initial enthusiasm for the mine, the Queensland Government, for its part, was galvanised to action as well, notably spending AUD15 million to upgrade and seal the road from Nardoo west to Gregory, and initiating the training that was intended to overcome low skill levels and lead to mining employment. It has been suggested, though, that investments in infrastructure and vocational training benefited the mine and Queensland as a whole, as much as advancing prospects of local Indigenous people (Howlett 2010). The local land council later observed that the interest groups they represented presumed that agreement commitments were additional to the government's responsibility for funding human capital development of its citizens:

> What needs to be highlighted is what proportion of the State's commitments constituted normal government service delivery ... which should have been provided in any event ... [and] commitments by the State which have been more about serving the primary interests of developing the Century Mine rather than benefitting the Gulf Communities generally. (CLCAC 2002: 49)

In fact, the Queensland Government itself observed 'an apparent trade-off' in targeting their education spending between the dual goals of their commitments to both supporting Century Mine by maximising trained labour available to them, and enhancing the broader human capital and skills in the Aboriginal population of the region (QSG 2002). It remained clear, five years into the GCA, that both the company and the state maintained a, possibly naïve, vision that 'The GCA can be a significant stimulus to regional development and the creation of a more diversified and robust post mining economy in which Aboriginal people have a much higher participation rate' (Pasminco et al. 2002).

The GCA anticipated early vigorous engagement by identifying some immediate 'wins' while other initiatives needed staged progression. Among the latter were the outstation commitments—and notably development of an outstation resource centre. Development of this centre and a policy on outstations were to be coordinated by Queensland's Department of Family, Youth and Community Affairs (Office of Aboriginal and Torres Strait Islander Affairs). However, there was criticism that there had been no action beyond a report and no attempt to align with federal (ATSIC) initiatives on outstations (CLCAC 2002). Such long-term commitments are always vulnerable to changes, not least to shifts in government policies, contrasting state and national priorities and turnover of personnel.

As a consequence of such changes, the GCA did not maintain a prominent status with the Queensland Government, and the government's active presence in this region, 2,000 km away from the state capital and locus of most development, faded. The initial road works, training and other investments in infrastructure and service provision (whether as Queensland Government obligations or under the GCA) were rarely delivered through an approach that favoured local enterprise development or boosted regional capacity. The benefits of some initiatives dissipated, and others never gained traction. In part, this related to insufficient attention to institutional capacity building and to the state relying on the Gulf Aboriginal Development Corporation as the voice of the various Indigenous rights holders. This was despite the fact that the draft document reporting the initial review of the GCA in 2002, supposedly produced according to Terms of Reference drafted by the Queensland Government, the company, and the Gulf Aboriginal Development Corporation itself, as representative of the Native Title signatory groups, described this body as 'no longer effectual' (Pasminco et al. 2002: 9). The draft report of that review also noted:

> Implementation of the Queensland commitments is substantially complete except for land transfers at Karumba, the Gulf Area Social Development Fund (GASDF) Trust implementation as part of the Social Impact Assessment initiative, ongoing initiatives in training support, and environmental information provision (following on from the Multiple Use Strategic Plan). Approximately 90 percent of the $30.29 million commitment has been allocated or committed, with substantial additional funding provided to training initiatives. (Pasminco et al. 2002: 4)

These generally positive conclusions were contested by many in the region. For instance, the Carpentaria Land Council, in response, and claiming to represent the interests of the Waanyi people, was scathing in its criticism of government and company. Their report, prepared as part of the five-year review of the GCA, claimed:

> The State's failure to conduct the SIA [social impact assessment] has been a glaring failure in the agreement. Not only is it a failure of implementation, but it has been a breach of the Agreement which has had serious ramifications for the Waanyi Native Title Group whose members are among the most disadvantaged in the Gulf Region. (CLCAC 2002: 1)

The Carpentaria Land Council further claimed that this meant that:

> [S]trategies to prevent the corruption of indigenous cultures, strategies to re-affirm cultures and enhance lifestyles and initiatives in relation to health, employment, education and other social indicators have lagged behind and that this has been especially disadvantageous for the Waanyi. (CLCAC 2002: 24)

Concerns were that various specifics were not complied with in the spirit or in the substance of the agreement. Queensland's initiatives are classified as commitments under the GCA, intended to benefit the native title holders and the Aboriginal communities and 'to demonstrate its commitment to the project and its good faith to the Native Title Groups' (GCA 1997: Schedule 1, Clause 11). However, some initiatives, such as road upgrades and investment in power supply served commercial interests of the mine and native title holders were only 'peripheral' beneficiaries.

A related concern was that the Queensland Government's undertaking to conduct a social impact assessment (SIA) was to serve as the basis for a framework for monitoring, assessing and responding to social change in the region. The native title parties to the agreement understood the Queensland Government to be committing to a major exercise over three years (1998–2001) costing AUD1.8 million to assess the social situation and potential impacts of the Century project, so as to develop a community development strategy to mitigate potential negative effects and enhance potential positive impacts (GCA 1997: Schedule 1, Clause 24). Though the draft five-year review reported that half of the allocation was spent, local stakeholders were adamant that a study had not occurred. This perceived breach of the GCA obligations with respect to the SIA (Schedule 1 Clauses 23–30), also led the native title groups to reject the establishment of a Gulf Area Community Social Development Trust Fund without an SIA since, 'Targeted, planned economic development according to a sound strategy cannot be undertaken without the SIA' (CLCAC 2002: 39). A major complication was that AUD3.5 million of the AUD5.76 million earmarked in Schedule 1 of the GCA for 'social impact projects' was channelled, in 2001, through the Northwest Queensland Community Benefits Strategy (as the first distribution of funds meant to address shortcomings identified in the promised SIA). The grants went to projects in Mt Isa, Cloncurry and other parts of what the Queensland Government regarded as the northwest minerals province that were outside of the affected lower gulf communities. This reflected the government's position from the outset that this was a 'regional agreement', with their definition of 'regional' being the whole northwest minerals province centred on Mt Isa rather than being confined to the lower gulf region (The Right Mind 2008). However, in the eyes of local people the spending had been strongly influenced by the Mt Isa-based local Member of the Legislative Assembly and benefited the core of his electorate rather than the periphery in the lower gulf (Anonymous roundtable participant, personal communication, November 2014). There was strong opposition to him having any future role in distribution of trust funds and to the funds having a broader geographical focus than the affected native title groups (CLCAC 2002). The vexed question of the extent to which Indigenous people have benefited from mining in Australia has been frequently raised in the context of the Century Mine and in other cases (Martin et al. 2014).

Cracks had begun to appear in the tripartite relationship as well. These were particularly evident in November 2002 when Waanyi people and other Aboriginal supporters from the region occupied the worker's camp and canteen at the mine site, staging a multi-day protest. Triggering issues included the lack of employment or other economic advancement opportunities for local Aboriginal people afforded by mining on their traditional land, and implementation of the GCA including the Queensland Government's commitments in the agreement. Other concerns were focused on the storage of stone tools and artefacts that had been displaced by the mine and a general lack of sensitivity about cultural issues on site (Martin 2019: 157). Then Queensland premier, Peter Beattie, is quoted as suggesting native title was part of the problem:

> Native title and all those issues has produced bugger all for Indigenous people and frankly not a lot for the mining industry. And that's why we have embraced a number of strategies involving negotiated outcomes to get results both for the mining industry, the State of Queensland and for Indigenous people. (Townsend 2002)

This reveals the Queensland Government's intent for the mining company and itself to benefit, and also suggests a preference for bilateral partnerships and contracts over the multilateral arrangements of the GCA. One notable memorandum of understanding (MOU) between the company and the government (specifically the education provider, TAFE Queensland) related to the employment and training obligations. This sort of issue-specific, bilateral arrangement in preference to working through the agreement and its structures and processes aligns with notions of the 'contractual state' (Plant 2012). Once the mine was in production, governance structures formed and immediate obligations fulfilled, the Queensland Government distanced itself from the mine and region and was described as passive and extraneous (The Right Mind 2008). Having established what it saw as the enabling conditions, it assumed a background role leaving GCA mechanisms and the market to proceed with implementation. Both the Queensland and federal governments engaged in strong, cross-government, pro-mining advocacy (Howlett 2010). Through policies such as 'practical reconciliation' and the Community Development Employment Program adopted in 1998, they expressed consistent propositions about responsibilities of government, the private sector and citizens. These favoured mainstreaming service delivery to Indigenous Australians and encouraging their economic participation. Such policies resonated with sections of the Indigenous communities of remote North Queensland (e.g. see Pearson 2000).

Phase 3 (2002–12): The 'Big Zinc' Years

In ensuing years, governments maintained an emphasis on 'steering' development (Peters 2000) to maximise the opportunities mining provides to Indigenous people, especially through employment, often in partnership with the private or non-government sectors. For example, there was an MOU between the Australian Government and the Minerals Council of Australia, reached in 2005, to increase Indigenous employment and training (Martin et al. 2014). The Queensland Government and the Queensland Resources Council formed a similar MOU in 2007. As well as public–private partnerships, partnerships between levels of government or various government departments also burgeoned. During this phase, a shift occurred as the focus of government partnerships, joint initiatives and agreements extended beyond the initial emphasis on employment creation and training to encompass areas of community development such as health, education, infrastructure and governance, as well (Martin et al. 2014). Most prominently, this related to the National Indigenous Reform Agenda when, in 2007, the Council of Australian Governments committed to 'Closing the Gap' in opportunities and outcomes between non-Indigenous and Indigenous Australians. This program originally had six targets and eight strategic areas for action that were subject to a suite of multi-year partnership agreements between the state and federal governments. From 2008–09, the Federal Government channelled AUD8 billion in funding over 10 years into these bilateral partnerships, which especially targeted remote areas like the northwest of Queensland where the gaps were most pronounced (QSG 2009).

Just before such reforms provided new impetus, the Ten-Year Review of the GCA had very similar messages to those surfaced five years earlier in the first review of the implementation of the GCA. The recommendations highlighted the need to improve communication between parties, especially from the government (The Right Mind 2008). As well, the reviews emphasised the critical requirement for effective Aboriginal institutions—eligible bodies, the Gulf Aboriginal Development Corporation, and committees including Century Liaison Advisory Committee—observing that there was insufficient investment in the relevant organisations to ensure they were strong and their capacity developed. While not underestimating the complexities of achieving such capacity development, and without minimising efforts particularly among Indigenous employees and committee members over the years, for much of this phase some institutions were dormant if not dysfunctional. In the state's eyes, these governance failures were a broader issue. This was illustrated in prime minister Howard's justification for dismantling ATSIC by asserting: 'We believe very strongly the experiment in elected representation for Indigenous people has been a failure' (Anon. 2004). With the abolition of ATSIC, the Indigenous supporters of the mine had lost a key federal supporter. The Queensland Government's attention similarly waned. Reflecting on the government's activity during the period 2002–07, the 10-year review of the GCA, halfway through this phase, noted, with respect to Schedule 1:

> Whilst the [Queensland] Government might have been engaged at the beginning of this agreement … The Government corporate memory, commitment and will to leverage the Agreement to build leadership and management, to improve regional development through carefully planned linkages from mining activities to regional development have been all but forgotten. (The Right Mind 2008: 22)

The review noted the stalling of key government commitments under the GCA—to the SIA, the Bidunggu outstation and fostering a Men's Business Association—as well as neglect of its responsibility to provide infrastructure and undertake regional coordination of training and other regional development initiatives. These issues had ramifications for the tripartite relationships. For instance, Bidunggu became reliant on the goodwill of the mining company for asset maintenance and some operational needs (such as diesel generator fuel).

In fact, over this third phase, there was no further progress reported on any of the government's obligations under Schedule 1 of the GCA that had already been highlighted as lagging at the five-year review. While this neglect may indicate challenges with less feasible elements in the GCA, it also encourages the inference that the Queensland Government directed attention to new objectives (such as Closing the Gap targets). Hence, the economic self-sufficiency and community development initially envisaged as accompanying the mine were not fully realised and programs outlined in the GCA languished (The Right Mind 2008; Everingham et al. 2013). Most parties to the GCA appeared to become defensive and reactive during this phase. For instance, the company reverted to a more internal focus and to appeasement and opportunistic tactics rather than former proactive, strategic approaches. Even the 2009 determination of Waanyi native title to the Lawn Hill and Riversleigh pastoral properties resulted in only cautious advancing of the phased transfer of control (Everingham et al. 2013). One view is that reactive approaches did not encourage strong Aboriginal organisations with the capacity to leverage full advantage from state-sanctioned mining, but encouraged a self-interested approach by the native title groups as responsible for their own participation in the economy and their own social circumstances rather than the state bearing any collective responsibility for their well-being (Trebeck 2007). However, we also note research across the region that has depicted internal drivers of self-interest and competition among Indigenous parties (Trigger 1988, 1997). Outcomes from mining then are best understood as a complex mix driven by customary norms and the circumstances of relations with the state and the company.

Significant changes in policies of the Queensland and federal governments reinforced the state's arm's-length position with respect to service provision and greater reliance on contracted service providers, including corporate actors, to fulfil governance requirements. For example, after a review found significant flaws and ineffectiveness in the Community Development Employment Program, that program was discontinued from July 2009 and replaced by the Remote Jobs and Communities Program. This more recent program aspired to greater emphasis on job readiness and genuine employment outcomes. However, judging by 2011 Census figures and 2015 Closing the Gap reports, it was slow to evidence any improvement in the lower gulf region (Brereton and Everingham 2016; QSG 2017).

This was unfortunate since education and employment were key targets set in 2008 in the National Indigenous Reform Agreement. Mid-2008, the Rudd Government established the National Indigenous Health Equality Council, and in subsequent years additional Closing the Gap targets were added to strengthen the focus on equity with respect to health, education, employment and community safety. Many of the national partnership agreements supporting these action areas, arguably 'marketise' the delivery of services to remote Indigenous communities at the same time as supporting Indigenous aspirations for participation in the market economy. This ambivalence in the policy agenda undermines state support for the kinds of benefits envisaged in the GCA.

Other developments over this decade similarly affected the relationships between company, communities and government and reveal the Queensland Government increasingly acting at a distance, through others, to achieve its goals while itself taking a 'hands off' approach to governance in the lower gulf. However, it was no longer feasible to rely on the corporate party to the GCA for continuity and implementation, since this period saw a sequence of changes of ownership and staff turnovers with seven general managers and four companies between 2002 and 2012.[16] As well, the mining operation faced technical challenges that consumed its attention. For instance, during a storm in 2007, the Wununa barge, used to tranship ore from the Karumba processing plant to the bulk export carriers offshore, had to be temporarily abandoned at sea, threatening the company's licence to operate. While that was an unwelcome prospect in 2007, by 2012 viable reserves were becoming depleted and the company sponsored a closure SIA with a view to ceasing production within five years, when they anticipated the mine would reach the end of its 'economic life'. This was accompanied by a scaling down

---

16 Corporate owners of Century Mine: CRA/Rio Tinto: 1990–97; Pasminco: 1997–2004; Zinifex: 2004–08; OzMinerals (a merger of Oxiana and Zinifex) 2008–09; MMG 2009–12.

(and some redirecting) of training and employment activities and eventual elimination of the 'community liaison officer' roles which had primarily supported those efforts in relation to Indigenous residents of the region.

The third phase, in sum, was characterised by gathering momentum of the federal rhetoric of practical reconciliation, the championing of economic liberalism and market-based economics as the route to regional development, and strategic partnerships with industry and service providers. This was particularly evident between 2005 and 2007, when the Howard Government controlled both houses of federal parliament. This less present, more distant, state was variously embraced and criticised by some regional residents and some within all parties to the GCA.

Discussion and Conclusion

The Queensland Government viewed mining as the ideal avenue to development and closing the gap in the remote northwest, a region where the state's capacity to directly achieve economic outcomes for the Indigenous majority is limited. Accordingly, Century Mine was a key driver of development, environmental impacts and sociocultural circumstances of residents in the lower gulf for 20 years. It was, for some years, the largest private employer of Indigenous people in Queensland (Miles et al. 2005). Perhaps its most striking achievements were maintaining 15–20 per cent Indigenous employment through direct employees and the contractor workforce, providing training opportunities in a region where these are in short supply and significantly increasing revenue to the region. For example, average household and individual incomes improved, and household material assets such as white goods and vehicles became more widespread. As the 15-year review of the GCA concluded, these were essentially narrow, individual benefits so 'the main regret of native title groups is that there have not been more such changes' (Everingham et al. 2013: 41).

In this region, as elsewhere, national economic growth and Queensland economic prosperity have provided temporary benefits for locals but have historically failed to deliver broad and lasting benefits to mining-affected areas (Eriksen 2018). Mining-affected communities now demand of the state a new balance between state economic priorities and local interests. In Century Mine's case, the finale is yet to come, with the years since 2012 seeing dramatic shifts in the leverage and the aspirations of the major parties all of whom have seized upon a 'stay of execution' as the closure of the mine has been averted. The mine ownership transferred in early 2017 to a smaller company with 'economic rehabilitation' plans. The Indigenous parties see renewed hope of active participation in environmental aspects of this economic lifeline to the region, while the Queensland Government grapples with the spectre of large mine closures in northwest Queensland and the need for regional resilience and revitalisation (Macguire 2019). However, in the period under review, initial expectations were not met. The very real achievements such as employment and financial benefits were unevenly distributed and the aspiration that two decades of mining could overcome decades of regional disadvantage proved overly ambitious.

Century Mine proceeded after a highly contentious negotiation process, in a historically troubled and disadvantaged region (Blowes and Trigger 1999). The GCA, to an extent, reflects Indigenous and regional dissatisfaction with various levels of government as much as it is an expression of hope for a better future. The negotiations about the mine came to represent the initial trial of the Commonwealth Government's 'right to negotiate' procedure contained in the *Native Title Act 1993* (Howlett 2010). Hence, there was unusually intensive government engagement in the pre-approvals negotiation and planning about the mine, and the Queensland Government was involved as a party to the GCA. However, within five years of the mine receiving the 'green light', a group of Waanyi people and their peers staged a sit-in to express their discontent with the progress and outcomes of the GCA. This was, in part, a consequence of the state leaving market forces to deliver its commitments, rather than adopting a more engaged role as a party to the agreement. Relevant sectors of government became less present and proactive soon after the mine began operating, and adopted a more arm's-length strategy with limited government intervention and institution-building in line with national regional policy (Tonts and Haslam-McKenzie 2005). As elsewhere in Australia, this region was losing jobs, services, public amenities and infrastructure rather than developing a thriving regional economy and tackling socioeconomic disadvantage. The economic injection stimulated by the mine provided a partial distraction from this underlying trend, but was not a sustainable antidote.

The state monopolises regulatory powers with respect to mining—but with Century mine, as in other cases, such regulations have at times encouraged a minimalist compliance approach. Governments have proven to be a weak link in moving from the historical, exploitative model of mining development to delivering the promise offered by the sustainable development approach that mining peak bodies advocate (IIED 2002). Australian governments' overriding rationality appeared to regard economic growth as a precursor to social justice and they demonstrated primary responsiveness to politico-economic imperatives at state and national levels while maintaining a distance from local socioeconomic, structural and cultural issues.

There are regular examples in the literature of forms of private regulation and governance—with both direct and arm's-length involvement of the state (Moon et al. 2011; Raschke et al. 2013). This has been observed to result in non-government actors, including mining companies, filling a governance void where state capacity is weak (Cheshire 2006, 2010; Eversole and Scholfield 2006; Cheshire et al. 2011). Wilson and colleagues even conceptualise private actors as 'meta-governors', steering the various actors (Wilson et al. 2018). Their analysis of the role of corporate actors in such arrangements suggests they shape social and economic agendas. Corporations operating in Australia may not experience the same level of pressure as corporations operating in countries like Papua New Guinea and similar developing economies to take on traditional state responsibilities (see, for example, Bainton and Macintyre, Chapter 4; Skrzypek, Chapter 2; and Burton and Levacher, Chapter 10, all this volume); nevertheless, where states are not fulfilling regional development mandates, the boundaries between state and private roles are blurred. It is difficult to distinguish government from company responsibilities with the state's steering being more assumed than scrutinised. What has been termed 'governance without government' (Rhodes 1996) can also be interpreted as a strategic withdrawal and implicit delegation of the state's active agency to self-managing, non-government actors largely free of state interference. Rather than these being 'stateless' spaces, with a passive state playing a limited role, they can be understood as spaces where the state takes 'action at a distance' to pursue its ends indirectly, through the actions of others, in this case a mining corporation. This has sometimes been described as cost-shifting from the state to mining companies (Martin 2009). The approach can also be criticised on the grounds of 'steering' and small government being an inadequate response to the policy challenges in regions like the lower gulf—where one corporate development project is insufficient to drive the transformation and reform needed to ensure desirable long-term socioeconomic outcomes.

There are multiple narratives of Century Mine during the 1992–2012 period. As Ballard and Banks (2003) argue, the state, community, company and 'fourth estate' of NGOs, each with their own matrix of relationships, interact in a network of roles and interests. The uneven distribution of benefits is but one of the imbalances evident in this case. The state's story is a celebratory one of economic development. In contrast, for many residents and traditional owners of the lower gulf, the salient threads of the story relate to rights, relationships to land and unmet expectations as state and private sector interests and roles converged. The Century Mine case illustrates conflicting notions of the extent to which the state should actively intervene in regional development, which reflects unresolved questions about the state's role in an era of governance (Rhodes 1996). Likewise, there are multiple ways to interpret the contestations over development in remote Indigenous Australia, and persistent tensions characterising resource relations in far northwest Queensland. The case also illustrates the significance of temporal dimensions, with the state alternating between presence and absence in the affected region. Periods of 'absence' frequently involved an indirect 'presence' through the mining company's actions. As well as the fluctuations over time, there were varying spatial limits to the state's presence, with the remotely located Century Mine communities experiencing mostly absence. These spatial and temporal influences varied as the mine moved through its project life cycle, as political changes brought new policy imperatives for Indigenous affairs, regional development and economic prosperity and as diverse actors within or beyond the state exerted an influence. The GCA proved to be an ambitious instrument, yet in some respects flawed in that it was unable to carry the sole responsibility for ensuring all parties contributed to regional advancement. Nevertheless, the case demonstrates that, in peripheral mining localities, agreements such as the GCA can play a critical role in engaging the state and compelling its presence.

References

Altman, J., 2009. 'Indigenous Communities, Miners and the State in Australia.' In J. Altman and D. Martin (eds), *Power, Culture, Economy: Indigenous Australians and Mining*. Canberra: ANU E Press (Centre for Aboriginal Economic Policy Research Monograph 30). doi.org/10.22459/caepr30.08.2009.02

————, 2014. 'Indigenous Policy: Canberra Consensus on a Neoliberal Project of Improvement.' In C. Miller and L. Orchard (eds), *Australian Public Policy: Progressive Ideas in the Neoliberal Ascendancy*. Bristol: University of Bristol, Policy Press. doi.org/10.2307/j.ctt1ggjk39.13

Anon., 2004. 'Clark Vows to Fight as ATSIC Scrapped.' *Sydney Morning Herald*, 15 April.

Arndt, H.W., 1983. 'The "Trickle Down" Myth.' *Economic Development and Cultural Change* 32: 1–10. doi.org/10.1086/451369

Ballard, C. and G. Banks, 2003. 'Resource Wars: The Anthropology of Mining.' *Annual Review of Anthropology* 32: 287–313. doi.org/10.1146/annurev.anthro. 32.061002.093116

Blowes, R. and D. Trigger, 1999. 'Negotiating the Century Mine Agreement: Issues of Law, Culture and Politics.' In M. Edmunds (ed.), *Regional Agreements: Key Issues in Australia—Volume 2: Case Studies*. Canberra: Australian Institute of Aboriginal and Torres Strait Islander Studies.

Brereton, D. and J. Everingham, 2016. *Making and Implementing Agreements with Indigenous Communities: A Case Study of the Gulf Communities Agreement*. St Lucia: University of Queensland, Centre for Social Responsibility in Mining.

Cheshire, L., 2006. *Governing Rural Development: Discourses and Practices of Self-Help in Australian Rural Policy*. Aldershot: Ashgate. doi.org/10.4324/ 9781315585598

————, 2010. 'A Corporate Responsibility? The Constitution of Mining Companies as Governance Partners in Remote Mine-Affected Localities.' *Journal of Rural Studies* 26: 12–20. doi.org/10.1016/j.jrurstud.2009.06.005

Cheshire, L., J. Everingham and C. Pattenden, 2011. 'Examining Corporate-Sector Involvement in the Governance of Selected Mining-Intensive Regions in Australia.' *Australian Geographer* 42: 123–138. doi.org/10.1080/0004918 2.2011.569986

CLCAC (Carpentaria Land Council Aboriginal Corporation), 2002. 'Response of the CLCAC on behalf of the Waanyi Native Title Group to the draft report by the State of Queensland [Department of State Development], PCML [Pasminco, Century Mine Ltd], and GADC [Gulf Aboriginal Development Corporation] and David Martin.' Unpublished report prepared for the first five-year review of the 1997 Gulf Communities Agreement.

Eriksen, T.H., 2018. *Boomtown: Runaway Globalisation on the Queensland Coast*. London: Pluto Press. doi.org/10.2307/j.ctv3mt8xn

Evans, R., K. Saunders and K. Cronin, 1975. *Exclusion, Exploitation, and Extermination: Race Relations in Colonial Queensland.* Sydney: Australia and New Zealand Book Company.

Everingham, J., R. Barnes and D. Brereton, 2013. *Gulf Communities Agreement 2008–2013: 15-year review.* St Lucia: University of Queensland, Centre for Social Responsibility in Mining.

Everingham, J., R. Barnes, J. Parmenter and D. Brereton, 2014. *Social Aspects of the Closure of Century Mine: Social Impact Assessment Stage 2.* St Lucia: University of Queensland, Centre for Social Responsibility in Mining.

Everingham, J., L. Cheshire and G. Lawrence, 2006. 'Regional Renaissance? New Forms of Governance in Nonmetropolitan Australia.' *Journal of Environment and Planning* 24: 139–155. doi.org/10.1068/c47m

Everingham, J. and J. Keenan, 2017. *Hindsight for Foresight: Lessons About Agreement Governance from Implementing the Gulf Communities Agreement.* St Lucia: University of Queensland, Centre for Social Responsibility in Mining. Viewed 13 October 2020 at: csrm.uq.edu.au/publications/hindsight-for-foresight-lessons-about-agreement-governance-from-implementing-the-gulf-communities-agreement-gca

Eversole, R. and K. Scholfield, 2006. 'Governance in the Gaps: Inter-Agency Action in a Rural Town.' *Rural Society* 16: 320–328. doi.org/10.5172/rsj.351. 16.3.320

GCA (Gulf Communities Agreement), 1997. Agreement between The State of Queensland and The Waanyi, Mingginda, Gkutharn and Kukatj Peoples and Century Zinc Limited.

Gunstone, A., 2008. 'The Failure of the Howard Government's "Practical" Reconciliation Policy.' In H. Babacan and N. Gopalkrishan (eds), *The Complexities of Racism: Proceedings of the Second International Conference on Racisms in the New World Order.* Caloundra: University of the Sunshine Coast.

Howlett, C., 2010. *The Politics of Mining on Indigenous Lands: The Story of Century Zinc Mine.* Saarbrücken: Lambert Academic Publishing.

ICMM (International Council of Mining and Metals), 2019. *Integrated Mine Closure: Good Practice Guide.* Viewed 13 October 2020 at: www.icmm.com/website/publications/pdfs/closure/190107_good_practice_guide_web.pdf. doi.org/10.36487/acg_rep/1915_63_brock

IIED (International Institute for Environment and Development), 2002. *Breaking New Ground: Mining, Minerals and Sustainable Development.* London: IIED, Mining Minerals and Sustainable Development Project. doi.org/10.4324/9781315541501-14

Langton, M., 2013. *The Quiet Revolution: Indigenous People and the Resources Boom*. Sydney: Harper Collins (The Boyer Lectures 2012).

Maguire, K., 2019. 'New Century Resources Signs Deal to be Mining Industry Leader for Indigenous Employment.' *Australian Broadcasting Corporation*, 10 July. Viewed 25 September 2020 at: www.abc.net.au/news/rural/2019-07-10/half-indigenous-employment-for-new-gulf-of-carpentaria-mine/11290912

Martin, D.F., 1998. 'Deal of the Century? A Case Study from the Pasminco Century Project.' *Indigenous Law Bulletin* 4(11): 4–7, 18.

———, 2009. 'The Governance of Agreements Between Aboriginal People and Resource Developers: Principles of Sustainability.' In J. Altman and D. Martin (eds), *Power, Culture, Economy: Indigenous Australians and Mining*. Canberra: ANU E Press (Centre for Aboriginal Economic Policy Research Monograph 30). doi.org/10.22459/caepr30.08.2009.05

Martin, D., D. Trigger and J. Parmenter, 2014. 'Mining in Aboriginal Australia: Economic Impacts, Sustainable Livelihoods and Cultural Difference at Century Mine, Northwest Queensland.' In E. Gilberthorpe and G. Hilson (eds), *Natural Resource Extraction and Indigenous Livelihoods: Development Challenges in an Era of Globalization*. London: Routledge. doi.org/10.4324/9781315597546-6

Martin, R., 2019. *The Gulf Country: The Story of People and Place in Outback Queensland*. Crows Nest: Allen and Unwin.

Miles, R.L., J. Cavaye and P. Donaghy, 2005. *Managing Post Mine Economies— Strategies for Sustainability*. Paper presented at the Sustainable Economic Growth for Regional Australia conference, Yeppoon, Queensland, September 2005.

Moon, J., A. Crane and D. Matten, 2011. 'Corporations and Citizenship in New Institutions of Global Governance.' In C. Crouch and C. MacLean. (eds), *The Responsible Corporation in a Global Economy*. Oxford: Oxford University Press. doi.org/10.1093/acprof:oso/9780199592173.003.0010

Palmer, V.J., 2008. 'Uneasy Terrains: Mapping Ethical Tensions in Corporate-Community Partnerships.' *Third Sector Review* 14: 51–66.

Pasminco, the State of Queensland and Gulf Aboriginal Development Company, 2002. 'Report of the First 5-Year Review of the Century Mine Gulf Communities Agreement in Accordance with Clause 63 of the Agreement.' Unpublished report.

Pearson, N., 2000. *Our Right to Take Responsibility*. Cairns: Noel Pearson & Associates.

Peters, B.G., 2000. 'Governance and Comparative Politics.' In J. Pierre (ed.), *Debating Governance: Authority, Steering and Democracy.* Oxford: Oxford University Press.

Plant, R., 2012. 'Foreword.' In J. Connelly and J. Hayward (eds), *The Withering of the Welfare State: Regression.* Basingstoke: Palgrave Macmillan.

QSG (Queensland State Government), 2002. 'Queensland Support for Training and Employment through the Gulf Communities Agreement and Century Mine: An Assessment of Impact.' Brisbane: Department of Employment and Training.

———, 2009. 'Queensland Closing the Gap Report 2008/09: Indicators and Initiatives for Aboriginal and Torres Strait Islander Peoples.' Brisbane: Department of Aboriginal and Torres Strait Islander Partnerships.

———, 2017. 'Know Your Community: Key Insights into Aboriginal and Torres Strait Islander Queenslanders.' Brisbane: Office of the Statistician and Department of Aboriginal and Torres Strait Islander Partnerships.

Raschke, A., F. de Bakke and J. Moon, 2013. 'Complete and Partial Organizing for Corporate Social Responsibility.' *Journal of Business Ethics* 115: 651–663. doi.org/10.1007/s10551-013-1824-x

Rhodes, R.A.W., 1996. 'The New Governance: Governing Without Government.' *Political Studies* 44: 652–667. doi.org/10.1111/j.1467-9248.1996.tb01747.x

Rowley, C.D., 1970. *The Destruction of Aboriginal Society.* Canberra: Australian National University Press.

Scambary, B., 2009. 'Mining Agreements, Development, Aspirations, and Livelihoods.' In J. Altman and D. Martin (eds), *Power, Culture, Economy: Indigenous Australians and Mining.* Canberra: ANU E Press (Centre for Aboriginal Economic Policy Research Monograph 30). doi.org/10.22459/caepr30.08.2009.08

———, 2013. *My Country, Mine Country: Indigenous People, Mining and Development Contestation in Remote Australia.* Canberra: ANU E Press (Centre for Aboriginal Economic Policy Research Monograph 33). doi.org/10.22459/CAEPR33.05.2013

St-Laurent, G.P. and P. Le Billon, 2015. 'Staking Claims and Shaking Hands: Impact and Benefit Agreements as a Technology of Government in the Mining Sector.' *Extractive Industries and Society* 2: 590–602. doi.org/10.1016/j.exis.2015.06.001

The Right Mind, 2008. 'Gulf Communities Agreement: Review of Performance 2002–2007.' Unpublished report.

Tonts, M. and F. Haslam-McKenzie, 2005. 'Neoliberalism and Changing Regional Policy in Australia.' *International Planning Studies* 10: 183–200. doi.org/10.1080/13563470500378861

Townsend, I., 2002. 'Indigenous Sit-In Continues at Century Zinc Mine.' Transcript, M. Colvin. Australian Broadcasting Corporation, 19 November. Viewed 13 October 2020 at: www.abc.net.au/pm/stories/s730483.htm

Trebeck, K.A., 2007. 'Tools for the Disempowered? Indigenous Leverage over Mining Companies.' *Australian Journal of Political Science* 42: 541–562. doi.org/10.1080/10361140701513604

Trigger, D., 1988. 'Equality and Hierarchy in Aboriginal Political Life at Doomadgee, Northwest Queensland.' *Anthropological Forum* 5: 524–544. doi.org/10.1080/00664677.1988.9967386

———, 1997. 'Reflections on Century Mine: Preliminary Thoughts on the Politics of Indigenous Responses.' In D.E. Smith and J. Finlayson (eds), *Fighting over Country: Anthropological Perspectives*. Canberra: The Australian National University, Centre for Aboriginal Economic Policy Research (Monograph 12).

———, 1998. 'Citizenship and Indigenous Responses to Mining in the Gulf Country.' In N. Peterson, and W. Sanders (eds), *Citizenship and Indigenous Australians: Changing Conceptions and Possibilities*. Cambridge: Cambridge University Press. doi.org/10.1017/CBO9780511552243.009

Trigger, D., J. Keenan, K. de Rijke and W. Rifkin, 2014. 'Aboriginal Engagement and Agreement-Making with a Rapidly Developing Resource Industry: Coal Seam Gas Development in Australia.' *Extractive Industries and Society* 1: 176–188. doi.org/10.1016/j.exis.2014.08.001

Williams, I., 1996. 'The Century Project: The Early Years: Engineering or Anthropology?' Unpublished reflections reported in D. Brereton and J. Everingham, 2016. *Making and Implementing Agreements with Indigenous Communities: A Case Study of the Gulf Communities Agreement*. St Lucia: University of Queensland, Centre for Social Responsibility in Mining.

Wilson, C.E., T.H. Morrison and J. Everingham, 2018. 'Multi-Scale Meta-Governance Strategies for Addressing Social Inequality in Resource Dependent Regions.' *Sociologia Ruralis* 58: 500–521. doi.org/10.1111/soru.12189

10

The State That Cannot Absent Itself: New Caledonia as Opposed to Papua New Guinea and Australia

John Burton and Claire Levacher

Introduction

The other contributors to this volume have come to broadly consistent views of the state in the two jurisdictions that are the focus of this book, Papua New Guinea (PNG) and Australia. The writers on PNG have profiled what community members aspire to: variously, a locally realised Utopia unhindered by the fact that 'the developers are foreigners and the State is only a concept' (Lihir: Bainton and Macintyre, Chapter 4); an end to 'waiting for the economic development' (Frieda: Skrzypek, Chapter 2); an Engan version of 'development, law and order, and prosperity', where the 'furious contest' for 'jobs, cash, houses, vehicles' can continue without restraint (Porgera: Golub, Chapter 3); and an end to the immorality of the 'corruption and neglect of an absent state' to save the world from entropic decline and, of course, bring the worldly rewards of development (PNG Liquefied Natural Gas (LNG) project: Main, Chapter 5). All the Indigenous voices represented capture a desire for something better than what national independence has so far brought them, but perhaps only the Frieda River people, in their feeling of simply being forgotten about (as opposed to the Huli feeling of being *immorally* neglected), expect the state to be the means of delivering it to them. The rest express a desire for their projects to trigger various versions of self-actualisation which, to judge from their various ways of expressing dissatisfaction, has yet to happen. Certainly, the overall picture is of an absent state. At Porgera, the executive part of the state, in the form of prime minister James Marape, suddenly woke up on 24 April 2020 to announce that Barrick Niugini Limited's lease renewal had been rejected. Was this a presence, or just feeble hand-waving? No one can be sure quite yet.

The writers on Australia portray a different, but similarly consistent picture of the Australian state. At the McArthur River Mine, the historic chicanery of the state and the company replicates the nineteenth-century dispossession of vast expanses of the Northern Territory for private gain (Lewis, Chapter 8). In the Northern Territory generally, poor 'recognition and regulatory guidance' means that the state is 'selectively absent' for Aboriginal people, has a 'fiscal absence' in the mining sector because of tax concessions and rebates to companies, and—as is notoriously the case in PNG—has a tendency to cut expenditure on Aboriginal communities to which mining benefits flow (Holcombe, Chapter 7). In Queensland's Western Downs region, the extremely rapid uptake of coal seam gas developments since 2010 highlights the state's 'organised irresponsibility' to approve resource developments without evaluating them effectively, leaving private land holders 'responsibilised' to do as best they can (Espig, Chapter 6). At the Century Mine, also in Queensland, the state is framed as taking 'action at a distance', with the ambitious Gulf Communities Agreement delivering unevenly distributed benefits to disadvantaged Aboriginal people whose land was needed for mining (Everingham et al., Chapter 9). All the above backs the decades-old finding that Indigenous economic status in remote parts of Australia is barely changed by the presence of long-life mines and with whose operators they have entered into seemingly favourable agreements (e.g. Taylor and Scambary 2005).

The consistency here is that, in all the areas described, local 'host communities' have an expectation that extractive industry projects will summon distant governments to be present to assist them with complex negotiations and afterwards to bring services which urban dwellers, in both jurisdictions, take for granted. Alas, this imagined emanation from nothingness disappoints time and again. In each case the kinds of state encountered, when stripped down to basics, cast themselves in the role of investor-enablers, and appear to be in retreat from (or never attained) the ideal of giving first priority to the human needs of citizens. The state is either remote or irrelevant (Skrzypek, Chapter 2; Bainton and

Macintyre, Chapter 4), or it can be irresponsible, immoral or malevolent (Espig, Chapter 6; Main, Chapter 5; Lewis, Chapter 8). In other words, to a greater or lesser extent, local communities consistently view the state as looking after someone else's interests.

What if there was a different kind of state entirely?

This chapter responds to this question by comparing experiences of PNG and Australia with those of New Caledonia. The aims of this chapter are twofold. First, we use Jeffrey Wilson's model of different state forms to analyse resource nationalism across the three nation states and explore whether, when compared to PNG and Australia, New Caledonia is, indeed, a different kind of state. Second, we explore if and how the state is absent or present in resource contexts in these countries and ask whether the notion of 'absent presence' can be applied to the study of the state in resource settings in New Caledonia where, unlike in the cases from PNG and Australia, it would seem that in some ways at least *the state cannot absent itself.*

Wilson's Kinds of States

Wilson (2015: 399, 403–4) identifies three forms of resource nationalism that can be summarised as follows. A first form is associated with *rentier states*, where governments often fashion loyal societal coalitions, finance repressive apparatuses and engage in neopatrimonialism, exemplified by the Gulf states of the Middle East. A second form is associated with *developmentalism*, where governments maintain their standing through an 'ability to deliver industrialisation, economic modernisation and high rates of growth', exemplified by states in East Asia. And a third form is associated with *liberal market economies*, where the role of the state is (theoretically) limited to that of an arm's-length regulator, exemplified by the USA, Canada and Australia.

Of the three countries considered in this chapter, only Australia, having recently overtaken Qatar as the world biggest exporter of LNG (Owen 2020), could become a rentier state. However, as we discuss below, its resource rental regime is not capturing enough of the benefits to make this possible. PNG is not in the running, while New Caledonia has a mixed outlook but, here too, no one economic sector is poised for an East Asian–style boom (Bouard et al. 2016). All three places aspire to some kind of developmentalism, but none has achieved the desired goal of economic modernisation. Australia has de-industrialised in recent years and seems to have abandoned an industry policy. PNG is not doing at all well at translating resources incomes into development outcomes (UNDP 2014), and, again, New Caledonia has a mixed outlook, although no single economic sector is placed for rapid take-off. Elements of the last of Wilson's forms of the state, the liberal market economy, can be found in each but, as we will show, the hand of the state intervenes to varying degrees.

For our purposes, Wilson's model provides a useful way of conceptualising the question of state presence or absence in PNG and Australia, using New Caledonia as a point of comparison. To aid this objective, we have structured our analysis along three policy axes, based on three core sets of policies identified by Wilson in his modelling of resource nationalism. We have further clarified the three sets and define them as follows:

- Axis 1. Concerning ownership or equity. The key question here, regardless of who invests the initial capital, is who ends up owning resource projects and what political project this matches.

- Axis 2. Concerning the exercise of mining, oil and gas powers, other aspects of regulation, and non-regulatory influences on the conduct of the resource industries. The exercise of mining powers ranges from full state control of extractive resources (Bebbington 2014) to various degrees of delegation to authorities at the subnational level (Childs 2016). Whether regulation is or is not effective is likely to be a proxy for the presence or absence of the state at different levels.

- Axis 3. Concerning resource rents and benefit distribution. Thinking on this ranges from the position that states must attract investors on generous terms to the opposite idea that if foreigners want national resources they should pay as much for them as the market will bear. Once received, how should economic returns be shared out: should the state put them into its coffers, or how much should it share with other levels of government, local communities and landowners? Both dimensions of this problem raise difficult economic and political questions.

These three axes now provide the basis for comparison between PNG, Australia and New Caledonia. In the sections that follow, we draw upon these different dimensions or axes to consider how the state may be more or less present (or absent) in resource extraction contexts.

New Caledonia: A Different Kind of State

New Caledonia became a French territory in 1853 and is now engaged in the process of decolonisation. A violent period in the 1980s known as 'the events' (*Les Événements*), centred on Kanak[1] struggles for independence, ended in a hostage siege on the island of Ouvéa. This triggered a first set of political agreements between loyalists and separatists, the Matignon-Oudinot Accords, negotiated in Paris in May and June 1988. The accords led to new initiatives for Kanak economic development and administrative decentralisation, seen as the first steps in a process of decolonisation. A territorial government was created; the previously wide powers of the High Commissioner, a delegate of Paris, were curbed; and provincial governments were set up for new North, South and Loyalty Islands Provinces. A second political agreement was signed in 1998, the Nouméa Accord, resulting in further transfers of powers from France, more formalised economic financing, and with provisions for up to three referendums on the question of independence from France. Two have now been held: the first, on 4 November 2018, resulted in a 56.7 per cent 'no' vote while the second, on 4 October 2020, resulted in a 52.3 per cent 'no' vote (HCRNC 2019, 2020). In 2020, the split between the Kanak-dominated Islands and North Provinces, where 84.3 per cent and 78.3 per cent voted 'yes' respectively, and the loyalist-dominated South Province where 70.1 per cent voted 'no', was manifest. While the results are narrowing, the heavier population of the South Province, where voters cast over 100,000 votes in 2020 as against 53,000 in the other two provinces, means that the result in 2022's third referendum will be extremely tight.

The nature of government in New Caledonia is very distinctive and differentiates it from the other two countries explored here. In both Australia and PNG, governments are formed in winner-takes-all systems of government, but this is not the case in New Caledonia. Here government is formed in a tiered way, from a starting point of five-yearly general elections for 76 provincial seats in the assemblies of the South (40), North (22) and Loyalty Island (14) Provinces. Without further elections, a quota of the members of each assembly is selected by party list proportional representation to also sit in the Congress (South: 32 seats; North 15 seats; Loyalty Islands: 7 seats). The 54 members of the

---

1   Indigenous Melanesian inhabitants of New Caledonia.

newly formed Congress then vote to determine the 11 members of the executive government of New Caledonia. From among their number, the 11 members of the government elect a president and vice-president. The government is termed a *collegial* government because it incorporates what in other jurisdictions would be opposition members. Its key role is to propose legislation for forwarding to the Congress for discussion and voting. An unwritten rule is that once the president is elected, the members of the losing bloc choose the vice-president.

A supra-governmental body, the Committee of the signatories of the Nouméa Accord (*Comité des signataires de l'accord de Nouméa*) meets annually in Paris to monitor progress on the implementation of the Nouméa Accord, nowadays bringing together an enlarged delegation of members of the Congress and mayors as well as surviving signatories from 1998.[2] This is supported by the GTPS, the working group of the presidents and signatories of the Nouméa Accord (*Groupe de travail des présidents et des signataires de l'accord de Nouméa*) which meets locally. Technically there are further superior bodies, such as the *Conseil d'État* ('Council of State') to which appeals can be made, but these are part of the French state.

The implication for critical legislation, such as that concerning resource extraction, is that the make-up of the government has always reflected the spectrum of political positions held by the members of the three provincial assemblies and, in turn, among the electorate of New Caledonia. Critical legislation cannot therefore be changed easily by the biggest party to fit with its electoral platform, in disregard of the electoral platforms of its rivals. While a majority of the members of the government is all that is technically required to make a decision, in practice a broader consensus must be achieved because of the complexity of party positions in the Congress. If the minority were to be disregarded in the government, it would run the risk of chaos in the Congress and of joint positions collapsing in relation to other matters.

Because of the unusual manner in which its institutions of democracy are interlaced, it is possible to suggest that the state, in its territorial form in New Caledonia, *cannot absent itself.* The institutions were designed in the post-conflict situation in which France and New Caledonia were locked

---

2   The first meeting was held on 2 May 2000. The nineteenth meeting, chaired by the French prime minister, Edward Philippe, was held in Paris on 10 October 2019.

together in May 1988 and have evolved over three decades into the form they have today. Whether it is due to the special design of the institutions, or that a kind of post-conflict vigilance on the part of all parties affects the way they operate, or that living people made the original agreements and they are still politically active, is not especially relevant to the present discussion. What is true is that the expression of the state in terms of policies, laws and governmental decrees *must* follow the political templates laid down in the foundational agreements—Matignon, Nouméa, Bercy, the Organic Law of 1999, etc.—and that deviations from them have not so far been possible.

This does not mean that there are no conflicts among lower-level entities; indeed, there have been bitter ones concerning which shareholder group should control this or that enterprise and which financing entity should or should not exercise influence or should yield its capital to local interests. But it does show the difference that the structures of the state can make to its relative absence or presence in resource extraction contexts. The value of this point of comparison is that it helps to highlight some of the structural similarities between PNG and Australia where they might otherwise be rendered invisible or masked by other more seemingly obvious differences, such as state capacity to use mineral wealth for the benefit of the population, as we discuss below.

A Short History of Mining in New Caledonia

New Caledonia's mineral wealth is linked to its nickel mining industry. Nickel was discovered there early in the colonial period, with exports starting in 1874. Today mining is carried out, in the first place, by three large companies that operate their own refineries. The first is Société Le Nickel (SLN), founded in 1880, from time to time partly state-owned, and now majority-owned by the French multinational group Eramet (see Newbury 1955; Bencivengo 2014; Black 2014). SLN's Doniambo nickel refinery on the outskirts of Nouméa has operated since 1910. The second is Koniambo Nickel SAS (KNS), a joint venture between the Société Minière du Pacifique Sud (SMSP) and Glencore plc. KNS has built a refinery in the North Province. The third is Vale Nouvelle-Calédonie SAS (Vale NC), a subsidiary of the Brazilian energy and mining group Vale SA. It has built a refinery at Goro in the South Province.

Twenty or more second-line companies that are only engaged in parts of the process of mining, and do not do their own refining, operate across the territory and are loosely known as '*petits mineurs*' (or 'small miners'). The fact that some of these 'small miners' are well capitalised, hold large concessions and directly export nickel (Bouard et al. 2019), only seems to reinforce Sarah Holcombe's observation (Chapter 7, this volume) that there is very limited consensus around the definition of a 'junior miner'.

The French state has asserted a presence in mining in various ways from 1859 when French mining law was first extended to New Caledonia. However, the hand of the state in mining was weakly felt to start with and a local decree in 1873 authorised mining on pegged claims ('*possession par invention*'), similar to what is seen in Anglophone countries (Le Meur and Mennesson 2011). The state remained a spectator while the Paris-based bank Rothschild Frères owned the main producer SLN from 1883, and after its merger in 1931 with its main rival, André Ballande, during the Great Depression. When the Second World War started, trade with Germany and Japan, which made up 95 per cent of exports before the war, came to a stop, halting the economy until the US entered the war and took all of New Caledonia's nickel production.

In the 1960s, in promoting the autonomy of the territory, the *Union Calédonienne*,[3] expressed interest in attracting foreign capital. But when the global price of nickel boomed in 1967, the French Secretary of State, Pierre Billotte, declared to the National Assembly in Paris that 'there can be no question of delivering New Caledonian nickel to a foreign trust' (Brou 1982: 112). Whereas previously licencing had been handled at the territorial level, the 1969 'Billotte Laws' classified nickel, chromium and cobalt as strategic raw materials and transferred control of them to the Ministry for Industry. This period also saw French interest in creating a company with French investment majority to face off foreign investors, but nothing eventuated before a slump in prices in 1972 threatened the entire industry with collapse. To prevent this from happening, a 50 per cent stake in SLN was taken out in 1974 by the (then) state-owned oil company, Elf Aquitaine. Further restructuring took place in 1985 when another majority state-owned enterprise, Eramet, boosted state-ownership to 85 per cent of SLN. Since this peak of state control, metropolitan

---

3   A political party founded in 1953 with the goal of promoting more autonomy for New Caledonia, the *Union Calédonienne* moved from its original multi-ethnic base to become part of the pro-Independence FLNKS in the 1970s.

public ownership of Eramet has been progressively diluted, then finally reduced to zero by offloading the last third to a New Caledonian entity STCPI (see below) in 2000.

Modernisation of the Mining Sector

Up to this point, in our exploration of the role of the state in resource contexts in New Caledonia, the state we are concerned with was exclusively conceived of as that of metropolitan France. After 1988, though, the political solutions of the metropolitan state pursued a new set of objectives directed at (a) averting violence and (b) redressing the injustice of the underdevelopment of Kanak communities, particularly through the doctrine of economic rebalancing (*rééquilibrage*). These objectives can be seen as aspiring to non-traditional, more progressive 'state effects' and creating new institutions able to define and enforce collectively binding decisions on members of society (Lund 2006: 685).

Mining had not been mentioned in the 1988 Matignon Accords, but it soon became integrated into pro-independence claims for sovereignty as a precondition for talks[4] in the following years. Simultaneously, the presence of the French state in the New Caledonian mining industry mutated from direct involvement—it was selling off most of the concessions it held globally at the time[5]—to indirect involvement, seen in its capital injection into the newly formed mining and metallurgical company, *Société de financement et d'investissement de la province Nord* (SOFINOR), a financing entity for the new Kanak-run North Province. The leader of the anti-independence groups in New Caledonia, Jacques Lafleur, was both a signatory to the accords and a major player in the nickel industry, through his ownership of another extractive company, *Société minière du Sud Pacifique* (SMSP). In a surprise move, Lafleur sold his shares to SOFINOR on 5 October 1990, giving it an 85 per cent holding. SMSP focused on the export of nickel ore for the next eight years.

---

4   The catchphrase *préalable minier* ('mining precondition') expressed the specific political position that talks could not proceed if mining issues were not included at the outset.
5   Between 1990 and 1995, the *Bureau de Recherche géologique et minière* divested much of its global mining portfolio; in New Caledonia its concessions were mostly acquired by SLN and Inco (Pitoiset and Wéry 2008: 101).

The first strictly mining-related agreement was the Bercy Accord, signed on 1 February 1998.[6] Two months later, a joint venture was entered into between SMSP (51 per cent) and Falconbridge of Canada (49 per cent) to form Koniambo Nickel SAS (KNS), which began to plan a USD5.3 billion nickel processing plant at Vavouto. The project was incentivised, along with the Goro project in the South Province, with a tax holiday for 15 years in 2001, with further exemptions granted in 2004. Construction started in 2007 after Falconbridge was bought by Xstrata, and the plant entered production in 2013 as Xstrata was bought by Glencore plc. This was a new form of state intervention—via the politics of providing a financing vehicle for a subterritorial provincial government—which has resulted in the North Province having majority control of both the Koniambo massif and its linked processing plant.

The Koniambo project was conceived from the start as an economic engine around which the North Province as a whole might develop. Far from the state being absent-malevolent, as seen in Australia, and absent-irrelevant, as seen in PNG, the Koniambo project is the projection of the presence of *two levels* of the state in an industry: foremost as part of a long-term political project of the Kanak-run North Province and its mining company, SMSP; and financing mechanisms ultimately devised by the French state in the shadows behind SOFINOR.

The North Province and its mining arm SMSP have been able to expand into offshore processing of ore unsuitable for processing at Koniambo (Bouard et al. 2016: 121–2):

- Investing in a new processing plant in Gwangyang, South Korea, in 2006, creating a Caledonian offshore company, the SSNC (*Société du Nickel de Nouvelle-Calédonie et de Corée*).

- Forming the New Caledonia–based Nickel Mining Company (NMC) with 51 per cent ownership by SMSP and 49 per cent by the Korean steelmaker POSCO, to feed the Gwangyang plant from NMC concessions at Oauco, Poya, Nakéty and Kouauoa.

- Concluding an agreement for a nickel processing plant in Guanxi Province, China, to also have 51 per cent ownership by SMSP (Jinchuan project) and for the offshore processing of New Caledonian ore.

---

6   Signed by SMSP, Eramet and its subsidiary SLN, and three French ministries (Economy and Finance, Industry, Overseas Territories), in the presence of the French Agency for Development (CFD, now AFD).

At the same time, the Goro project, located in the communes of Mont-Dore and Yaté, came into being outside the template involving Kanak interests that the Matignon and subsequent agreements provided for. It began with the acquisition of mining titles at Goro by Inco of Canada in 1992. Inco tested a pilot plant in 1999 and began construction of a refinery in 2002. Technical problems resulted in the suspension of construction until 2005. Inco was acquired by Vale of Brazil to form Vale Nouvelle-Calédonie (Vale NC), but projected costs rose from USD1.4 billion to an eventual USD4.3 billion when it began operating four years after the Vale takeover, in 2010. The project has most recently been described as having cost USD9 billion (Mainguet 2019). Ownership was transferred to Prony Resources, a 51 per cent locally owned consortium, in March 2021.

Goro faced stiff opposition from both local communities and civil society organisations between 2001 and 2008. Loyalist forces dominate in the South, and the manner of developing the project was seen as a way to assert the old way of doing things. Three chieftaincies in Yaté formed the Rhéébu Nùù ('Eye of the Country') committee, which was successful in widening the circle of stakeholders to include local interest groups (see Demmer 2007; Horowitz 2009; Levacher 2016, 2017). This led to the signing of a Pact for the Sustainable Development of the Far South (*Pacte pour un Développement Durable du Grand Sud*) by Vale and local stakeholders, including Rhéébu Nùù, in 2008 (Le Meur, Horowitz and Mennesson 2013; Horowitz 2015).

Three Axes of Resource Policy: Comparisons with PNG and Australia

So far in this chapter we have introduced Wilson's model of resource nationalism and provided information necessary to establish New Caledonia as a point of comparison with PNG and Australia. For the remainder of this chapter, we turn our attention to the three policy axes, based on core sets of policies identified by Wilson, along which we make comparisons between the three countries. The axes are drawn along the critical issues of ownership or equity; exercise of mining powers, other aspects of regulation and non-regulatory influences; and resource rents and benefits distribution, respectively. We use them here to consider ways in which the state is absent and/or present in resource extraction contexts in PNG, Australia and New Caledonia across the identified policy areas.

Axis 1: Concerning Ownership or Equity

Papua New Guinea

The PNG government has frequently expressed an intention to own equity in resource projects. At the Ok Tedi copper and gold mine in the Western Province, the state started as a 20 per cent owner when the mine opened in 1984. Wounded by the environmental record of the mine, the majority owner BHP placed its 52 per cent share ownership in an offshore company structure, the PNG Sustainable Development Program (SDP), in 2001, and left the country. In the 2000s, SDP's and the state's shares grew to about 65 per cent and 35 per cent respectively as minor partners divested, but a long-standing feud between the 2011–19 prime minister, Peter O'Neill, and the 1999–2001 prime minister, Mekere Morauta, changed things when Morauta became chairman of SDP. O'Neill, who hails from Southern Highlands Province, pushed through legislation to cancel SDP's shares, nationalise the mine and replace Morauta as chairman, saying, 'I ask, who is he? He's not from Western Province. He has done enough damage to this country and he has protected his foreign friends enough'. He added, 'the people of Western Province must run the company' (Nalu 2013). But when the new executive team was announced, it did not noticeably include anyone from Western Province, and O'Neill appointed a prominent Australian as the CEO.

At Porgera the state initially took a 10 per cent stake in the mine, which opened in 1990. When the mine produced a million ounces of gold each year, the government of Paias Wingti demanded a greater stake (Davis 1993; Jackson and Banks 2002: 191–8). Seemingly a textbook example of a state wanting to take a lead role in managing national mineral wealth, the next five prime ministers gradually lost interest, and the government's shareholding in Porgera was diluted from a high of 25 per cent in the mid-1990s to 0 per cent in 2007. Most of the shares ended up back in the hands of the operator to 2020, the Porgera Joint Venture, 95 per cent owned by Barrick Gold and Zijin Mining. In 2020, James Marape's government refused an application for a 20-year mine lease extension, under the policy slogan 'Take Back PNG' (see Afterword, this volume). Marape prepared poorly for the announcement, with no obvious plan as to what new entity would run the mine. Meanwhile, 2,700 workers were laid off just as the gold price passed USD2,000/oz, a record high.

In the oil and gas sector, the O'Neill government borrowed AUD1.2 billion from the Union Bank of Switzerland (UBS) to buy a 10 per cent stake in Oil Search, a partner in the PNG LNG project, on the advice of consultants. Commentators deprecated the move (Yala et al. 2014), and an official report concluded that O'Neill and several others had broken the law (GPNG 2018). When the oil price slumped in 2016, PNG abandoned the loans at a loss of about a third of the amount borrowed (Grigg et al. 2019). Kumul Petroleum Holdings, the state-owned oil and gas company, and the beneficiary of the loans, made losses of PGK481 million in 2016 and PGK1.8 billion in 2017, despite being a 16.6 per cent partner in the PNG LNG project which has been exporting gas since 2014.

Australia

Taking a different approach, Australia does not seek government equity in the extractive industries. A short-lived Petroleum and Minerals Authority was created in 1974 but was struck down by the High Court over states' rights (Bambridge 1979). However, Australia invests in other ways. It has recently taken up a long-standing commitment to the International Energy Agency to maintain a 90-day oil reserve (Taylor 2020), and its 'critical minerals strategy' incentivises various kinds of research funding allocations to universities and the CSIRO (the Commonwealth Scientific and Industrial Research Organisation) (AFG 2019, 2020). The Northern Australia Infrastructure Facility offers concessional loans to resource companies to upgrade ports, airports, and road and rail links, while many states own or part-own energy retailers like Energy Queensland (Queensland), Aurora Energy (Tasmania) and Horizon Power (Western Australia).

A further 'other way' concerns the Australia–Timor Leste sovereignty dispute. In 2004, Australian secret agents bugged Timor Leste's delegation during talks over oil and gas sharing in the Timor Sea leading up to the signing of a treaty providing for 50 per cent of the income to be allocated to each country. When the spying operation was exposed, Timor sued Australia in the Hague and negotiated a new agreement giving 80 per cent of the income to Timor Leste and 20 per cent to Australia. Agreement on the maritime boundary between the two countries, and therefore the absolute ownership of the oil and gas, was settled in 2018 (Rothwell 2018).

New Caledonia

We have already discussed the long-standing political project for New Caledonia to assume majority ownership of resource projects. In the South, both Indigenous and environmental movements favoured the South Province investing in the Goro Nickel project as the North Province had done at Koniambo. The idea was endorsed politically at a meeting of the *Comité des signataires* and in 2005 a new entity, the SPMSC (*Société de participation minière du Sud calédonien*), acquired a stake in Goro Nickel, though a small one. In 2000 the three provincial governments created yet another equity vehicle, STCPI (*Société territoriale calédonienne de participation industrielle*), as a step towards joint involvement in mining. By means of various transfers from the French state, via each province's financing arms—SOFINOR, PROMOSUD, SODIL,[7] etc.—STCPI had acquired 34 per cent of the historically largest nickel company SLN by 2007. As shown in Figure 10.1, all three onshore refineries are part-owned by provincial governments, with the North Province's 51 per cent of Koniambo being the highest. The same and/or linked entities, like Nord Avenir in the North Province, also offer *financing* for mining contractors.

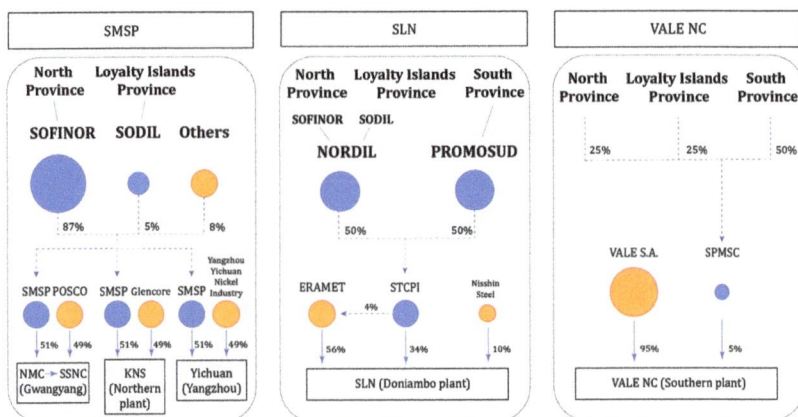

Figure 10.1 Provincial shareholding in New Caledonian nickel refineries
Source: C. Levacher (data to 2020)

---

7   PROMOSUD, *Société De l'Financement et de Développement de la Province Sud*; SODIL, *Société de Développement et d'Investissement des Îles*.

Axis 2: Concerning the Exercise of Mining, Oil and Gas Powers, Other Aspects of Regulation, and Non-Regulatory Influences on the Conduct of the Resource Industries

Papua New Guinea

The first piece of mining legislation to be passed after PNG's independence in 1975 was the *Mining (Ok Tedi Agreement) Act 1976* (Jackson 1982; Burton and Haihuie 2017: 22–5). A working group drawn from the major economic ministries in Port Moresby was the point of consultation for key decisions. Reference to local viewpoints and interests was almost non-existent (Burton 1997). Ok Tedi is well known for the environmental damage done to the Ok Tedi and Fly Rivers below the mine, and landowners in the impact area, in alliance with international lawyers, mounted litigation so damaging that it eventually forced BHP out of PNG, as recounted above (Banks and Ballard 1997; Kirsch 2001, 2007). All later extractive industry projects have been governed by national mining and environmental legislation: currently, the *Mining Act 1992*, the *Oil and Gas Act 1998*, and the *Environment Act 2000* (Filer and Imbun 2009). However, each project to date has been characterised by localised agreements (Burton and Haihuie 2017) emerging from Development Forums (Filer 2008) where the state takes second place to negotiations between companies and landowner representatives (Bainton 2010; Main, Chapter 5, this volume).

In reality, the state is barely visible in regulatory matters. Just to cite waste disposal problems, the government has watched on while landowners have rebelled, sued, lobbied ministers, engaged with international NGOs, or got their (opposition) MPs to complain about damage to five separate river systems: Jaba River (Bougainville mine: May and Spriggs 1990), Ok Tedi-Fly river system (Ok Tedi mine: Banks and Ballard 1997), Porgera-Lagaip river system (Porgera mine: CSIRO 1996; Bainton et al. 2020), Angabanga River (Tolukuma mine: OCAA 2004), and Watut River (Hidden Valley mine: Ketan and Geita 2011).

From the 1990s, international mining companies operating in PNG have turned to sustainability reporting to assure investors about their social and environmental performance. Close examination, though, reveals a patchy approach to applying standard reporting criteria (Burton and Onguglo 2017) and selective amnesia over past failings that poor government alertness allows operators to get away with. A quarter of a century later, mining, oil and gas policy is out of step with several cycles of national development plans (e.g. GPNG 2009, 2015) and has a puzzling fit with PNG's international commitments to initiatives like the Sustainable Development Goals and the Paris Agreement. Worst, the Department of Petroleum Energy is unfit for purpose: its ministers have failed to take up external offers of help and the department has not published an annual report since 2009 (PNGEITI 2019).

Australia

Unlike PNG, in Australia responsibility for the regulation of the extractive industries is located at the subnational level in state and territory governments and governed by subnational legislation. This means that the expectations placed upon extractive companies differs between states and territories. For example, as Holcombe has observed, there is a large gulf between the requirements and guidance for environmental and social impact assessment for new extractive projects in New South Wales compared with the Northern Territory. As a result, we also find a good deal of variation in the capacity and willingness of state and territory governments to closely regulate the activities of extractive companies to minimise social and ecological impacts, and to harness the economic benefits of extraction to provide broad-based development for local project-affected communities and the wider population. This partly helps to explain the incomplete or uneven nature of the state, where different actors experience different incarnations of the state in stronger or weaker ways, or as a more or less consequential force supporting some interests and not others. These subnational differences are particularly apparent when it comes to unconventional gas extraction and development, which is sometimes known as 'fracking'. While some states have placed a moratorium on unconventional gas extraction, others have actively encouraged these developments. But the discrepancy between the state's sovereign power to approve extractive projects and its ability to oversee and responsibly regulate the sector can be problematic, particularly for local residents who are caught up in this disparity (Espig, Chapter 6, this volume).

The influence of the coal seam gas industry, and the feeble regulatory capacity of the Queensland Government, point to larger considerations around the degree of state capture by industry and a 'growth-first' political economy in Australia. The considerable political influence and lobbying power of the extractive industries has found a favourable reception among successive federal, state and territory-level governments—especially in Queensland, Northern Territory and Western Australia. Nowhere is this more evident than in regards to Indigenous interests. Since at least the 1970s, the mining and petroleum industries have become increasingly influential in Australian politics and policy making. As Jon Altman has previously noted, historically the mining industry's relationship with the state in terms of land rights and native title has been fundamentally oppositional, seeing consent or negotiation provisions where the state has supported land rights as just another regulatory hurdle in the way of unimpeded access to Aboriginal land for mineral exploitation (Altman 2009: 25). In the 1970s, the Australian Mining Industry Council, now known as the Minerals Council of Australia, strongly opposed land rights, and mounted an effective campaign against these rights in Western Australia. These diminished rights are now reflected in Western Australia's *Aboriginal Heritage Act 1972*, specifically Section 18 which provides a legal basis for extractive companies to destroy specified Aboriginal heritage objects and places. This piece of legislation has served the interests of the extractive industries over Aboriginal people, most recently witnessed in the apparently legal destruction of the Juukan Gorge by Rio Tinto in 2020 (see Afterword, this volume). So while many multinational extractive companies working in Australia have embraced the rhetoric of corporate social responsibility—including Rio Tinto, who was once at the forefront of this corporate shift (Kemp and Owen 2018)—it would seem that in practice these higher corporate standards are trumped by lower legislative standards.

New Caledonia

A key institutional effect of the 1998 Bercy Accord was the transfer of mining powers from Paris to New Caledonia. An organic law[8] followed that prescribed work on a Framework for the Development of Mineral Resources (*Schéma de Mise en Valeur des Richesses Minières*) and decreed that future decisions relating to mining must follow the principles set out in it. It was meant to be ready by 2004, when a local Mining and Energy Department was indeed created—DIMENC, the *Direction de l'Industrie, des Mines et de l'Energie de la Nouvelle-Calédonie*—but rounds

---

8   Organic law n°99-209 of 19 March 1999 in relation to New Caledonia.

of consultations across New Caledonia took five more years until the Framework, and a territorial mining law, were adopted concurrently by the Congress of New Caledonia.

The Framework and the mining law not only set out the rules by which the mining should be run, as standard mining policy and legislation might be expected to do but, backed by the organic law, they also act as almost inviolable constitutional documents of a longer-term view of the mining industry in New Caledonia. That is why the adoption in 2015 of a 'nickel doctrine' by the territory-wide pro-independence coalition FLNKS (*Front de Libération Nationale Kanak et Socialiste*) raised strong opposition especially from the smaller producers and 'truckers' (*rouleurs*) who wanted to retain the freedom to export raw nickel to lower-cost overseas processors (Demmer et al. 2018). This policy advocates: better use of nickel resources by seeking efficiency gains in local mining operations and by striking economic agreements more advantageous to local interests; ceasing exports of unprocessed nickel ore (except those destined for New Caledonian offshore processing plants); and promoting the territorially owned STCPI to reach a 51 per cent shareholding in SLN (Bouard et al. 2016: 122) up from 34 per cent at present (Figure 10.1). The question of unprocessed mineral exports and the balance prescribed by the Framework between exporting nickel and doing onshore processing continues to lie at the heart of political contestation over the nickel industry (Demmer 2017; Anon. 2020b).

The combination of the political agreements (Matignon-Oudinot and Nouméa), the increased importance of mining issues in politics since the 1990s and the rise of environmental concerns has led to an evolution in the relations between mining companies and local populations. Newer negotiated agreements aim to recognise the special Kanak link to the land and to compensate for the impacts of mining in a variety of ways, notably in undertakings regarding employment and contracting by local entities.

Another point of evolution in regulations in the last two decades is the rise in awareness of New Caledonia's status as a biodiversity hotspot and increasing local concerns over environmental impacts combined with calls for environmental compensation (Levacher and Le Meur 2019). This is illustrated in the fact that protests around the Goro Nickel mining project in the south played a key role in the inscription of the New Caledonian lagoon as a UNESCO World Heritage Site in 2008 (Merlin 2014; Horowitz 2016). At the institutional level, both North and South Provinces introduced new environmental legislation in 2008–09, which include compulsory environmental impact assessments, with a special emphasis on impacts on ecosystems with heritage value and protected species, provisions for compensation, action plans for mitigating impacts, and long-term monitoring and reporting of results.

Axis 3: Concerning Resource Rents and Benefit Distribution

Papua New Guinea

Despite a perpetually renewed national rhetoric of reaping benefits from the exploitation of the country's endowment of natural resources, the latest version of which is James Marape's mantra of 'Take Back PNG', the effectiveness of the state in collecting fair returns has waned over time. The Australian colonial administration included tax incentives in its 1967 agreement with the mining company Conzinc Riotinto of Australia (CRA) to develop the Bougainville copper mine. A normal tax regime prevailed by the time of independence and an additional profits tax (Garnaut and Clunies Ross 1975) was brought into use, exemplifying Wilson's third policy axis: the use of tax mechanisms to capture economic rents for public purposes. This was when the state's powers to control income from mining were strongest. In the last year of full operations, the government's share amounted to 66 per cent of revenues from the Bougainville mine, after operating costs, while CRA's shareholders received 34 per cent (Bougainville Copper Limited 1988: 3–4).

As in New Caledonia, it was assumed that people in Bougainville would benefit from business and employment; compensation payments to traditional landowners were small. Local grievances escalated during the 1980s (Filer and Le Meur 2017: 18–9) until landowners demanded PGK10 billion (approximately USD11.6 billion) in compensation and 50 per cent of future profits to be shared by local stakeholders (AGA 1989). As is well known, rebel landowners fought a decade-long civil war against the PNG Defence Force (Filer 1990; May and Spriggs 1990; Braithwaite at al. 2010). A referendum on independence from PNG was held in 2019 at which over 98 per cent of the electorate voted 'yes'.

Not surprisingly, later mining agreements made greater provisions for benefits to be made to local communities (Filer 1997). Royalty had been set at 1.25 per cent of production at Bougainville and Ok Tedi, of which the host province and landowners took 95 per cent and 5 per cent shares respectively. In 1989, landowners at Porgera negotiated a balance of 77 per cent (province) and 23 per cent (landowners and local institutions). In 1995, the Lihir landowners negotiated a balance of 50 per cent (province), 50 per cent (landowners and local institutions) and an increase in the rate of royalty to 2 per cent of production (Bainton 2010: 168). In turn, this caused the Porgera landowners to renegotiate their agreement, and from mid-1995, they too shared royalty at the rate of 2 per cent and the breakdown became 50 per cent (province) and 50 per cent (landowners and local institutions) (Banks 2001: 44).

On the side of compensation and business development grants, where in the 1970s landowners reckoned in the tens of thousands of PNG Kina, in the 1980s they began reckoning in hundreds of thousands of Kina (e.g. Banks 2001), in the 1990s in millions of Kina (e.g. Bainton 2010: Table 5-1; Golub 2014), in the 2000s in tens of millions of Kina (e.g. Johnson 2012) and in the 2010s in hundreds of millions of Kina (Nicholas 2011, 2019).

All this points to the state exhibiting a weaker and weaker capacity to control resource revenues: (a) its local presence, manifesting in its bargaining power with citizens, has faded and (b) its ability to collect mining and petroleum taxes from corporations has faltered. Tax receipts averaged AUD650 million/year 2010–12, but only AUD85 million/year 2014–17 (Figure 10.2).

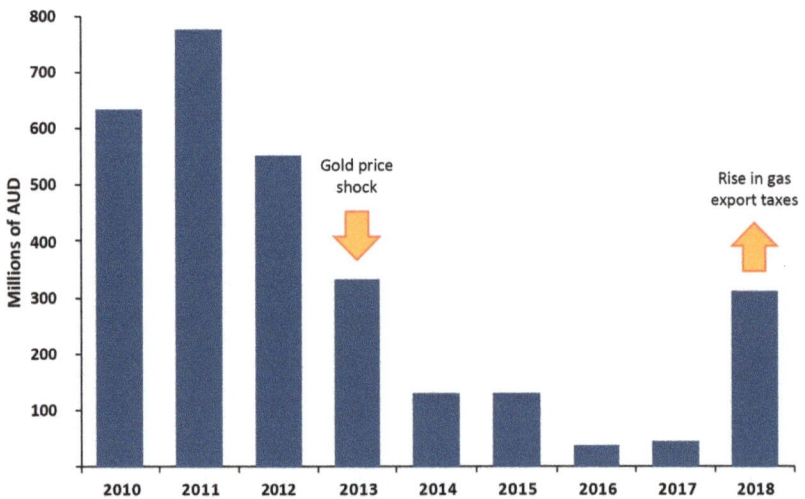

Figure 10.2 Mining and petroleum taxes collected in PNG during the 2010s

Source: PNG Department of Treasury data 2010–18

A brief gold price shock in mid-2013 was a local trigger for gold miners to restructure their tax affairs, but this was merely an acceleration of worldwide trend for multinationals to more aggressively minimise tax. It should be noted that Australia (below) has its own troubles with resource rents, but as a developing country PNG has fewer means of redress and PNG's long-awaited sovereign wealth fund, intended to capture LNG revenues and stabilise the economy (Osborne 2014) and supposed to start accumulating from 2014, has still received no deposits (Anon. 2020a). As Banks and Namorong (2018) note, something has gone significantly wrong with PNG's investment settings.

Australia

Extractives-friendly politicians in Australia constantly boast about the number of jobs created in the key producing states, with a multiplier effect generating benefit in the wider economy. However, the tax take from resources is acknowledged as inefficient.

The 2010 'Henry Tax Review' (AFG 2010) recommended that the government should replace the existing regime of state-based mining and petroleum royalties with a resource rent tax. As is well known, the attempt by the then Labor Government to introduce a Resource Super Profits Tax (RSPT)—matching the existing petroleum resource rent tax (PRRT)—was met with strident industry protests: the announcement was 'shocking' (Rio Tinto), 'highly regrettable' (Xstrata) and 'a surprise attack on us' (Fortescue Metals). The Minerals Council of Australia (MCA)—the peak lobby group representing the mining industry—ran an advertising blitz over the next two months, a key factor in the Labor Party replacing its leader, Kevin Rudd, in mid-2010 (Davis 2011). His successor, Julia Gillard, introduced a weaker Minerals Resource Rent Tax (MRRT) which was initially forecast to raise AUD3.7 billion a year, but net revenue was only AUD88 million in its two years of operation, 2012–14. Reasons for its failure have been split among weak commodity prices, 'industry capture' of the process of designing the legislation—BHP, Rio Tinto and Xstrata had a hand in this—and insufficient technical expertise in the Treasury (Valle de Souza et al. 2016). The situation is no better in the energy sector, where it could be decades before the PRRT brings significant returns from new gas projects coming online in the north of Australia (AFG 2016: 4).

The MCA says its members 'dig deep' to pay royalties and tax, contributing AUD39.2 billion to public finances in 2018–19 (DAE 2020; Zakharia 2020). But data from the Australian Taxation Office (ATO) shows that many of the largest extractive companies, with collective revenues of AUD54.5 billion, paid no tax at all in 2017–18, including Chevron, ExxonMobil, Woodside, Shell, Santos, Peabody Coal, Yancoal and BHP.[9] In recent years, the ATO has taken to the courts to recover unpaid tax. In 2018, Chevron settled a dispute for AUD866 million concerning loans between subsidiaries in different countries. BHP, which reduced tax by using the 'Singapore Sling', a method of transfer pricing via Singapore, settled claims for AUD529 million in 2018 (Chenoweth 2020). Rio Tinto, which does pay tax, faces action over AUD500 million in the transfer pricing of aluminium and iron ore (TP News 2020).

New Caledonia

The preferred method of obtaining returns from the mining industry in New Caledonia is through dividends arising from provincial equity-holding in mining companies, rather than from production levies or royalties, and by adding as much economic value as possible by maximising onshore processing. In the current absence of a royalty regime, benefits to local communities arise largely as government spending on infrastructure and services, and through opportunities for employment and subcontracting, especially when agreements were signed. However, it seems inevitable that some form of production levy or royalty[10] will be introduced eventually; both a recent agreement at Kouaoua (Mainguet 2020) and the proposals currently before the Congress to permit mineral exports are believed to contain provisions for a levy as a form of compensation (Lalande 2020).

Discussion

Wilson's three policy axes provide an informative way to see the presence or absence of the state. It is suggested that two particular phenomena are shared between the three country contexts explored here.

---

9   See: data.gov.au/data/dataset/corporate-transparency
10   The French term *redevance* is frequently heard in demands by local communities. While *redevance* is the standard translation of 'royalty', the French term has additional meanings that royalty does not in English.

First, each version of the state has problems collecting resource rental incomes. PNG appears addicted to attracting new projects to repair the budget problems caused by the old ones, only to again fall into the trap of offering overly generous tax holidays and other incentives. Australia has a huge resource industry, but it is not doing a lot better at getting good returns on exporting the national wealth. New Caledonia is at the mercy of the fluctuating price of a single commodity and, as a high-cost producer, struggles to match its political commitments to onshore processing with the reality of lower-cost competition in China, South Korea and elsewhere. Sometimes, different dimensions of the same state present a very different face to different parts of the same issue. For example, political leaders can see themselves as *enablers* of the extractive industries, while tax collectors like Australia's ATO are locked in combat *against* large corporations— an observation which points to the multivalence of the state.

Second, if the domain of the feasible lies in the hands of transnational corporations, putatively regulated by each state in its own way, it is increasingly the case that their activities are moderated by local actors with their own points of view and which none of the states (or corporations) has found a way to consult with in an optimal way. We see numerous disputes with landowners in PNG; blockades and vandalism in New Caledonia, where the label '*conflit des coutumiers*' (approximately: 'conflict with clans in the local communes') conceals much complexity; and polarising political stand-offs in Australia forming part of the so-called 'Culture Wars' (Wallace 2019) between conservative and progressive political actors.

Issues of this type have been widely discussed elsewhere in terms of 'glocalisation'—the simultaneous movement of influence away from states, 'upward' to globalised international capital and 'downward' to local citizens' groups (Swyngedouw 2004; Le Meur, Ballard, Banks and Sourisseau 2013). While the processes of glocalisation are evident in all three places, it is only in PNG and Australia that they are accompanied by a series of 'cut-offs' of the democratic will of voters. The two countries have electoral systems where citizens elect representatives to parliaments, but where the range of viewpoints canvassed in elections is greatly simplified by the process of forming governments. Only the viewpoints of the winning parties prevail; those of the opposition parties, and the electors who voted for them, fail a first cut-off and are discarded. The process of forming a cabinet of ministers from the members of the winning parties further forms a second cut-off. Only ministers and their assistants, a small handful of winning party representatives in a parliament, translate the policies of their parties into the actions of government. In this way, governments can adopt courses of action that citizens did not broadly vote for and that ultimately distance them from their electorate and election promises—though governments have long sought to persuade citizens after the fact that particular policies are what they wanted all along. The absence of politically alternative voices in the executive can mean that decision makers are susceptible to industry capture; if so, 'successful' legislation may be that which is smiled on by interests outside mainstream politics, not by a consensus of the parliament.

Conclusions: What a Difference the State Makes

In the introduction to this chapter we noted that in PNG and Australia the state is largely seen as 'the enemy' in the eyes of customary owners seeking some good to come out of permitting mining or oil and gas extraction from their land. This, in turn, informed peoples' experiences and perception of state absence and/or presence and created space for experience of the 'absent presence' phenomena, where the presence of the state is experienced through its perceived absence. We asked: is there an alternative where the state is *not* the enemy? Is there a different kind of a state? And what impact would such difference have on the role and experience of the state in resource contexts?

As we have demonstrated above, New Caledonia's 'collegial' government is of a different form where electors vote once and their representatives are filtered by party list into the Congress, and then 11 are chosen to form the territorial government, which will always be from a mix of parties. While parties may differ in their support for broad platforms hammered out over years of consultation, such as the Framework for the Development of Mineral Resources discussed earlier, when voters back them their positions are carried all the way through to the formation of the territorial government so that alternative voices remain intact among the executive. This makes New Caledonia a different kind of state when compared to PNG and Australia.

There are also similarities between PNG, Australia and New Caledonia, which emerge from the comparative material presented above, and our findings demonstrate the value of the comparative approach adopted here. Most notably, our analysis shows that crisis plays a central role in shaping—or rather, periodically resetting—public and industry policies in all three places. Even when all the tools to control and exercise powers are present in principle, crisis appears to act as a policy brake, forcing perspectives to remain local and focused on particular projects, political movements and communities. What sets New Caledonia—and notably its North Province—apart is that, when things are democratically in step, the policy brake reaches directly from communities into the heart of government. When things are out of step, of course, conflicts jolt those in power and bring them to the table. In PNG and Australia, the first scenario is rare. In many cases, the only recourse that communities have is to go on treating the state as *the enemy*, experienced through its irrelevant presence or malevolent absence, or a phenomenon described in this volume as 'absent presence'—which is especially evident when the state fails to meet expectations and fulfil its role in resource contexts, when it is present for some but absent for others, and where its presence (and absence) fluctuate over time.

In their work on the resource curse theory, Emma Gilberthorpe and Dinah Rajak (2017) discuss the roles of marginalised or subaltern actors on the one hand, and the elite agencies of actors with power on the other. Their aim is to shift the focus away from revenue management and to make the social and political relations of extraction 'front and centre'. We can agree with this and suggest that it is not the problem of resource rents in the three jurisdictions that exposes them to a resource curse pathology, but the degree to which decisions made at the top fall in line with the interests of citizen stakeholders. In this respect, New Caledonia holds an advance over Australia and PNG in that, as we said above, the state cannot absent itself.

This may be a counter-intuitive proposition to make for a jurisdiction so divided over the politics of territorial sovereignty, and on other matters, and we are certainly not waving away the deep fissure that has existed for so long between loyalists and separatists. What we observe is that exactly the same balance of these political forces as exists in society as a whole is mirrored in the Congress, in the government of New Caledonia, and indeed in the review body, the *Comité des signataires*. Citizens may not obtain satisfaction that the positions they debate among themselves can be resolved easily, but at least they can be sure they are *the same ones* that are fought out in the Congress, and in the executive government above that. At the same time, as we show here, interpreted in the context of the historical as well as contemporary relationship between New Caledonia and the French government, and with consideration of Kanak struggles for independence, the questions of absence and presence of the state, or indeed its 'absent presence', gains significant traction. So while the distinctive form of the state in New Caledonia means that, in terms of political representation and forms of governance *the state cannot absent itself*, exploring the role of the state in resource contexts in terms of an interplay between its presence and absence, and how they demonstrate at different levels of discourse and experience, offers new insights and reveals the dynamic and uncertain nature of the state in resource contexts—in PNG and Australia, as well as New Caledonia.

References

AFG (Australian Federal Government), 2010. *Australia's Future Tax System: Overview*. Canberra: Department of Treasury.

———, 2016. 'Petroleum Resource Rent Tax Review: Final Report.' Canberra: Department of Treasury.

———, 2019. *Australia's Critical Minerals Strategy*. Canberra: Productivity Commission.

———, 2020. *Trade and Assistance Review 2018–19*. Canberra: Department of Industry, Innovation and Science.

AGA (Applied Geology Associates), 1989. 'Environmental, Socio-economic and Public Health Review of Bougainville Copper Mine, Panguna.' Unpublished report to PNG Department of Minerals and Energy.

Altman, J., 2009. 'Indigenous Communities, Miners and the State in Australia.' In J. Altman and D. Martin (eds), *Power, Culture, Economy: Indigenous Australians and Mining*. Canberra: ANU E Press (Centre for Aboriginal Economic Policy Research, Research Monograph 30). doi.org/10.22459/CAEPR30.08.2009

Anon., 2020a. 'Dreams of a Better Future.' *Post-Courier*, 25 May.

————, 2020b. 'Export de minerai: la province Nord opposée à la dilapidation du patrimoine calédonien' [Mineral Exports: The North Provinces Opposes the Squandering of Territorial Resources]. *Les Nouvelles Calédoniennes*, 9 July.

Bainton, N.A., 2010. *The Lihir Destiny: Cultural Responses to Mining in Melanesia*. Canberra: ANU E Press (Asia-Pacific Environment Monograph 5). doi.org/10.22459/ld.10.2010

Bainton, N.A., J.R. Owen, S. Kenema and J. Burton, 2020. 'Land, Labour and Capital: Small and Large-Scale Miners in Papua New Guinea.' *Resources Policy* 68: 101805. doi.org/10.1016/j.resourpol.2020.101805

Bambridge, S., 1979. *Australian Minerals and Energy Policy*. Canberra: Australian National University Press.

Banks, G.A., 2001. 'Papua New Guinea Baseline Study.' London: International Institute for Environment and Development, Mining Minerals and Sustainable Development Project (Working Paper 180).

Banks, G.A. and C. Ballard (eds), 1997. *The Ok Tedi Settlement: Issues, Outcomes and Implications*. Canberra: The Australian National University, National Centre for Development Studies (Pacific Policy Paper 27).

Banks, G.A. and M. Namorong, 2018. 'Papua New Guinea's Disappearing Resource Revenues.' DevPolicy blog, 15 April. Viewed 22 September 2020 at: www.devpolicy.org/papua-new-guineas-disappearing-resource-revenues-20180815/

Bebbington, A., 2014. 'Governing Natural Resources for Inclusive Development.' In S. Hickey, K. Sen and B. Bukenya (eds), *The Politics of Inclusive Development: Interrogating the Evidence*. Oxford: Oxford University Press. doi.org/10.1093/acprof:oso/9780198722564.003.0004

Bencivengo, Y., 2014. *Nickel : La naissance de l'industrie calédonienne* [*Nickel: The Birth of the Industry in New Caledonia*]. Tours: Presses Universitaires François Rabelais. doi.org/10.4000/books.pufr.12654

Black, P., 2014. 'Green Gold': The Contribution of New Caledonia's Nickel Industry to the 'Age of Steel' 1870–1920. Canberra: The Australian National University (PhD thesis).

Bouard S., C. Levacher, Y. Bencivengo, L. Decottigny, C. Demmer, P.Y. Le Meur, S. Blaise, J Burton, F. Enjuanes and S. Grochain, 2019. 'PME Minières en Nouvelle-Calédonie: Petites et Moyennes Entreprises Minières en Nouvelle-Calédonie' [Mining SMEs in New Caledonia: Small and Medium Mining Enterprises in New Caledonia]. Nouméa: Institut de Recherche pour le Développement.

Bouard, S., J.-M. Sourisseau, V. Geronimi, S. Blaise and L. Ro'i (eds), 2016. *La Nouvelle-Calédonie face à son destin. Quel bilan à la veille de la consultation sur la pleine souveraineté?* [*New Caledonia Facing its Destiny: What is the State of Things on the Eve of the Referendum on Independence?*] Paris: Éditions Karthala.

Bougainville Copper Limited, 1988. *Annual Report 1988*. Melbourne: Bougainville Copper Limited.

Braithwaite, J., H. Charlesworth, P. Reddy and L. Dunn, 2010. *Reconciliation and Architectures of Commitment: Sequencing Peace in Bougainville*. Canberra: ANU E Press. doi.org/10.22459/RAC.09.2010

Brou, B., 1982. *30 Ans d'Histoire de la Nouvelle-Calédonie: 1945/1977* [*30 Years in the History of New Caledonia: 1945–77*]. Nouméa: Société d'Études Historiques de la Nouvelle-Calédonie.

Burton, J., 1997. 'Terra Nugax and the Discovery Paradigm: How Ok Tedi was Shaped by the Way it was Found and how the Rise of Political Process in the North Fly took the Company by Surprise.' In G.A. Banks and C. Ballard (eds), *The Ok Tedi Settlement: Issues, Outcomes and Implications*. Canberra: The Australian National University, National Centre for Development Studies (Pacific Policy Paper 27).

Burton, J. and Y. Haihuie, 2017. 'Corruption Risks in Mining Awards: Papua New Guinea Country Report.' Port Moresby: Transparency International Papua New Guinea.

Burton, J. and J. Onguglo, 2017. 'Disconnected Development Worlds: Responsibility towards Local Communities in Papua New Guinea.' In C. Filer and P.-Y. Le Meur (eds), *Large-Scale Mines and Local-Level Politics: Between New Caledonia and Papua New Guinea*. Canberra: ANU Press (Asia-Pacific Environment Monograph 12). doi.org/10.22459/LMLP.10.2017.09

Chenoweth, N., 2020. 'And the Winners from BHP's Singapore Sling are ...' *Australian Financial Review*, 3 June.

Childs, J., 2016. 'Geography and Resource Nationalism: A Critical Review and Reframing.' *Extractive Industries and Society* 3: 539–546. doi.org/10.1016/j.exis.2016.02.006

CSIRO (Commonwealth Scientific and Industrial Research Organisation), 1996. 'Review of Riverine Impacts.' Unpublished report to Porgera Joint Venture.

DAE (Deloitte Access Economics), 2020. 'Estimates of Royalties and Company Tax Accrued in 2018–19.' Canberra: DAE for Minerals Council of Australia.

Davis, M., 1993. 'How PNG pulled off the Porgera Coup.' *Australian Financial Review*, 23 April.

———, 2011. 'A Snip at $22m to Get Rid of PM.' *Sydney Morning Herald*, 2 February.

Demmer, C., 2007. 'Autochtonie, nickel et environnement. Une nouvelle stratégie kanake' [Independence, Nickel and the Environment: A New Strategy for the Kanaks]. *Vacarme* 39: 43–48. doi.org/10.3917/vaca.039.0043

———, 2017. 'L'export du nickel au cœur du débat politique néo-calédonien' [The Export of Nickel at the Heart of Political Debate in New Caledonia]. *Mouvements* 91: 130–140. doi.org/10.3917/mouv.091.0130

Demmer, C., P.-Y. Le Meur, J.-M. Sourisseau, 2018. 'Quelle stratégie nickel après le référendum d'indépendance?' [What Strategy for Nickel After the Referendum on Independence?]. *Le Monde*, 30 October.

Filer, C., 1990. 'The Bougainville Rebellion, the Mining Industry and the Process of Social Disintegration in Papua New Guinea.' In R.J. May and M. Spriggs (eds), *The Bougainville Crisis*. Bathurst (NSW): Crawford House Press.

———, 1997. 'The Melanesian Way of Menacing the Mining Industry.' In B. Burt and C. Clerk (eds), *Environment and Development in the Pacific Islands*. Canberra: The Australian National University, National Centre for Development Studies (Pacific Policy Paper 25).

———, 2008. 'Development Forum in Papua New Guinea: Upsides and Downsides.' *Journal of Energy & Natural Resources Law* 26: 120–149. doi.org/10.1080/02646811.2008.11435180

Filer, C. and B. Imbun, 2009. 'A Short History of Mineral Development Policies in Papua New Guinea, 1972–2002.' In R.J. May (ed.), *Policy Making and Implementation: Studies from Papua New Guinea*. Canberra: ANU E Press. doi.org/10.22459/PMI.09.2009.06

Filer, C. and P.-Y. Le Meur. 2017. 'Large-Scale Mines and Local-Level Politics.' In C. Filer and P.Y. Le Meur (eds), *Large-Scale Mines and Local-Level Politics: Between New Caledonia and Papua New Guinea*. Canberra: ANU Press (Asia-Pacific Environment Monograph 12). doi.org/10.22459/LMLP.10.2017.01

Garnaut, R. and A. Clunies Ross, 1975. 'Uncertainty, Risk Aversion and the Taxing of Natural Resource Projects.' *Economic Journal* 85: 272–287. doi.org/10.2307/2230992

Gilberthorpe, E. and D. Rajak, 2017. 'The Anthropology of Extraction: Critical Perspectives on the Resource Curse.' *Journal of Development Studies* 53: 186–204. doi.org/10.1080/00220388.2016.1160064

Golub, A., 2014. *Leviathans at the Gold Mine: Creating Indigenous and Corporate Actors in Papua New Guinea.* Durham (NC): Duke University Press.

GPNG (Government of Papua New Guinea), 2009. *Papua New Guinea Vision 2050.* Port Moresby: National Strategic Plan Taskforce.

———, 2015. *National Strategy for Responsible Sustainable Development.* Port Moresby: Department of National Planning and Monitoring.

———, 2018. *An Investigation into the Alleged Improper Borrowing of AU$1.239 Billion Loan from the Union Bank of Switzerland.* Port Moresby: Ombudsman Commission.

Grigg, A., J. Shapiro and L. Murray, 2019. 'Swiss Regulator Probes UBS Australia's $1.2b PNG Loan.' *Australian Financial Review*, 15 March.

HCRNC (Haut-Commissariat de la République en Nouvelle-Calédonie), 2019. 'Les résultats du référendum 2018: Résultats définitifs' [Outcome of the 2018 Referendum: Final Figures]. Nouméa: HCRNC.

———, 2020. 'Référendum du 4 octobre 2020—Résultats définitifs' [Referendum of 4 October 2020: Final Figures]. Nouméa: HCRNC.

Horowitz, L.S., 2009. 'Environmental Violence and Crises of Legitimacy in New Caledonia.' *Political Geography* 28: 248–258. doi.org/10.1016/j.polgeo.2009.07.001

———, 2015. 'Culturally Articulated Neoliberalisation: Corporate Social Responsibility and the Capture of Indigenous Legitimacy in New Caledonia.' *Transactions of the Institute of British Geographers* 40: 88–101.

———, 2016. 'Rhizomic Resistance Meets Arborescence Assemblage: UNESCO World Heritage and the Disempowerment of Indigenous Activism in New Caledonia.' *Annals of the American Association of Geographers* 106: 167–185. doi.org/10.1080/00045608.2015.1090270

Jackson, R., 1982. *Ok Tedi: The Pot of Gold.* Port Moresby: University of Papua New Guinea.

Jackson, R. and G.A. Banks, 2002. *In Search of the Serpent's Skin: The Story of the Porgera Gold Project.* Port Moresby: Placer Niugini.

Johnson, P., 2012. 'Lode Shedding: A Case Study of the Economic Benefits to the Landowners, the Provincial Government, and the State from the Porgera Gold Mine: Background and Financial Flows from the Mine.' Port Moresby: National Research Institute (Discussion Paper 124).

Kemp, D. and J.R. Owen, 2018. 'The Industrial Ethic, Corporate Refusal and the Demise of the Social Function in Mining.' *Sustainable Development* 26: 491–500. doi.org/10.1002/sd.1894

Ketan, J. and M. Geita, 2011. *An Assessment of the Combined PNG Government-Morobe Mining Joint Venture (Hidden Valley) Environmental Audit Consultations and Awareness Program in the Watut and Markham valleys, Morobe Province, Papua New Guinea.* Port Moresby: Tanorama Ltd.

Kirsch, S., 2001. 'Lost Worlds, Environmental Disaster, "Culture Loss," and the Law.' *Current Anthropology* 42: 167–198. doi.org/.10.1086/320006

———, 2007. 'Indigenous Movements and the Risks of Counterglobalization: Tracking the Campaign Against Papua New Guinea's Ok Tedi Mine.' *American Ethnologist* 34: 303–321. doi.org/10.1525/ae.2007.34.2.303

Lalande, C., 2020. 'Vie des usines et après-nickel: des avancées majeures' [Outlook for the Refineries and the Post-Nickel Economy: Significant Developments]. *Les Nouvelles Calédoniennes*, 5 August.

Le Meur, P.-Y., C. Ballard, G. Banks and J.-M. Sourisseau, 2013. 'Two Islands, Four States: Comparing Resource Governance Regimes in the Southwest Pacific.' Paper presented at the 2nd International Conference on Social Responsibility in Mining, Santiago, 5–8 November.

Le Meur, P.-Y., L.S. Horowitz and T. Mennesson, 2013. '"Horizontal" and "Vertical" Diffusion: The Cumulative Influence of Impact and Benefit Agreements (IBAs) on Mining Policy-Production in New Caledonia.' *Resources Policy* 38: 648–656. doi.org/10.1016/j.resourpol.2013.02.004

Le Meur, P.-Y. and T. Mennesson, 2011. 'Le cadre politico-juridique minier en Nouvelle-Calédonie. Mise en perspective historique' [The Political and Legal Framework for Mining in New Caledonia]. Nouméa: Centre National de Recherche Technologique, Nickel et Son Environnement (Document de Travail 3).

Levacher, C., 2016. De la terre à la mine? Les chemins de l'autochtonie en Nouvelle-Calédonie [From the Earth to the Mine? Indigenous Pathways in New Caledonia]. Paris: École des Hautes Études en Sciences Sociales (PhD thesis).

————, 2017. 'Contesting the Goro Nickel Mining Project, New Caledonia: Indigenous Rights, Sustainable Development and the Land Issue.' In C. Filer and P.-Y. Le Meur (eds), *Large-Scale Mines and Local-Level Politics: Between New Caledonia and Papua New Guinea*. Canberra: ANU Press (Asia-Pacific Environment Monograph 12). doi.org/10.22459/lmlp.10.2017.06

Levacher, C. and P.-Y. Le Meur, 2019. 'The Compensation Arenas in South New Caledonia: Minescape Management, Governmentality and Politics.' Paper presented at a workshop on 'The Micropolitics of Mining Capitalism', University of Liège, 11–13 September.

Lund, C., 2006. 'Twilight Institutions: Public Authority and Local Politics in Africa.' *Development and Change* 37: 685–705. doi.org/10.1111/j.1467-7660.2006.00497.x

Mainguet, Y., 2019. 'Vale abaisse sérieusement la cote de l'usine du Sud' [Vale Writes Down the Southern Plant]. *Les Nouvelles Calédoniennes*, 29 November.

————, 2020. 'Kouaoua: que contient le protocole signé?' [Kouaoua: What does the Agreement Say?] *Les Nouvelles Calédoniennes*, 6 July.

May, R.J. and M. Spriggs (eds), 1990. *The Bougainville Crisis*. Bathurst (NSW): Crawford House Press.

Merlin, J., 2014. 'L'émergence d'une compétence environnementale autochtone?' [The Emergence of an Indigenous Capacity for Environmental Monitoring]. *Terrains and travaux* 24: 85–102. doi.org/10.3917/tt.024.0085

Nalu, M., 2013. 'Govt to Debate Mine Takeover.' *The National*, 18 September.

Newbury, C., 1955. 'La Société "Le Nickel", de sa fondation à la fin de la deuxième guerre mondiale, 1880–1945.' ['Société Le Nickel': From its Foundation to the end of the Second World War]. *Journal de la Société des Océanistes* 11: 97–123. doi.org/10.3406/jso.1955.1847

Nicholas, I., 2011. 'Gas Owners Fight over K115mil.' *The National*, 23 February.

————, 2019. 'Landowners in Line to Get K300m Royalty.' *Post-Courier*, 2 February.

OCAA (Oxfam Community Aid Abroad), 2004. 'Mining Ombudsman Case Report: Tolukuma Gold Mine.' Melbourne: OCAA.

Osborne, D., 2014. 'What has happened to Papua New Guinea's Sovereign Wealth Fund?' DevPolicy blog, 28 October. Viewed 22 September 2020 at: devpolicy.org/what-has-happened-to-papua-new-guineas-sovereign-wealth-fund-20141028/

Owen, W., 2020. 'Australia Officially the World's Largest Exporter of LNG.' LNG Industry blog, 6 January. Viewed 25 September 2020 at: www.lngindustry. com/liquid-natural-gas/06012020/australia-officially-the-worlds-largest-exporter-of-lng/

Pitoiset, A. and C. Wéry, 2008. *Mystère Dang* [*The Mystery of André Dang*]. Nouméa: Le Rayon Vert.

PNGEITI (PNG Extractive Industries Transparency Initiative), 2019. 'Papua New Guinea 2018 EITI Report.' Port Moresby: PNGEITI Secretariat.

Rothwell, D., 2018. 'Australia and Timor Leste Settle Maritime Boundary after 45 Years of Bickering.' *The Conversation*, 7 March.

Swyngedouw, E., 2004. 'Globalisation or "Glocalisation"? Networks, Territories and Rescaling. *Cambridge Review of International Affairs* 17: 25–48. doi.org/ 10.1080/0955757042000203632

Taylor, A., 2020. 'Australia to Boost Fuel Security and Establish National Oil Reserve.' Media Release from the Minister for Energy and Emissions Reduction, April 22.

Taylor, J. and B. Scambary, 2005. *Indigenous People and the Pilbara Mining Boom: A Baseline for Regional Participation.* Canberra: ANU E Press (Centre for Aboriginal Economic Policy Research Monograph 25). doi.org/10.22459/ IPPMB.01.2006

TP News, 2020. 'Mining Group Rio Tinto in new $86 Million Dispute with ATO Over Pricing of Aluminium.' TPcases blog, 9 April. Viewed 25 September 2020 at: tpcases.com/mining-group-rio-tinto-in-new-86-million-dispute-with-ato-over-pricing-of-aluminium/

UNDP (United Nations Development Programme), 2014. *From Wealth to Wellbeing: Translating Resource Revenue into Sustainable Human Development.* Port Moresby: UNDP and PNG Department of National Planning and Monitoring (National Human Development Report).

Valle de Souza, S., B.E. Dollery and M.A. Kortt, 2016. 'A Critical Evaluation of Australian Mineral Resources Rent Tax.' *International Journal of Public Administration* 40: 472–480. doi.org/10.1080/01900692.2015.1136944

Wallace, C., 2019. 'After Years of Vicious Culture Wars, Hope May Yet Triumph over Hate in Australian Politics.' *The Conversation*, 8 March. Viewed 29 September 2020 at: theconversation.com/after-years-of-vicious-culture-wars-hope-may-yet-triumph-over-hate-in-australian-politics-110887

Wilson, J.D., 2015. 'Understanding Resource Nationalism: Economic Dynamics and Political Institutions.' *Contemporary Politics* 21: 399–416. doi.org/10.1080/13569775.2015.1013293

Yala, C., O. Sanida and A. Make, 2014. 'The Oil Search Loan: Implications for PNG.' DevPolicy blog, 21 Mach. Viewed 22 September 2020 at: devpolicy.org/the-oil-search-loan-implications-for-png-20140321-2/

Zakharia, N., 2020. 'Minerals Industry Digs Deep for Tax Payments.' Australian Mining blog, 11 March. Viewed 25 September 2020 at: www.australianmining.com.au/oil-gas/news-oil-gas/minerals-industry-digs-deep-for-tax-payments/

Afterword: States of Uncertainty

Nicholas Bainton, John R. Owen
and Emilia E. Skrzypek

Estragon: Godot?
Pozzo: You took me for Godot.
Estragon: Oh no, sir, not for an instant, sir.
Pozzo: Who is he?
Vladimir: Oh, he's a ... he's a kind of acquaintance.
Estragon: Nothing of the kind, we hardly know him.
Vladimir: True ... we don't know him very well ... but all
the same ...

Waiting for Godot (Samuel Beckett 1956)

Introduction

This volume opened with a set of questions about the relationship between absence and presence and what this might tell us about the nature of contemporary states from the perspective of resource arenas: how the state may be experienced as more or less present for different actors in different times and circumstances, and how its presence can be experienced through its absence—an absent presence. As the dialogue between Beckett's troubled characters attests, persons, things and processes that exhibit an absent presence are often experienced as ambiguous and indeterminate phenomena. Standing on the broken ground of resource extraction settings, the state is sometimes like a chimera: its appearance and intentions are misleading, and for some actors, it is unknowable and incomprehensible. It may be easily mistaken for someone or something else, like a mining company, for example. As a partial and incomplete project, the state is experienced in

Papua New Guinea (PNG) and Australia as both a form of uncertainty and a progenitor of uncertainty. If this condition of uncertainty is slightly tempered in places like New Caledonia where the state assumes a somewhat different form, as Burton and Levacher claim in their final comparative chapter for this volume, this reinforces the point that the absence or the presence of the state is never absolute.

Two significant events have recently occurred in the period since we compiled the bulk of this volume. These events directly challenge our thinking on absence and presence and happen to have occurred in Australia and PNG: the destruction of the Juukan Gorge rock shelters in Western Australia by Rio Tinto, and the announcement by the PNG prime minister James Marape that his government would not renew Barrick Nuigini Limited's mining lease at the Porgera Gold Mine. In closing out this volume, these events give pause to reflect further on the meaning and the effects of the absence and the presence of the state. These events evidence the uncertainties that constitute the modern state and serve as a kind of postscript to the cases we have considered in this volume— but they do so in unexpected ways and open up new perspectives on the presence and absence of the state, and the relationship between the 'state-idea' and the 'state-system' (Abrams 1988). We briefly describe these events here, necessarily glossing their immense historical complexity, followed by some concluding comments. Each of these events will likely reshape future resource relations and encounters in these nations, and beyond. As the ramifications of these actions reverberate past the publication of this volume, it will be important to keep in mind ideas about absence, presence and absent presence as we try make sense of ongoing state and corporate effects at resource extraction projects.

The Destruction of the Juukan Gorge Rock Shelters

In the lead up to the July 2020 NAIDOC[1] week, which celebrates the history, culture and achievements of Aboriginal and Torres Strait Islander peoples in Australia, the Puutu Kunti Kurrama People and

---

1   NAIDOC originally stood for 'National Aborigines and Islanders Day Observance Committee'. This committee was once responsible for organising national activities during NAIDOC week. The acronym has since become the name of the week itself. See: www.naidoc.org.au/

Pinikura People (PKKP) requested permission from Rio Tinto to visit their ancestral lands encompassed by the lease for the Brockman 4 iron ore mine in Western Australia's Pilbara region. The mine, one of 16 in the region owned by Rio Tinto, was opened in 2010 at an initial cost of more than AUD1.5 billion, with an estimated mine life of 20 years, constituting a major boost to the already mining-dependent state economy. The PKKP wanted to visit the rock shelters at Juukan Gorge, which had been confirmed as a site of Aboriginal occupation dating back some 46,000 years before the present, and was now threatened by expanding mining operations.

Rio Tinto had commenced negotiations with the PKKP over access to their lands in 2003, and in 2011 this relationship was formalised through an Indigenous Land Use Agreement, or what Rio Tinto prefer to call a 'Participation Agreement'.[2] The Juukan Gorge area lay within the proposed mine expansion footprint, and as such, it has been the subject of considerable archaeological and ethnographic investigation to assess its heritage value. In 2013, Rio Tinto obtained state permission to destroy the rock shelters—known as Juukan 1 and Juukan 2—for mining purposes. Under the terms of Western Australia's *Aboriginal Heritage Act 1972*, ministerial consent was granted for the destruction of the Juukan Gorge rock shelters. This authorisation occurred under Section 18 of the Act, commonly referred to as a 'Section 18 approval' or 's18' for short. This set in train a major program of salvage archaeology, sponsored by Rio Tinto, which confirmed that the site was of the highest archaeological importance in Australia, and of global significance. Archaeological surveys conducted in 2014 provided evidence of human life during the Pleistocene and continued human presence in that place going back tens of thousands of years.

Undeterred by these findings, and satisfied that the site was 'fully salvaged' and that they were legally compliant, Rio Tinto continued with a mine plan that progressively encroached upon the site. By 2020 the company had been blasting for two years within the vicinity of the rock shelters, and in May 2020 a sequence of explosives was loaded into 382 blast holes to access an estimated AUD135 million worth of ore located around Juukan 1 and Juukan 2. When the PKKP approached Rio Tinto for access to their country to 'celebrate Juukan' as part of NAIDOC week, they were

---

2   In 2003, Rio Tinto was present in the negotiations through its wholly owned subsidiary, Hamersley Iron Pty Ltd.

informed of the impending blast and reminded via email correspondence that 'the sites are within the current mine pit design, and RT was granted s18 approval for that activity in 2013' (Rio Tinto 2020: 32). Last-minute appeals from the PKKP and their representatives were not enough to stop the destruction of the rock shelters and on 24 May Rio Tinto proceeded with the blast on the basis that it was no longer feasible to remove the explosives.

The timing of this event has been critical to the broader response. In addition to the upcoming NAIDOC week, Australia, like much of the global North, was gripped by the rapid escalation of the Black Lives Matter movement. Local, national and international condemnation ensued. In June 2020 the Australian Senate referred the matter to the Joint Standing Committee on Northern Australia, and a parliamentary inquiry was launched.[3] By September, outraged investors had forced the board of Rio Tinto to sack its chief executive along with two senior executives partially responsible for the destruction of the rock shelters (Hopkins and Kemp 2020). While many observers have applauded this move, Marcia Langton (2020) was quick to remind the public that the decision evidenced the power of shareholder interests, not Aboriginal interests.

In a statement submitted to the inquiry, Rio Tinto expressed their belief that, under the terms of the 2011 Participation Agreement, it had secured 'Free Prior and Informed Consent' to conduct mining operations on PKKP land including the destruction of the rock shelters. Reflecting on the archaeological evidence amassed over the past decade, Rio Tinto described these reports as 'missed opportunities' to re-evaluate the mine plan (2020: 3). In a submission to the same inquiry, the Government of Western Australia stated that the destruction of the rock shelters was 'devastating for all parties involved and was clearly avoidable' (2020: 1). As Langton later lamented:

> Rio Tinto had four opportunities to stop the destruction of the Juukan Gorge caves. There were alternatives that would have allowed mining but lessened the impact on the site. The company deliberately and consciously failed to share these possibilities with the traditional owners and instead chose the most profitable and expedient option. (Langton 2020)

---

3   At the time of writing, the inquiry had received 161 submissions, and remained open.

The PKKP have rejected Rio Tinto's claim that they provided their free, prior and informed consent for the destruction of Juukan 1 and Juukan 2. In their damning submission to the inquiry, the PKKP stated that while Rio Tinto agreed to their request to delay the blast (to assess options for removing the explosives) the company kept loading charges at the site without informing them. In the words of the PKKP, 'The Juukan Gorge disaster tells us that Rio Tinto's operational mindset has been driven by compliance to minimum standards of the law and maximisation of profit' (PKKP 2020: 8). They described destruction of the rock shelters as a 'yet another example of the low importance accorded to Aboriginal people and Aboriginal culture' (ibid.: 60) and argued that 'Rio Tinto's submission ignores the grossly unequal negotiating position of the parties, a matter Rio Tinto was acutely aware of' (ibid.: 30). Ultimately, 'the destruction of Juukan 1 and Juukan 2 has caused immeasurable cultural and spiritual loss and profound grief to the PKKP People' (ibid.: 7).

Many questions have been asked from many different quarters, including 'where is the state in all of this that such an outcome can occur?' These types of questions lie at the heart of the parliamentary inquiry—which essentially involves the state looking at the state as much as, if not more so, than the state looking at the mining company.

Reclaiming the Porgera Mining Lease

In late August 2020, the Government of PNG announced that it had granted the Special Mining Lease (SML) for the Porgera mine to Kumul Minerals Holdings Limited (KMHL), a state-owned mining company.

This follows a three-year period of uncertainty over Barrick Niugini Limited's tenure in Porgera, after their lease expired in 2019 and their application for a further 20-year extension was denied by the incoming Marape government as part of his strategy to 'take back PNG'. Barrick has challenged the decision domestically in the National Court to no immediate avail. One commentator, who goes by the name Vailala on The Australian National University's DevPolicy blog, reported Deputy Chief Justice Kandakasi as stating the following on 10 July 2020 as he denied Barrick's right to appeal:

> Counsel for the defendants led by the learned Solicitor General's arguments are that the State is under no obligation to give reasons and one of his colleagues in this case, joined in to say it is like a lease situation. When a landlord decides to terminate a lease, the landlord is not required to give reasons. Whether that is a correct analogy or not, I am not getting into that space except to say in this case that a decision has been made and that, there is no expressed statutory provision for disclosure of reasons. (Vailala 2020)

For observers tracking the project, the lead up to the lease renewal (or refusal) has been at least a decade in the making. Much of the observable commentary has been dominated by the project and its discontents. For example, former prime minister Sir Michael Somare went on the parliamentary record in 2005 over 29 alleged killings involving the mine's security forces, announcing his intention to establish a committee to investigate the matter: 'we want to know why they are killing those people, and whether the law allows them to do that' (Anon. 2005). A report presented at the time by the Akali Tange Association, a local Porgera organisation, alleged that the company was directly involved in extrajudicial killings going back as far as 1993 (ATA 2005).

Similarly, allegations of sexual assault by company personnel have circulated around Porgera for well over a decade. In response to claims that Barrick personnel were responsible for the gang rape of local women, Barrick's chief executive Greg Wilkins issued a letter to Porgeran leaders stating that the allegations were 'most distasteful, to say the least as you know these allegations to be untrue.' Three years later in 2011, after several investigative reports from international organisations, including MiningWatch Canada, Human Rights Watch and Amnesty International, Barrick finally admittedly publicly that there was a problem (Anon. n.d.a).

The proposed solution invoked further issues, and more international scrutiny. Barrick's approach to resolving what were effectively criminal activities was to construct a series of direct financial settlements as a means to expunge its future liability to the victims. During the first half of 2013, MiningWatch Canada issued at least two letters to the United Nation's High Commissioner for Human Rights, alleging that the company's remediation framework ran contrary to the United Nations Guiding Principles on Business and Human Rights (UNOHCHR 2013). In April 2015, the General Counsel for EarthRights International said that 'Porgera presents one of the worst cases we've seen of human rights abuse associated with extractive industry' (Anon. n.d.b). EarthRights International had been central in supporting the Porgera Landowners Association (PLOA) in its complaint to the Canadian government and had been actively exploring a case against Barrick in the United States.

By this stage, it looked as if Barrick was trying to get itself out of the Porgera Joint Venture (PJV). In early June of 2015, the PLOA wrote what became an open letter (courtesy of PNG Mine Watch) to Barrick raising concerns about the recent 50 per cent acquisition of the project by Chinese developer Zijin, noting in particular, a concern that the company under Zijin's management would not honour its commitments to the landowners (PLOA 2015).

In the same year, the company finally proposed to resettle households from the villages surrounded by the Anawe dump after the PNG Mineral Resources Authority (the state regulator of the industry) indicated that a new national policy guideline would require developers to resettle all people living on an SML. This commenced a four-year pilot project in which Barrick actively sought to demonstrate progress against a very long and very overdue set of commitments to resettle villages impacted by the operation (Kemp and Owen 2015). Among a raft of sticking points, the lack of a clear position by either the national government or PJV on who would take responsibility for law and order post-relocation, off-lease in the Porgera Valley, was a major hurdle that none of the parties could see beyond.

In April 2020, the decision by several of the leaders of the PLOA to issue a joint press release with Barrick, defending the company's right to continue its operations would indeed seem curious—especially in light of the project's colourful past (PLOA and PJV 2020). Some commentators have suggested that this move by factional leaders was not supported by the broader community, or indeed by Zijin. This may well be the case; however, following the formal announcement that the national government had granted the SML to Kumul Consolidated Minerals Limited (KCML), the leadership of the PLOA issued yet another media release in which Mr Maso Mangape, the Chairman of the PLOA Negotiating Team, was quoted as saying:

> Our interest in reaching an agreement with BNL was fuelled by the desire to have certainty about issues such as resettlement and closure, as well as increased benefits for our community. Will KCML and the National Government now meet these obligations? (Anon. 2020)

Roles and Interests, Absences and Presences

On the surface, both of these events reflect common patterns of absence and presence. They also raise critical questions about the nature of the state. In the case of Juukan Gorge, an absent state allowed Rio Tinto and other extractive companies to operate almost unilaterally. At the Porgera mine, an otherwise absent state has suddenly presented itself, initially proposing to displace Barrick Niugini Limited. The latest development, as this volume goes to press, suggests that Marape has cut an eleventh-hour deal with Barrick's chief executive Mark Bristow, which will give the state a majority interest while retaining Barrick Niugini Limited as the mine operator (PJV 2020).

To a large extent, public commentary on these events, and the analysis of other cases in this volume, has drawn upon and reinforced traditional assumptions about the roles of different actors: states are supposed to act in the best interests of their citizens, and it is generally accepted that corporations are simply out to make a profit on behalf of their shareholders. From these 'role assumptions' we easily find corresponding absences and presences as states fail to act in expected ways and regulate the excesses of the industry. At the same time, corporations appear to colonise these absent spaces. But if we direct our assumptions at the 'interests' of specific actors, then the absence or the presence of the state begins to look very different—and from a longer-term perspective it seems that neither Barrick nor Rio Tinto have really acted in their own interest.

As noted in the introductory chapter, and reiterated throughout the volume, the interests of the Australian state have long been captured by the extractive industries at all jurisdictional levels. In the case of Western Australia, there is a structural tension between the state's heavy reliance upon extraction, and its executive role in issuing extractive licences and safeguarding Aboriginal heritage. These apparently contradictory functions draw attention to the interests of the state, or what motivates the state to act. From the vantage point of interests, one could argue that the state's presence is clearly felt—in ways that privilege extractive capital. While there may be a certain truth to this observation, perhaps the question is not 'whether the state has acted to protect or promote', but whether the permissive approach to the extractive industries is an active and total representation of the state's interests.

The uncertainty created by this proposition reinforces the value of an ethnographic approach to the state. As Alex Golub reminds us in his chapter, an accurate understanding of history cannot take collective actors like 'the state' or 'the company' for granted. If we are going to grasp the interests of the state then we must engage the networks of people who emerge in different times and situations as personating these kinds of entities, promoting particular interests. From this perspective, states may well appear both concrete and uncertain. For example, it cannot be assumed that the current Minister for Aboriginal Affairs in Western Australia necessarily shares the same interests as the Minister for Mines and Petroleum or the Minister for Environment, or that these office holders have any interests in common with previous incumbents or the minster who originally provided the Section 18 approval to destroy the Juukan Gorge caves. At this level, the 'interests of the state' appear far less certain.

On another level, regardless of the interests pushed by politicians enacting the state at any single point in time, the state has long exerted a hegemonic presence in structural and ideological ways through the Aboriginal Heritage Act, specifically the Section 18 terms.[4] Once ministerial consent has been granted to destroy heritage sites there are very few legal pathways to oppose the decision, and the destruction of Aboriginal heritage and the erasure of Aboriginal interests appears to become *unavoidable*. More importantly, other options, like not proceeding with existing mine plans, appear to be *unthinkable*. The logic of the Section 18 terms reflects a historical coalition of state and corporate interests that licences a dominant social order that governs 'common sense' by shaping ideas about what is 'acceptable', feasible and necessary, and what is unthinkable, unreasonable and unworkable. The captains of the industry and various state actors have used these forms of ideological, structural and instrumental power to set the 'rules of the game' and suppress the political power or effectiveness of their opponents (Lukes 1974).

It could be argued that this particular form of state presence is manifest in a certain level of resignation about the inevitability of heritage destruction and the power of extractive interests. Recognising or acknowledging

---

4   The 1972 Aboriginal Heritage Act was already under review by the Western Australian government prior to the destruction of the Juukan Gorge caves. This event has highlighted both the need for urgent legislative reform, and the historical influence of the extractive industries over state processes.

resignation does not imply that people actually consent to the destruction of their heritage, or that they necessarily agree with the terms of any contractual arrangements that have been struck. Rather, attention to resignation can highlight the structural limitations that impede the ability to bring about change. Or in this case, the structural presence and the power of the state that supresses more active forms of opposition—particularly through its legislation and the so-called 'gag clauses' embedded in land use agreements which limit traditional owners' ability to publicly object to specific mining activities.

The appearance of interests over roles and responsibilities is likewise front and centre in the unfolding narrative over the fate of the Porgera Gold Mine. In the quote provided to the media explaining their preference for Barrick, the PLOA make it amply clear that the known presence of the developer (for all its faults) was more desirable than an absent state (for all its faults). Over months of negotiations the PLOA representatives demonstrated an acute awareness of this critical difference between state roles and state interests. The begging question from the above quote is this: what responsibilities can landowners expect the state to step into under these new arrangements? Will this newly acquired role as majority shareholder translate into a presence of another kind?

There is, in addition, the looming question of whether the state's 'role absence' from Porgera across the decades, and especially in relation to the more egregious issues that have unfolded around the project, has not in fact worked to the advantage of the state as it now exerts its interests. This is almost an exercise in hypothetical history, but the question remains: if the state had positively intervened in the many serious environmental, social, legal and governance matters in the last 30 years, would the project have reached the stage where onlookers see appropriation by the state as the only responsible course of action?

Marape's expressed enthusiasm for taking back PNG on behalf of the country's eight million 'shareholders' represents a curious turn in language, a quasi-corporate utilitarianism that offers self-contained justifications for otherwise quite challenging decisions. One supposes that corporations habitually act in their self-interest. Marape is suggesting that nation states ought to do the same, and with a similar resolute sentiment in the communication of the outcomes. Acquiring majority ownership of the Porgera Gold Mine, according to Marape, could see a substantial windfall for the state's coffers, and a marked expansion in the country's share of resource development projects. This assumes that together, Barrick Niugini Limited, Kumul Minerals and the state will be able to restart the operation after these developments and avoid the 'time bomb' scenario, hinted at by Filer, Burton and Banks, which is created by the conflict within the mine-affected communities over the distribution of the burdens and benefits of mining (Filer et al. 2008: 165).

Of course, there is a fine line between this type of corporate utilitarianism and good old-fashioned opportunism. The time at which the SML expired certainly worked in Marape's favour. Striking a balance between his nationwide shareholders and the many legitimate demands of the project's most immediate stakeholders will be key for the state in this new venture, and any slippage between its commercial interests and its local responsibilities in the Porgera Valley will be imminently visible and could have devastating effects.

In both the Australian and the Porgeran case, it has taken a series of disastrous events to remind the state of its obligations to its citizens. While the Australian state, after notable absence, has suddenly appeared for the PKKP in the form of a parliamentary inquiry into the destruction of their heritage, assuming a more critical (even hostile) stance towards Rio Tinto, the PNG state now claims to be acting in the interests of the Porgera landowners. But this sudden appearance of the PNG state seems to have only created more uncertainty, and not everyone is convinced that the prime minister, or the directors of Kumul Minerals, have acted with the interests of the landowners foremost in mind.

While we might conclude that the presence of the state of Australia and PNG appears uncertain or indeterminate from the vantage point of resource extraction settings, the existential presence of the state cannot be passed over. As the material in this volume reveals, and the Juukan Gorge and Porgera cases confirm, the consequences that follow in the wake of state interests are, in fact, highly visible if not patently palpable.

References

Abrams, P., 1988. 'Notes on the Difficulty of Studying the State'. *Journal of Historical Sociology* 1: 58–89.

Anon., 2005. 'Probe Looms for Mine Deaths'. *The National*, 6 May.

————, 2020. 'PNG PM Marape's Secret Porgera SML Grant to KCHL Tramples Mine Landowners Rights.' Pacific Mining Watch blog, 28 August. Viewed 2 October 2020 at: mine.onepng.com/2020/08/png-pm-marapes-secret-porgera-sml-grant.html

————, n.d.a. 'Abuse by Barrick Gold Corporation.' EarthRights International factsheet. Viewed 2 October 2020 at: earthrights.org/wp-content/uploads/documents/barrick_fact_sheet_-_earthrights_international_1.pdf

————, n.d.b. 'Survivors Who Alleged Rape and Killing at Papua New Guinea Mine Pleased with Barrick Gold Settlement.' EarthRights International press release. Viewed 2 October 2020 at: earthrights.org/media/survivors-who-alleged-rape-and-killing-at-papua-new-guinea-mine-pleased-with-barrick-gold-settlement/

ATA (Akali Tange Association), 2005. 'The Shooting Fields of Porgera Joint Venture: Now a Case to Compensate and Justice to Prevail.' Unpublished submission to Porgera Joint Venture.

Beckett, S., 1956. *Waiting for Godot: A Tragicomedy in Two Acts*. London: Faber & Faber.

Filer, C., J. Burton and G. Banks, 2008. 'The Fragmentation of Responsibilities in the Melanesian Mining Sector.' In C. O'Faircheallaigh and S. Ali (eds), *Earth Matters: Indigenous Peoples, The Extractive Industries and Corporate Social Responsibility*. Sheffield: Greenleaf Publishing.

Government of Western Australia, 2020. 'Western Australian Government Submission to the Joint Standing Committee on Northern Australia Inquiry into the Destruction of Indigenous Heritage Sites at Juukan Gorge.' Canberra: Parliament of Australia (Submission 24).

Hopkins, A. and D. Kemp, 2020. 'Corporate Dysfunction on Indigenous Affairs: Why Heads Rolled at Rio Tinto.' The Conversation blog, 11 September. Viewed 2 October 2020 at: theconversation.com/corporate-dysfunction-on-indigenous-affairs-why-heads-rolled-at-rio-tinto-146001

Kemp, D. and J.R. Owen, 2015. 'A Third Party Review of the Barrick/Porgera Joint Venture Off-Lease Resettlement Pilot: Operating Context and Opinion on Suitability.' St Lucia: University of Queensland, Centre for Social Responsibility in Mining.

Langton, M., 2020. 'The Destruction of the Juukan Gorge Caves.' *The Saturday Paper*, 19 September.

Lukes, S., 1974. *Power: A Radical View*. London: Macmillan.

PJV (Porgera Joint Venture), 2020. 'Barrick (Niugini) Limited Update on Porgera Negotiations.' News release, 17 October. Viewed 29 October 2020 at: www.porgerajv.com/Company/Media/Press-Release

PKKP (Puutu Kunti Kurrama People and Pinikura People), 2020. 'Submission to the Joint Standing Committee on Northern Australia Inquiry into the Destruction of 46,000-Year-Old Caves at the Juukan Gorge in the Pilbara Region of Western Australia.' Canberra: Parliament of Australia (Submission 129).

PLOA (Porgera Landowners Association), 2015. 'Sale of 50 per cent Share on Barrick's Operation Porgera Mine to Zijin Mining Group.' Letter to Barrick Gold Corporation, 2 June. Viewed 2 October 2020 at: www.porgeraalliance.net/wp-content/uploads/2015/06/Barrick-Sales.pdf

PLOA (Porgera Landowners Association) and PJV (Porgera Joint Venture), 2020. Letter to James Marape, 14 April. Viewed 2 October 2020 at: miningwatch.ca/sites/default/files/ltr_from_ploa_bnl_to_pm_marape_-_14april2020_signed_copy.pdf

Rio Tinto, 2020. 'Submission to the Joint Standing Committee on Northern Australia Inquiry into the Destruction of 46,000-Year-Old Caves at the Juukan Gorge in the Pilbara Region of Western Australia.' Canberra: Parliament of Australia (Submission 25).

UNOHCHR (UN Office of the High Commissioner for Human Rights), 2013. 'Re. Allegations Regarding the Porgera Joint Venture Remedy Framework.' Geneva: UNOHCHR.

Vailala, 2020. Reply to J. Burton and G. Banks, 'The Porgera mine in PNG: some background'. DevPolicy blog, 1 September. Viewed 2 October 2020 at: devpolicy.org/the-porgera-mine-in-png-some-background-20200507-2/